TOURIST BEHAVIOR

Past, Present, and Future

Advances in Hospitality and Tourism Series

TOURIST BEHAVIOR

Past, Present, and Future

Edited by

Narendra Kumar, PhD

Bruno Barbosa Sousa, PhD

Swati Sharma, PhD

AAP APPLE ACADEMIC PRESS

First edition published 2023

Apple Academic Press Inc.
1265 Goldenrod Circle, NE,
Palm Bay, FL 32905 USA

760 Laurentian Drive, Unit 19,
Burlington, ON L7N 0A4, CANADA

CRC Press
6000 Broken Sound Parkway NW,
Suite 300, Boca Raton, FL 33487-2742 USA

4 Park Square, Milton Park,
Abingdon, Oxon, OX14 4RN UK

Library and Archives Canada Cataloguing in Publication

Title: Tourist behavior : past, present, and future / edited by Narendra Kumar, PhD, Bruno Barbosa Sousa, PhD, Swati Sharma, PhD.
Names: Kumar, Narendra (Lecturer in travel and tourism), editor. | Sousa, Bruno Barbosa, editor. | Sharma, Swati (Lecturer in travel and tourism), editor.
Series: Advances in hospitality and tourism book series.
Description: First edition. | Series statement: Advances in hospitality and tourism book series | Includes bibliographical references and index.
Identifiers: Canadiana (print) 20220280649 | Canadiana (ebook) 20220280851 | ISBN 9781774910245 (hardcover) | ISBN 9781774910252 (softcover) | ISBN 9781003282082 (ebook)
Subjects: LCSH: Tourists—Psychology—Case studies. | LCSH: Consumers' preferences—Case studies. | LCSH: Tourism—Case studies. | LCGFT: Case studies.
Classification: LCC G156.5.P79 T68 2023 | DDC 910/.019—dc23

Library of Congress Cataloging-in-Publication Data

CIP data on file with US Library of Congress

ISBN: 978-1-77491-024-5 (hbk)
ISBN: 978-1-77491-025-2 (pbk)
ISBN: 978-1-00328-208-2 (ebk)

ABOUT THE ADVANCES IN HOSPITALITY AND TOURISM BOOK SERIES

Editor-in-Chief:
Mahmood A. Khan, PhD
Professor, Department of Hospitality and Tourism Management,
Pamplin College of Business,
Virginia Polytechnic Institute and State University,
Falls Church, Virginia, USA
Email: mahmood@vt.edu

This series reports on research developments and advances in the rapidly growing area of hospitality and tourism. Each volume in this series presents state-of-the-art information on a specialized topic of current interest. These one-of-a-kind publications are valuable resources for academia as well as for professionals in the industrial sector.

BOOKS IN THE SERIES:

Food Safety: Researching the Hazard in Hazardous Foods
Editors: Barbara Almanza, PhD, RD, and Richard Ghiselli, PhD

Strategic Winery Tourism and Management: Building Competitive Winery Tourism and Winery Management Strategy
Editor: Kyuho Lee, PhD

Sustainability, Social Responsibility and Innovations in the Hospitality Industry
Editor: H. G. Parsa, PhD
Consulting Editor: Vivaja "Vi" Naraparreddy, PhD
Associate Editors: SooCheong (Shawn) Jang, PhD,
Marival Segarra-Oña, PhD, and Rachel J. C. Chen, PhD, CHE

Managing Sustainability in the Hospitality and Tourism Industry: Paradigms and Directions for the Future
Editor: Vinnie Jauhari, PhD

Management Science in Hospitality and Tourism: Theory, Practice, and Applications
Editors: Muzaffer Uysal, PhD, Zvi Schwartz, PhD, and Ercan Sirakaya-Turk, PhD

Tourism in Central Asia: Issues and Challenges
Editors: Kemal Kantarci, PhD, Muzaffer Uysal, PhD, and Vincent Magnini, PhD

Poverty Alleviation through Tourism Development: A Comprehensive and Integrated Approach
Robertico Croes, PhD, and Manuel Rivera, PhD

Chinese Outbound Tourism 2.0
Editor: Xiang (Robert) Li, PhD

Hospitality Marketing and Consumer Behavior: Creating Memorable Experiences
Editor: Vinnie Jauhari, PhD

Women and Travel: Historical and Contemporary Perspectives
Editors: Catheryn Khoo-Lattimore, PhD, and Erica Wilson, PhD

Wilderness of Wildlife Tourism
Editor: Johra Kayeser Fatima, PhD

Medical Tourism and Wellness: Hospitality Bridging Healthcare (H_2H)©
Editor: Frederick J. DeMicco, PhD, RD

Sustainable Viticulture: The Vines and Wines of Burgundy
Claude Chapuis

The Indian Hospitality Industry: Dynamics and Future Trends
Editors: Sandeep Munjal and Sudhanshu Bhushan

Tourism Development and Destination Branding through Content Marketing Strategies and Social Media
Editor: Anukrati Sharma, PhD

Evolving Paradigms in Tourism and Hospitality in Developing Countries: A Case Study of India
Editors: Bindi Varghese, PhD

The Hospitality and Tourism Industry in China: New Growth, Trends, and Developments
Editors: Jinlin Zhao, PhD

Labor in Tourism and Hospitality Industry: Skills, Ethics, Issues, and Rights
Abdallah M. Elshaer, PhD, and Asmaa M. Marzouk, PhD

Sustainable Tourism Development: Futuristic Approaches
Editor: Anukrati Sharma, PhD

Tourism in Turkey: A Comprehensive Overview and Analysis for Sustainable Alternative Tourism
Editor: Ahmet Salih İkiz

Capacity Building Through Heritage Tourism: An International Perspective
Editor: Surabhi Srivastava, PhD

Medical Travel Brand Management: Success Strategies for Hospitality Bridging Healthcare (H2H)
Editors: Frederick J. DeMicco, PhD, RD, and Ali A. Poorani, PhD

Strategic Management for the Hospitality and Tourism Industry: Developing a Competitive Advantage
Editor: Vincent Sabourin, PhD

Tourist Behavior: Past, Present, and Future
Editors: Narendra Kumar, PhD, Bruno Barbosa Sousa, PhD, and Swati Sharma, PhD

The Hospitality and Tourism Industry in ASEAN and East Asian Destinations: New Growth, Trends, and Developments
Editors: Jinlin Zhao, PhD, Lianping Ren, D.HTM, and Xiangping Li, PhD

Brand Co-Creation Tourism Research: Contemporary Issues and Challenges
Editor: Raouf Ahmad Rather, PhD

Small Island and Small Destination Tourism: Overcoming the Smallness Barrier for Economic Growth and Tourism Competitiveness
Robertico Croes, PhD

Event Tourism in Asian Countries: Challenges and Prospects
Editor: Shruti Arora, PhD

ABOUT THE SERIES EDITOR

Mahmood A. Khan, PhD, is a Professor in the Department of Hospitality and Tourism Management, Pamplin College of Business at Virginia Tech's National Capital Region campus. He has served in teaching, research, and administrative positions for the past 35 years, working at major U.S. universities. Dr. Khan is the author of several books and has traveled extensively for teaching and consulting on management issues and franchising. He has been invited by national and international corporations to serve as a speaker, keynote speaker, and seminar presenter on different topics related to franchising and services management. Dr. Khan has received the Steven Fletcher Award for his outstanding contribution to hospitality education and research. He is also a recipient of the John Wiley & Sons Award for lifetime contribution to outstanding research and scholarship; the Donald K. Tressler Award for scholarship; and the Cesar Ritz Award for scholarly contribution. He also received the Outstanding Doctoral Faculty Award from Pamplin College of Business. He has served on the Board of Governors of the Educational Foundation of the International Franchise Association, on the Board of Directors of the Virginia Hospitality and Tourism Association, as a Trustee of the International College of Hospitality Management, and as a Trustee on the Foundation of the Hospitality Sales and Marketing Association's International Association. He is also a member of the several professional associations.

ABOUT THE EDITORS

Narendra Kumar, PhD

Narendra Kumar, PhD, is currently associated with the Amity Institute of Travel and Tourism, Amity University, Uttar Pradesh, India. He holds a total of 20 years of industry and academic experience, out of which four years is industry experience and 16 years is academic experience. He has been associated with many universities in various capacities and is well known in the academic fraternity. Dr. Kumar also is well connected in the travel and hospitality industry and has helped many young students in setting up their start-ups in India. He is also associated with many journals as an advisory member and regularly holds interactive sessions on various issues related to the tourism and hospitality industry. Recipient of many intradepartmental recognitions for outstanding contributions to the institutional and academic development, Dr. Kumar believes in hard work and dedication toward achieving targets. He is the reviewer of many scientific conferences and international journals, including *Current Issues in Tourism*.

Bruno Barbosa Sousa, PhD

Bruno Barbosa Sousa, PhD, is Professor at the Polytechnic Institute of Cávado and Ave (IPCA, Portugal) and Head of the Master Program in Tourism Management and PhD Marketing and Strategy. He was a market analyst at Sonae Distribuição, Modelo e Continente, S.A., and he was a marketing assistant for the *Jornal O Jogo at Controlinveste*. He is the author or co-author of several papers. His research interests include tourism management, marketing, and strategy. He is an editorial board member and an ad-hoc reviewer for several peer-reviewed scientific journals. He is also a member of the scientific committees of several national and international congresses and conferences. His awards include a Teaching Award from the School of Economics and Management of the University of Minho (2015–2016), Best Thesis in Tourism Award from the International Conference on Innovation and Entrepreneurship in Marketing and Consumer Behaviour (2015), among others.

Swati Sharma, PhD

Swati Sharma, PhD, is currently associated with the Amity Institute of Travel and Tourism, Amity University, Uttar Pradesh, India. She holds 15 years of industry and academic experience and is well known in the academic fraternity. She has published many papers in journals of repute. She has been awarded her PhD under Prof. Manjula Chaudhary who is one of the key members of the Indian Tourism Academic fraternity. She has also been instrumental in organizing various career-related exhibitions for the students and has major interest in events and exhibitions.

CONTENTS

CONTRIBUTIONS

Muhammad Ashfaq
Institute of Agricultural and Resource Economics, University of Agriculture, Faisalabad, Pakistan

Cevdet Avcıkurt
Balıkesir University, Faculty of Tourism, Balikesir, Turkey

Arup Kumar Baksi
Department of Management and Business Administration, Alliah University, West Bengal, India; E-mail: baksi.arup@gmail.com

Gunjan Bhayana
Institute of Management Studies & Research, MD University, Rohtak, Haryana, India

César Lapa Barros
Universidad de Vigo, Spain; E-mail: cesarlapa83@gmail.com

Nilgün Demirel
Travel-Tourism and Leisure Services Department, Igdir Vocational School, Igdir University, Turkey; E-mail: nilgundemirel90@hotmail.com

Sawitri Devi
Institute of Management Studies & Research, MD University, Rohtak, Haryana, India

Dinesh Dhakhar
Department of Tourism & Hotel Management, Kurukshetra University, Kurukshetra, Haryana, India

Ana Rita Dias
Polytechnic Institute of Cávado and Ave (IPCA), Portugal

Athula Gnanapala
Department of Tourism Management, Sabaragamuwa University of Sri Lanka, Belihuloya, Sri Lanka; E-mail: athulatmsusl@gmail.com

Asad Afzal Humayon
COMSATS Institute of Information Technology, Vehari, Pakistan

Mustafa Kan
Kirsehir Ahi Evran University, Agricultural Faculty, Department of Agricultural Economics, Kirsehir, Turkey

Magno Angelo Kelmer
UFJF - Federal University of Juiz de Fora, Brazil

Ismail Kizillirmak
Faculty of Economics, Istanbul University, Turkey

Annaelise Fritz Machado
Polytechnic Institute of Cávado and Ave, Portugal; E-mail: annaelisefritz@yahoo.com.br

Muhammad Hassan Mahboob
Institute of Agricultural and Resource Economics, University of Agriculture, Faisalabad, Pakistan;
E-mail: m.hassan1695@gmail.com

A Apsara Saleth Mary
Centre for Tourism and Hotel Management, Madurai Kamaraj University, Madurai, India;
E-mail: prof.apsara@gmail.com

Richa Mishra
Amity Institute of Travel & tourism, Amity University Uttar Pradesh, India;
E-mail: richa1188@gmail.com

Ipseeta Nanda
K L University, Guntur, India

Meltem Altinay Özdemir
Faculty of Tourism, Mugla Sitki Kocman University, Turkey; E-mail: meltemozdemir@mu.edu.tr

Gizem Özgürel
Muğla Sıtkı Koçman University, Marmaris Tourism Vocational School, Muğla, Turkey;
E-mail: gizemozgurel@mu.edu.tr

Ritu Punhani
Amity University, Noida, Uttar Pradesh, India

Rashmi
Management Department, Delhi Global Institute of Management, Faridabad, Haryana, India;
E-mail: mbahr006@gmail.com

Miguel Ribeiro
Polytechnic Institute of Cávado and Ave, Portugal

J. A. R. C. Sandaruwani
Department of Tourism Management, Sabaragamuwa University of Sri Lanka, Belihuloya, Sri Lanka; E-mail: jarudmi@gmail.com

Fauzia Ashraf Shaikh
Amity University Dubai, United Arab Emirates

Vinod Kumar Shukla
Amity University Dubai, United Arab Emirates; E-mail: vinodkumarshukla@gmail.com

Lakhvinder Singh
Department of Tourism Management, Government College, Kaithal (Haryana), India;
E-mail: lakhvindersingh.kuk@gmail.com

Aiman Siraj
Amity University Dubai, United Arab Emirates

Bruno Barbosa Sousa
Polytechnic Institute of Cávado and Ave (IPCA), Portugal; E-mail: bsousa@ipca.pt

Ezgi Kirici Tekeli
Department of Tourism Guidance, School of Applied Sciences,
Karamanoğlu Mehmetbey University, Turkey

Bárbara de Vasconcellos
Polytechnic Institute of Cávado and Ave, Portugal

Adrian Lubowiecki-Vikuk
SGH Warsaw School of Economics, Poland

ABBREVIATIONS

ADNKS	address-based population registration system
AI	artificial intelligence
ANEW	affective norms of English
ANOVA	analysis of variance
AR	augmented reality
ASV	average shared variance
AVE	average variance extracted
BIAS Map	behaviors from intergroup affect and stereotypes map
CAGR	compound annual growth rate
CBBE	consumer-based-brand-equity
CBC	cross-border cooperation
CBR	cross-border region
CFA	confirmatory factor analysis
CFI	comparative fit index
CHA	Cultural heritage's attributes
CHT	cultural and heritage tourist
CMB	common method bias
CR	composite reliability
CVM	contingent valuation method
CVI	content validity index
CVR	content validity ratio
DNM	Department of National Museums
ECR	ethnocultural rootedness
EDP	expectancy disconfirmation paradigm
EFA	exploratory factor analysis
FIML	full information maximum likelihood
GFI	goodness of fit index
GPS	Global Positioning System
ICOM	International Council of Museums
IE	immersive experience
IoT	internet of things
KMO	Kaiser-Meyer-Olkin
MSV	maximum shared variance

NFI	normed fit index
OLS	ordinary least square
QDAS	qualitative data analysis software
QR	quick response
SCM	stereotype content model
SEM	structural equation modeling
SPSS	statistical package for social science
STL	SusTraveLux
TLI	Tucker-Lewis fit index
UGC	user-generated content
UNESCO	United Nations Educational, Scientific and Cultural Organization
UNWTO	United Nations World Tourism Organization
VR	virtual reality
WHO	World Health Organization
WTP	willingness to pay
WTTC	World Travel and Tourism Council
WWDC	Worldwide Developers Conference

PREFACE

Over the years, tourist behavior has been a major interest area not only for business managers but also for researchers as these behaviors give an insight into policymaking, designing new products, and developing marketing strategies. What tourists do at destinations and what are their preferences have fascinated many of us. We are especially more inclined to know what interests or influences tourists. Tourism is no doubt a global phenomenon now, and a large number of spots have become complete tourist destinations in themselves as they provide all facilities and attractions at one place. This is possible because of the understanding of tourist behavior over the years and subsequent product designing and development.

Tourism is a progressively segmented, fragmented, and geographically complex activity. There are many tourist motivations that highlight the importance of tourism (e.g., religious, shopping, health, business, nature, cultural, leisure, among others). This book is intended to be a support tool for readers who work directly or indirectly on the phenomenon of tourism, with special emphasis on consumer behavior and the purchase decision process (choice of a tourist destination).

Tourism is important for the economy, for marketing, and for regional and local development. This book is an important contribution to the tourism sector literature, with an interdisciplinary look at the past, the present, and new challenges for the future.

The book has chapters and cases of tourist behavior from different cultural and geographical backgrounds and is one of those few publications available that provides case studies from across the globe. Despite the fact that tourists are more knowledgeable nowadays, there is a need for understanding the changing preferences of tourists to provide them the best possible experience. This book will be a handy source for understanding tourist behavior and patterns, and scholars can use it for referring to their research pertaining to tourist behavior. Also, business managers can use it as a ready reckoner as it will help them design their marketing and promotional strategies. For policymakers, this book shall provide benefit by discussing the trends to develop future tourist destinations. Overall, this book will provide an insight on tourist behavior from the past, understand the present, and help to predict the future.

CHAPTER 1

RELATIONSHIP MANAGEMENT IN SPA TOURISM: AN APPROACH TO CROSS-BORDER REGIONS

CÉSAR LAPA BARROS[1,*] and BRUNO BARBOSA SOUSA[2]

[1]*Universidad de Vigo, Spain*

[2]*Polytechnic Institute of Cávado and Ave (IPCA), Portugal*

[*]*Corresponding author. E-mail: cesarlapa83@gmail.com*

ABSTRACT

The practice of healthcare and wellness management depends on successfully informing potential customers about procedure options, service excellence, treatment facilities, tourism opportunities, travel benefits, and destination choice. As a tourism segment example, health, wellness, and thermal tourism is becoming an increasingly popular option for visitors looking to access procedures that are seemingly unavailable to them in their home countries due to lack of affordability, lack of availability. or lengthy waiting lists. This research seeks to gather and present the main aspects related to marketing and relationship management process in specific tourism context, in particular the case of health, wellness and thermal tourism in cross-border destination Portugal–Spain. In specific, in the Galicia Northern Portugal Euroregion, this activity still constitutes a vector of sustainable development, serving as leverage for complementary activities like healthcare and thermal tourism. The study has as purpose to

Tourist Behavior: Past, Present, and Future. Narendra Kumar, PhD, Bruno Barbosa Sousa, PhD & Swati Sharma, PhD (Eds.)

provide a better understanding regarding the essence of innovation process and relational marketing: theoretical and practical implications, in specific contexts of health, wellness and thermal tourism in cross-border destinations. Regarding the methodology used here, this is a conceptual paper with a literature review that brings together the major components of innovation and its implications tourist perspective and relationship marketing (i.e., visitor perspective). The present manuscript intends to contribute theoretically to the management of innovation in specific tourism contexts (i.e., wellness, health, and thermal tourism) in cross-border tourist destinations (Portugal and Spain, specifically the Euroregion of Northern Portugal and Galicia). In the end, we intend to explore the relationship with thermal tourism consumers, suggesting a conceptual model. The model should be tested empirically in the cross-border tourist area of Northern Portugal and Galicia (with the partnership of four thermal tourism companies, as one of the main types of tourism common to the territorial space). A future study should include emotional factors. The new vogue of wellness, heatlh, and thermal tourism forces us to challenge and re-visit the power relationships that exist within contemporary tourism and the limitations imposed by the new coronavirus.

1.1 INTRODUCTION

Tourism is a dynamic, economic, and social phenomenon due to its ability to create cross-border and Euroregion development. According to Ratten et al. (2019), there has been increasing recognition of the way tourism can be innovative, particularly through creative means (e.g., wellness, health, and thermal tourism). This is important in specific services developed that integrate technology and sustainability trends. The subject of interest for several years, the role of innovation in tourism is growing in significance. There are two different main reasons for the dynamic growth. First of all, the tourism sector has been the booster for new technological innovations that have created new segments and niches (e.g., medical tourism, health tourism, wellness, and thermal tourism), and secondly, the tourism sector has a large number of small-sized enterprises which includes tourism providers, such as travel agencies, hotels, transport companies, or educational providers (Ratten et al., 2019). The creation process in tourism contributes to business success and increased competitive advantage, which is a theme that has clearly captured the attention of academics in the field of tourism management. The dynamic nature of this process, coupled with the constant need to invest in a highly distinctive and differentiating offering new researches, have provided in order to score market position and achieve tourism consumer satisfaction, including the level of emerging realities confirmed by new trends and desires (Costa, 2001; Buhalis and Costa, 2006).

Tourism innovation is the adoption of a change that results in something new to the organization and relevance to the environment, and the creative idea and its development are the seed germinated by innovative, effective for the tourism economy (Sousa and Alves, 2019). There is a large body of knowledge about tourism innovation and management but less is known from a specific wellness, health, and thermal perspective. This is unusual as thermal management has an abundance of innovation due to its emphasis on technology and change. In this sense, innovation has attracted the attention of several scholars and entrepreneurs together with this research topic. Therefore, relationship marketing in wellness and thermal tourism can be defined as creating, maintaining, and improving the mutually beneficial relations or possible termination of unfavorable relationships of healthcare management and thermal tourism with stakeholders. This is undertaken to satisfy the stakeholder's need and to seek consensus between guest and wellness management and thermalism interests in the development of tourism experiences (Sousa and Alves, 2019). The present manuscript intends to contribute theoretically to the management of innovation in specific tourism contexts (i.e., thermal tourism) in cross-border tourist destinations (Portugal and Spain, specifically the Euroregion of Northern Portugal and Galicia). In the end, we intend to explore the relationship with thermal tourism consumers, suggesting a conceptual model (to be tested empirically) that relates relational marketing and the result of innovation in tourism (i.e., relational marketing dimensions). The model should be tested empirically in the cross-border tourist area of Northern Portugal and Galicia (with the partnership of four thermal tourism companies, as one of the main types of tourism common to the territorial space). This makes it possible to open new doors and open new markets, enabling greater efficiency in social, economic, and regional growth and new ways of targeting an increasingly competitive market, in specific, the case of wellness and thermal tourism contexts in cross-border destinations.

1.2 WELLNESS, HEALTH, AND THERMAL TOURISM

Tourism is a phenomenon that moves millions of people around the world, taking as a major driver of the global economy (Ferreira and Sousa, 2020). Moreover, wellness, health, and thermal tourism are a segment of great importance to the economy, responsible for creating a large number of jobs. Every year, much due to the several changes in the tourism environment, fosters competition between and within tourist destinations (Bigné and Andreu, 2004; Sousa and Rocha, 2019; Sousa et al., 2020). Like other emerging sectors in a modern economy, thermal tourism is a dynamic and ever-changing industry (e.g., wellness and thermal centers or specific organizations).

As such, comes an increasing need to understand the consumer behavior of wellness, health, and thermal destinations. According to Lordkipanidze et al. (2005), the growth of the tourism sector and increasing demand with regard to new tourism segments, makes it even more visible and required its connection with the emerging trends in companies (e.g., thermalism, thalassotherapy, spas and quality of life, and senior thermalism). Since tourism is composed of many of small firms, they are constantly called upon to respond quickly to new trends and market demands (assuming as a source of innovation). The role of each actor in the distribution channel, within the operational management of tourism, will be critical to appreciate the range and nature of the emerging changes. According Costa et al. (2015), wellness, health, and thermal tourism have become an emerging market segment directed not only for those who pursue solely thermal treatments but also for those seeking illness prevention, physical improvement, and spiritual balance or even for those eager of cultural and relaxation programs. Figure 1.1 shows how the sector may be classified and segmented. Wellness, health, and thermal tourism offer a more traditional tourism experience, whereby the customer is pampered with the major motivation of relaxation and rejuvenation and fits within more established notions of tourism (Sousa and Alves, 2019). Cosmetic tourism, although involving a medical element and often an operation, commonly mixes the procedure with usual tourist behavior (Tresidder, 2011). According to Hudson and Li (2012), the literature to date has been fairly descriptive and many researchers have called for more in-depth research on the topic (Goldbach and West, 2010; Lunt and Carrera, 2010; Hopkins et al., 2010; Sousa and Alves, 2019).

Most of the research and studies have focused on the following themes: (i) the reasons for growth, (ii) conceptualization and definition, (iii) quality and accreditation, (iv) services offered and types of care, (v) health tourism destinations, and (vi) consumer attitudes toward heath tourism. For instance, and according to Lee et al. (2012), health tourism has existed since at least the 19th century when people started to travel to other areas for healing and relaxation, such as at spas and springs. Modern health, wellness, and thermal tourism are grounded in what is commonly referred to as push and pull factors (Sousa and Alves, 2019). Push factors include inflated healthcare costs (Hunter, 2007; Ye et al., 2008), a shortage of care personnel in the public healthcare system (DeMicco, 2017; Hunter, 2007), long wait times, and low quality care for the cost of treatment (e.g., DeMicco, 2017; Kang and Oh, 2008). Pull factors are favorable currency exchange rates, improvement of medical standards, and the development of air transportation (Lee et al., 2012).

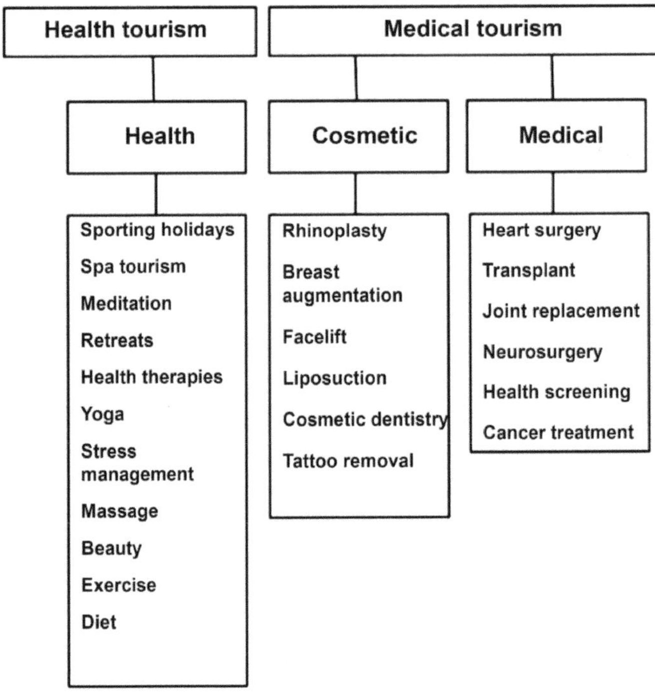

FIGURE 1.1 The shape of health and medical tourism (Henderson, 2004; Tresidder, 2011).

In this sense, this specific niche tourism has become more complex and creative, demanding high quality equipment and infrastructures, providing a wide range of products and services related not only with the diversity of spa treatments but also with complementary recreational activities regarding the connection to nature and to cultural patrimony as well as to other regional resources, creating effective links to the territory and becoming a significant force that will shape the regional development, by creating a positive pressure on local economic bases with spill over effects on the territory. For instance, and according to Araujo et al. (2015), thermal spa treatment is a touristic product that is highly recognized for its potential in the exploration of endogenous regional resources. Consequently, the development of new and competitive thermal tourism products can play an important role in the development of the inland regions, contributing effectively to addressing the economic asymmetries of the region. On top of this, the researchers Fleus et al. (2018) highlight the potential identified by governments and the tourism industry itself of wellness tourism—and thermal spa tourism in particular—to serve as a tool for empowering individuals and societies to adopt wellness. In a similar manner, according to Ladeiras et al. (2015), thermalism benefits the cultural industry as it helps boosting the demand for cultural products, promoting local identity and consolidating the feeling of belonging by transforming natural endogenous resources into differentiation elements for the destination, and consequently, improving the local communities' quality of life.

In fact, Roget et al. (2018) identify spas businesses as a framework of a company committed with sustainable development, because these companies make use of resources without compromising their future, create jobs and generate wealth, settle populations, and satisfy needs by improving people's quality of life. As thermalism goes, this activity underwent significant changes throughout the last centuries. It is referred by Del Río-Rama et al. (2018) that thermalism has moved from a tradition that focuses solely on the curative purposes for a variety of health conditions, such as respiratory, gastric or rheumatic, among others—to a wider approach that highlights the properties of thermal waters and their benefits for the contemporary society. The changes underwent by thermal tourism and the subsequent diversification of its offer echoed in the type of consumer of this industry. Del Río-Rama et al. (2018) also make reference to the objectives of this type of tourism: the improvement of the tourist's health status, rest and relaxation, contact with nature, aesthetics and beauty. This shift in the tourist's positioning forces the companies that provide thermal tourism services to give him/her special attention beyond technical quality, as is stated by Rodriguez Míguez (2005), since this tourist is not seen as a patient but as a consumer (Del Río-Rama et al., 2018). According to Del Río-Rama et al. (2018), this tourist seeks treatment for his/her physical and mental diseases, while at the same time, searching for a good level of service, adequate accommodation facilities, leisure programs, cultural and sports activities, and pleasant natural surroundings. According to Portuguese Thermal Association data, demand for wellness packages in 2005 registered a growth of 37.1% in relation to 2004, while in the same period, classical therapeutic thermalism showed negative growth with a 6% loss in clients. The same trend occurred in the following year, with a fall in the number of classical thermalism users, from 80,309 users in 2005 to 76,999 users in 2006, and growth in the number of wellness thermalism users, to a total of 22,049 users compared with the 17,730 seen in 2005. In 2008, there were 38 active hot springs, 19 of which were located in the center of Portugal (50% of the total), 16 in the North (42%) and 3 in the South (8%). Wellness has become an important topic of research, as well as a rapidly increasing business. There is demonstrable growth in health and wellness tourism across Europe, as a result of a number of social circumstances (Quintela et al., 2011: 2). Similarly, Loutraki has a long history in spa and Thermalism. The water derived from the Loutraki natural thermal springs was characterized as "The Water of Life" since antiquity. The geothermal springs derive from many faults of the coastal area in a 750-m-wide zone. The natural spring hot waters, at the bottom of Mount Geraneia, are discharged from a great mineral depth. Visitors enjoy balneology, spa treatments and wellness sessions in a luxurious environment in the modern Loutraki Thermal Spa (Fig. 1.2).

FIGURE 1.2 Loutraki Spa and Thermalism.

Similarly, wellness tourism is promoted through various services that are practiced in order to relax and rejuvenate. Hot springs have been one such effective method that has tremendously added to the dimensions of the wellness tourism. Naturally occurring hot springs and the use of their water is highly important in various wellness therapies as well as the procedures that are required to provide rejuvenating experience to the guest (Sousa and Alves, 2019). Geothermal energies and hot spring waters are the basis of the wellness therapies like balnoetherapy, thermalism, etc. The water of these naturally occurring hot springs usually possesses a very high quantity of mineral content that is used extensively for clinical purposes. Spas and other wellness centers run on the mercy of these waters for these waters are the basis for nearly all the wellness therapies (Jagyasi, 2010).

Also in thermal tourism context, the process of creation and innovation have been increasingly used to describe business behavior, the destinations, the tourism sector as well as all their planning (Dredge, 1999). However, tourism is going through significant changes, facing new challenges that require new perspectives and implementing ideas (Stamboulis and Skayannis, 2003).

1.3 RELATIONAL MARKETING AND DYNAMIC NATURE OF ENTREPRENEURIAL PROCESS

Relationship marketing aims to generate profitable, long-term relationships between partners (Miquel-Romero et al., 2014). Gupta and Sahu (2012: 63) argue that an "effective relationship marketing program delivers better results in terms of increasing customer base, sales, and profitability." Marketing literature refers some important dimensions of relationship marketing, such as commitment, trust, cooperation, among others, factors that are very important for any business to consider in their marketing program (Gupta and Sahu 2012). For Sousa and Alves (2019) satisfaction, trust, and commitment are the results of relationship marketing implementation. According to Bart et al. (2005), trust complacency of one of the actors in the relationship in accepting the vulnerability and dependence on the other party, with the expectation that you can believe achieve a particular goal. Trust is used as a basis for strengthening the commitment relationship in organizational sciences (Cambra-Fierro and Polo-Redondo, 2011). Trust is then defined as a belief that neither party of the relationship will exploit the other's vulnerability (Dwyer et al., 1987). Thus, several authors (e.g., Sousa and Alves, 2019) consider that the concept of trust encompasses three other important dimensions: the belief that the other will show benevolence in his actions; honesty and the belief that the other intervener will have competence to act in favor of the relationship between the two (Fynes et al., 2005; Moorman et al., 1992). There are several authors who explore the importance of commitment. Commitment has been pointed out as an important component of close relationships in B2B (business to business) marketing (Chang et al., 2012). Schmiedel et al. (2013) refer to commitment as the desire and the will to contribute to the achievement of the objectives of the process. Trust and commitment are fundamental elements for the success of the relationship marketing strategy (Siguaw et al., 1998). Trust decreases perceived risk and vulnerability in a relationship, which leads to a greater commitment to the relationship (Ganesan, 1994). Inter-company relations are guided by relational factors, such as commitment standards, which are based on trust (Riel et al., 2011). Many researchers (e.g., Sousa and Alves, 2019) consider cooperation as a central element in relational exchanges and use this construct to evaluate the quality of a company's relationship (Leonidou et al., 2006, Palmatier et al., 2007). In relations between exporters and distributors, cooperation is essential because it helps companies cope with the complexity and multiplicity of distribution tasks across borders. Several authors propose the following definitions for the construct cooperation. Also, given the above, and since the supply of health and thermal tourism can be designed to specific markets, we suggest that commitment, trust and cooperation, can be assumed as a relevant variable to visitor satisfaction and consequent loyalty. This situation is strongly related to the innovative and entrepreneurial spirit in health and well-being activities. Entrepreneurial activity is seen as a process: includes a set of steps, is subject to tourism management, continuous and applicable in many contexts (Stevenson and Jarillo 1990). This process becomes a source of competitive advantage on a global scale. This entrepreneurial process is, as a rule, inextricably linked to certain factors that allow initial "take the leap" (circumstances, triggering events). Shindehutte et al. (2000) present a conceptual model that summarizes the dynamic nature of the triggering process in an attempt to summarize the various contributions on this topic.

The entrepreneurial process is conceptualized as a response to the awakening of the event. However, the nature of the trigger, relevance, and impact depend on the dynamic interaction between the characteristics of managers/employees (e.g., personal life), firm characteristics (e.g., size, culture) and external developments of the environment (e.g., competitiveness). The different types of entrepreneurial activities are probably the result of different types of triggers.

Shindehutte et al. (2000) present a summary which reflect the main differences between conventional marketing and entrepreneurial marketing. Therefore, relationship marketing in wellness and thermal tourism can be defined as creating, maintaining, and improving the mutually beneficial relations or possible termination of unfavorable relationships of healthcare management and thermal tourism with stakeholders.

1.4 CROSS-BORDER DESTINATIONS: EUROREGION OF NORTH PORTUGAL AND GALICIA

The definition of the concept of cross-border region (CBR) depends, in the first instance, on what we understand to be cross-border cooperation (CBC). Regarding the concept of CBR, there are several definitions around it. However, the definitions used by the European Council, determine that these regions are characterized by homogeneous goals and functional interdependence, otherwise there would be no need for such cooperation (Perkmann, 2003) and the fact that the region is considered as potential, inherent in geography, history, ecology, ethnic groups, economic possibilities among other issues, but that was fragmented by the sovereignty of central governments responsible for each side of the border (CoE, 1995). On the other hand, these are spatial units in which there is a strong internal interaction when compared to external relations (Berry, 1969).

Thus, a cross-border region can be defined as "a territory on the border composed of local governments involved in cross-border cooperation agreements" (Perkmann, 2003: 156). For Perkmann (2007), cross-border cooperation can be divided taking into account its geographic reach (be it greater or lesser), the area where the region is inserted, the intensity of cooperation (namely the degree of autonomy) and even taking into account it counts the type of actors that go from the municipal (local) to the regional level. Contextualizing this study in terms of the Euroregion of North Portugal and Galicia requires its geographical definition. This Euroregion comprises the lands in the Iberian Peninsula delimited by Atlantic Ocean, the eastern mountains (Pardellas and Figueroa, 2017) and the Douro River. Far more important than its frontiers, for the Euroregion's identity are the resources, namely the natural resources therein, with unique common elements that can be found both in Galicia and Northern Portugal, such as an axis of thermal waters, as pointed out by Pardellas and Figueroa (2017). Particularly, in terms of natural resources and thermalism, this Euroregion is one of the most important in the Iberian Peninsula, with a large number of quality springs and the richest medicinal mineral waters (Amboage et al., 2019).

However, because this Euroregion is composed by two regions of two different countries, it is easily understandable that in political terms, these two regions live different realities. Cross-border regions may vary, for example, depending on the degree of cooperation (high or low intensity) or geographic extent (low or high). In particular, Euroregions tend to define themselves as micro cross-border regions, with a reduced geographical extension between contiguous spaces. In view of the above, it was intended to carry out fieldwork in the Euroregion North of Portugal and Galicia (as an integral part of the study of cross-border tourist regions), not only for the local convenience but, above all, for the contribution that will be given to it. This Euroregion belongs to two different states: In one of them, an autonomous administrative level (Galicia) is created, while in the other (Northern Portugal), an administrative region with no legislative capacity is maintained (Commission for the Coordination and Development of the Region do Norte), according to Pardellas and Padín (2005). Figure 1.3 shows the organization and geographic area that affects this cross-border region.

According to Sousa (2014), it is a Euroregion with more than 6,000,000 inhabitants, based on a space of strong social, economic, and cultural interaction. The Euroregion North of Portugal and Galicia shares a significant cultural and natural heritage, with enormous tourist potential. With the opening of national borders between Member States of the European Union, regional exchange registered a clear increase, both in economic and social terms, creating highly compensatory regional synergies. This region is formed by 12 NUTS III (common nomenclature of statistical territorial units) of which five constitute the basic border area (Pontevedra and Ourense in Galicia, Cávado, Minho-Lima and Alto Trás-os-Montes in Northern Portugal).

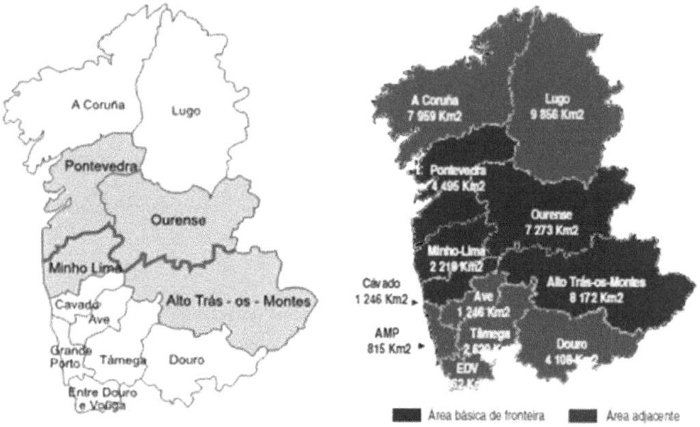

FIGURE 1.3 Euroregion of Galicia and Northern Portugal.
Source: http://regioes.blogspot.pt/

In a study on banding and strategy in the Euroregion of Galicia and Northern Portugal, Oliveira (2015) pointed out that as an autonomous region, Galicia has its own regional government with legislative and executive powers in such fields as infrastructure, environment and regional planning, while in contrast, the Northern Portugal region lacks this autonomy and is ruled by the central government and this limits the region's decision-making capabilities.

In terms of thermalism, Galicia stands out as being Europe's second largest thermal region and the city of Orense is only surpassed by Budapest in number of water reserves (Roget et al., 2018). According to Roget et al. (2018), Galicia possesses up to 300 water catchments and the largest part of the 116 spas in Spain.

It is underlined by Roget et al. (2018) that in Galicia, thermal tourism presents itself as a tourism alternative that comprises health, well-being, and leisure. These researchers point out that a large number of Galician spas are located in the inlands, in rural areas characterized by a low-income population and an economic landscape where the primary sector is dominant (Roget et al., 2018). Roget et al. (2018) add that the Galician spa play an important role of socioeconomic promoters of the municipalities where they are established. Particularly in the Galicia Northern Portugal Euroregion, this activity still constitutes a vector of sustainable development, serving as leverage for complementary activities like accommodation, restaurants, and transports (Ladeiras et al., 2015). Roger et al. (2018) state that these spas are generally small-sized, family-owned companies of Galician origins.

In the beginning of this century, Nóvoa (2010) found that one-fifth of all of Spanish spas are located in Galicia, with these spas generating approximately 126 million euros in wealth and being responsible for directly employing 2700 professionals in the the year of 2009, which served 140,000 tourists in the same year (Gândara et al., 2013).

The research of Solla (2012) shows us that by the beginning of the 21st century, Galician spas had a rate of occupancy around 80%. This researcher argues that although a large number of spas' customers benefit from public subvention programs, the popularity of spas and the services provided in spas has significantly increased throughout the years (Solla, 2012). This increase in popularity of spas translates into a growth of this market. Roget et al. (2018), referencing DBK (2016) show that in Spain thermalism is a 230 million Euromarket, with a growth trajectory.

Roget et al. (2018) noticed a change in the thermal tourist's demographic profile which is becoming younger. These researchers state that the most pro-eminent age interval of the thermal tourist is between the age of 46 and 60 years old which contains 46% of these consumers (Roget et al., 2018). Roget et al. (2018) add that the contemporary thermal tourist shows a greater interest for his/her health and well-being than before.

The amount of money spent by the thermal tourist in Galicia during his/her stay, according to Roget et al. (2018), lies between 251 and 500 Euros. Roget et al. (2018) noted that this amount is concomitant with the average amount spent by a user of the thermalism social program "IMERSO." Roget et al. (2018), making reference to Pilquiman (2014), highlight the importance of the thermalism social program regulated by the Ministry of Social Affairs decree of March 15th, 1989 which was responsible for dinamizing spas tourism activity. According to Roget et al. (2018), on the one hand, this program helped minimize the seasonal effect that can be observed in other types of tourism, and on the other hand, one can observe an indirect effect of this program on repeat visits to spas outside

IMERSO's program and on these tourists' recommendations and influence on other potential customers of these spas.

Over on the Portuguese side, Gustavo (2010) describes the country as a case-study for thermalism due to it abundance of hot springs and the popularity of thermalism. The work of this researchers shows that in the first decade of the 21st century there were approximately 40 active hot springs in Portugal, most of them located in Northern Portugal and specialized in the therapeutic component of thermalism (Gustavo, 2010). Gustavo (2010) adds that in recent years, Portugal has witnessed an increase in the number of modern spas, mostly located in urban areas and often integrated in hotel units.

According to Quintela et al. (2011), in 2005, the total number of visitors of portuguese thermal units was approximately 98,000, increasing to approximately 99,000 visitors in the following year. Spanish tourists accounted for a little over half of foreign visitors to Portuguese traditional thermal units. These researchers noted that if on the one hand, the demand for wellness packages increased dignificantly between 2004 and 2005, on the other hand, the demand for classic thermalism in the same period shortened 6% (Quintela et al., 2011). This trend continued in the following years, as is shown by Quintela et al. (2011), however, since the increase of wellness tourism was insufficient to offset the decrease in the demand for classical thermalism, the Portuguese thermal units suffered a decrease in business of 4.3% in that period.

Nevertheless, innovation in thermal tourism and its offer in the Euroregion of Galicia and Northen Portugal continued to thrive in the following years as is shown in the works of researchers such as Roget et al. (2018). These researchers state that in Galicia, innovative products are being introduced in the market, integrated in a strategy combined of the public and private sector to promote the thermal sector as an innovative tourism destination for well-being and thermalism seekers (Roget et al., 2018). The city of Ourense and its product "TrenBalnearios" is an example of innovation provided by Roget et al. (2018). This product comprises the mobility of tourists through seven of the province's spas (Roget et al., 2018). Another example provided in this article is the CERLAC—the rowing training center of Laias, Arnoia and Castrelo de Miño—which is a 2014 inniciative of Grupo Caldaria and INORDE—Ourense's Economic Development Institute—that combines indoor nautic sports and Laia's spa infrastructure and services (Roget et al., 2018).

1.5 PROPOSED RESEARCH MODEL

Preliminary research was conducted in order to develop the research instrument. Details of the preliminary research are given below. After the collection of the preliminary data, empirical data would be collected through fieldwork. This study discusses the increasing phenomena of thermal tourism and the relationship marketing perspective associated with innovation. Based on the literature review, we propose a model that connects the relationship marketing dimensions (i.e., commitment, trust, and cooperation) with the consequent satisfaction with thermal tourist and the behavioral intention of repeating the experience or reinforcing the experience with a Euroregion marketing and niche tourism perspective. Several studies in marketing and tourism contexts have discussed the association of relationship marketing to the site and consumer purchasing behavior, including the study of satisfaction, loyalty, or quality of service (Heung et al., 2011). An extensive review of the current literature reveals that no integrated theoretical framework for the holistic study of the innovation and thermal tourism industry exists. The proposed research model not only has implications for future research, but also provides useful information for practitioners in the thermalism. Several studies have shown that more brand commitment and brand trust leads to more positive satisfaction behavior (e.g., Belaid and Temessek Behi, 2011; Sousa and Alves, 2019). The theoretical framework highlights that satisfaction tends to favor the development of behavioral loyalty. These results are particularly evident in tourism contexts, as the example of the importance of satisfaction in revisiting or recommendation to family or friends (Bigné et al., 2005; Prayag, 2008; Sousa and Rocha, 2019). Based on this discussion, we propose that in a context of healthcare and thermal tourism, the increase in trust, cooperation, and commitment facilitates the development of behavioral satisfaction and loyalty. The proposed model intended to contribute theoretically to the management of innovation in specific tourism contexts (i.e., thermal tourism) in cross-border tourist destinations (Portugal and Spain, specifically the Euroregion of Northern Portugal and Galicia). The relationship with consumers of thermal tourism was explored, suggesting a conceptual model (to be tested empirically) that relates relational marketing and the result of innovation in tourism (favoring trust, commitment, quality of service, satisfaction, and loyalty). The model should be tested empirically in the cross-border tourist area of Northern Portugal and Galicia (with the partnership of four thermal tourism companies, as one of the main types of tourism common to the territorial space). The proposed model is presented in Figure 1.4. Hence, we propose that P1: cooperation has a positive effect on satisfaction of healthcare and thermal tourism; P2: commitment has a positive effect on satisfaction of healthcare and thermal tourism; P3: trust has a positive effect on satisfaction of healthcare and thermal tourism; P4: Satisfaction has a positive effect on the healthcare and thermal tourism.

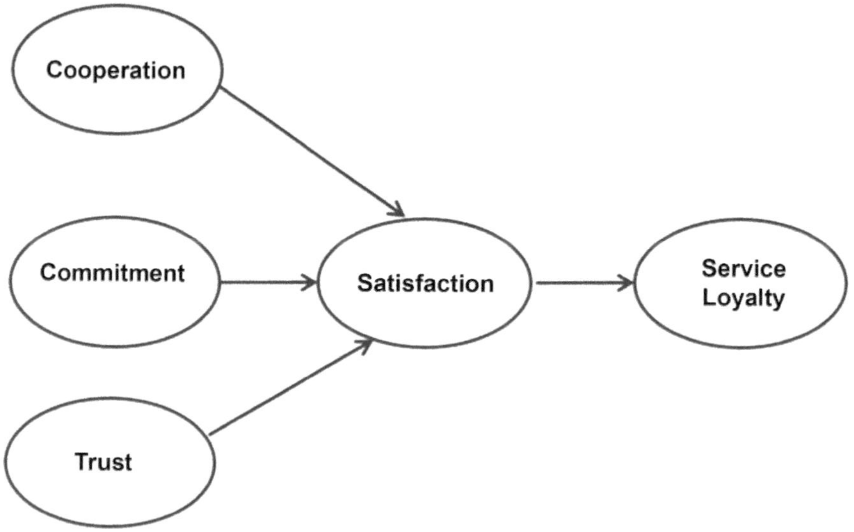

FIGURE 1.4 Relationship marketing and behavioral intentions in thermal tourism services.

This study should reveal the potentials of individual destination attributes to cause satisfaction and/or dissatisfaction, and it shed light on the most determinant and critical attributes in explaining the overall tourist experience, in specific, the case of thermal tourism contexts. It is well established in the literature that the tourist's assessment of different destination attributes influences his or her overall satisfaction, and subsequently, intentions to revisit a cross-border destination. Studies typically apply linear modeling techniques in analyzing these influences, such as multiple regression analysis or structural equations modeling and focus group analysis.

1.6 CONCLUSION AND NEXT STEPS

In an increasingly global world, which tends to predominate competitiveness and change, the difference is, so often, the ability to create discontinuities in the external environment. Many times the success is the ultimate goal, which focuses on searching for new products, new markets, new organizational forms and new sources of customer value. In this field, it inevitably highlights the thermal tourism as some of the examples mentioned in this paper, as vehicles of innovation, and that may prove suitable choices of context for the operationalization of the proposed study. For instance, the creation process associated with wellness and spa reservation systems, mechanisms and information sharing as a form of competitive advantage over other tourist destinations (social networks), the marketing of thermal and heatlh tourism products and services as well as other related activities that can leverage the increased business synergies.

Entrepreneurship factors that may trigger an entrepreneurial activity (particularly in thermal tourism contexts) and whose origin may be associated with internal, external factors, and characteristics of the organization should be noted. The external factors are more difficult to control such as political, economic, social, technological, environmental factors, or legal issues. However, the combination of those factors with the intrinsic nature will form the basis of an innovative event, along with the organizational characteristics (e.g., culture, values and business traditions). Finally, and this manuscript is primarily theoretical and reflection, is expected to arise in the future some work of a practical nature that bring greater robustness to support and perspectives presented.

In terms of thermalism, Galicia and Northern of Portugal stands out as being Europe's second largest thermal region and the city of Orense is only surpassed by Budapest in number of water reserves. Galicia possesses up to 300 water catchments and the largest part of the 116 spas is located in Spain. In the last years, Portugal has witnessed an increase in the number of modern spas, mostly located in urban areas and often integrated in hotel units.

The present manuscript is intended to contribute theoretically to the management of innovation in specific tourism contexts (i.e., thermal tourism) in cross-border tourist destinations (Portugal and Spain, specifically the Euroregion of Northern Portugal and Galicia). In the end, the relationship with consumers of thermal tourism was explored, suggesting a conceptual model (to be tested empirically) that relates relational marketing and the result of innovation in tourism (favoring trust, commitment, quality of service, satisfaction, and loyalty). The model should be tested empirically in the cross-border tourist area of Northern Portugal and Galicia (with the partnership of four thermal tourism companies, as one of the main types of tourism common to the territorial space). We intent to show how these valid questions can be answered by introducing causal modeling and discussing one approach to it, structural equation modeling (SEM), which is a rigorous technique for building and testing such models. It is an advanced statistical methodology that delivers models and results that are easy to understand and employ. We intent to confirm that a causal relationship exists (i.e., measurement over time should confirm that change in one variable precedes change in the other; a controlled experiment is required that changes a single variable; the direction of causality must be correctly specified; the relationship must not disappear when common causes of both are removed and the causal effect "makes sense" in domain knowledge) (Elliott, 2003).

This preliminary study has already identified that the global movement of tourism is seemingly showing an increased focus on the thermal tourism (niche product or niche service). In this case, the question seems to be whether the further growth in demand for healthcare management and thermal tourism. A future study should include emotional factors. The new vogue of health, wellness and thermal tourism forces us to challenge and re-visit the power relationships that exist within contemporary tourism and the host–guest relationship.

KEYWORDS

- **thermal tourism**
- **cross-border destination**
- **relationship marketing**
- **wellness**

REFERENCES

Amboage, E. S.; Fernández, V. A. M.; Boga, O. J.; Fernández, M. M. R. Redes sociales y promoción de destinos turísticos termales de la Eurorregión Galicia-Norte de Portugal. *Observatorio (OBS*)* **2019,** *13* (1), 137–152.

Antunes, V. M. P. *Marketing empreendedor no sector de turismo de saúde e bem-estar*; Doctoral dissertation, 2009.

Araujo, A. R.; Paiva, T.; Ribeiro, M. P.; Coutinho, P. Innovation in Thermalism: An Example in Beira Interior Region of Portugal. In *Health and Wellness Tourism*; Springer International Publishing, 2015; pp 165–180.

Belaid, S.; Temessek Behi, A. The Role of Attachment in Building Consumer-Brand Relationships: An Empirical Investigation in the Utilitarian Consumption Context. *J. Prod. Brand Manage.* **2011,** *20* (1), 37–47.

Berry, B. A Synthesis of Formal and Functional Regions Using a General Field Theory of Spatial Behavior. In *Spatial Analysis*, Berry, B. e Marble, D., org.; Prentice Hall: Englewood Cliffs, NJ, 1969.

Bigné, J.; Andreu, L. Emotions in Segmentation—an Empirical Study. *Ann. Tour. Res.* **2004,** *31* (3), 682–696.

Buhalis, D. *eTourism: Information Technology for Strategic Management*; Prentice Hall: London, 2003.

Buhalis D.; Costa, C. *Tourism Management Dynamics, Trends, Management and Tools*; Elsevier Butterworth Heinemann: Oxford, UK, International Edition, 2006.

CoE. *Manuel de coopération transfrontalière à l'usage des collectivités locales et régionales en Europe*. Council of Europe: Strasbourg, 1995.

Cole, A. *Business Enterprise in its Social Setting*; Harvard University, Press: Boston, 1959.

Costa, C. An Emerging Tourism Planning Paradigm? A Comparative Analysis Between Town and Tourism Planning. *Int. J. Tour. Res.* **2001,** *3* (6), 425–441.

Costa, C.; Quintela, J.; Mendes, J. Health and Wellness Tourism: A Strategic Plan for Tourism and Thermalism Valorization of São Pedro do Sul. In *Health and Wellness Tourism*; Springer International Publishing, 2015; pp 21–31.

Davenport, T. *Process Innovation, Reengineering Work through Information Technology*; Harvard Business School Press, 1993.

Del Río-Rama, M.; Maldonado-Erazo, de la C. C.; Álvarez-García, J. State of the Art of Research in the Sector of Thermalism, Thalassotherapy and Spa: A Bibliometric Analysis. *Eur. J. Tour. Res.* **2018,** *19*, 56–70.

DeMicco, F. J. (Ed.). *Medical Tourism and Wellness: Hospitality Bridging Healthcare (H2H)*; CRC Press, 2017.

Dosi, G. Sources, Procedures, and Microeconomic Effects of Innovation. *J. Econ. Literat.* **1988,** *26*, 1120–1171.

Dredge, D. Destination Place Planning and Design. *Ann. Tour. Res.* **1999,** *26* (4), 772–791.

Drucker, P. *Innovation and Entrepreneurship*, 1ª ed.; Collins, 1993.

Hudson, S.; Li, X. Domestic Medical Tourism: A Neglected Dimension of Medical Tourism Research. *J. Hosp. Market. Manage.* **2012,** *21* (3), 227–246.

Hunter, W. C. Medical Tourism: A New Global Niche. *Int. J. Tour. Sci.* **2007,** *7* (1), 129–140.

Kastenholz, E.; Davis, D.; Paul, G. Segmenting Tourists in Rural Areas: The Case of North and Central Portugal. *J. Travel Res.* **1999,** *37*, 353–363.

Ferreira, J.; Sousa, B. Experiential Marketing as Leverage for Growth of Creative Tourism: A Co-creative Process. In *Advances in Tourism, Technology and Smart Systems. Smart Innovation, Systems and Technologies*; Rocha, Á., Abreu A., de Carvalho J., Liberato D., González E., Liberato P., Eds., vol. 171; Springer, Singapore Systems, Smart Innovation, Systems and Technologies 171, 2020; pp 567–577. https://doi.org/10.1007/978-981-15-2024-2_49

Gândara, J. M. G.; Brea, J. A. F.; Manosso, F. C. Calidad de la experiencia en los hoteles termales de Galicia, España: Un análisis a través de la reputación online. *Estudios y perspectivas en turismo* **2013,** *22* (3), 492–525.

Gartner, W. A Conceptual Framework for Describing the Phenomenon of New Venture Creation. *Acad. Manage. Rev.* **1985,** *10* (4), 696–706.

Goldbach, A. R.; West, D. R. Medical Tourism: A New Venture of Healthcare. *J. Global Busi. Iss.* **2010,** *4* (2), 43–44.

Gupta, A.; Sahu, M. A Literature Review and Classification of Relationship Marketing Research. *Int. J. Customer Relat. Market. Manage.* **2012,** *3* (1), 56–81.

Gustavo, N. S. A 21st-Century Approach to Health Tourism Spas: The Case of Portugal. *J. Hosp. Tour. Manage.* **2010,** *17* (1), 127–135.

Gursoy, D.; McCleary, K. An Integrative Model of Tourists' Information Search Behavior. *Ann. Tour. Res.* **2004,** *31*, 353–373.

Heilman, M.; Chen, J. Entrepreneurship as a Solution: The Allure of Self-Employment for Women and Minorities. *Human Resour. Manage. Rev.* **2003,** *13* (2), 347–364.

Heung, V. C.; Kucukusta, D.; Song, H. Medical Tourism Development in Hong Kong: An Assessment of the Barriers. *Tour. Manage.* **2011,** *32* (5), 995–1005.

Hopkins, L.; Labonte, R.; Runnels, V.; Packer, C. Medical Tourism Today: What Is the State of Existing Knowledge? *J. Public Health Policy* **2010,** *31* (2), 185–198.

Jagyasi, P. *Medical Tourism: Research & Survey Report*; Dr Prem, 2010.

Kang, J. J.; Oh, S. H. An Exploratory Analysis of Demographic Differences on Choice of Medical Tourism Destination. *The Korean Academic Society of Tourism and Leisure* **2008,** *20* (4), 139–158.

Kim, W.; Lee, C.; Hiemstra, S. Effects of an Online Virtual Community on Customer Loyalty and Travel Product Purchases. *Tour. Manage.* **2004,** 25 (3), 343–355.

Ladeiras, A.; Mota, A.; Pardo, M. C. A Comparative Study of Thermal Legislation in the Galicia–North Portugal Euroregion. In *Health and Wellness Tourism*; Springer: Cham, 2015; pp 1–20.

Lee, M.; Han, H.; Lockyer, T. Medical Tourism—Attracting Japanese Tourists for Medical Tourism Experience. *J. Travel Tour. Market.* **2012,** *29* (1), 69–86.

Lordkipanidze, M.; Brezet, H.; Backman, M. The Entrepreneurship Factor in Sustainable Tourism Development. *J. Clean. Prod.* **2005,** *13*, 787–798.

Luchs, B. Quality as a Strategic Weapon. *Eur. Busi. J.* **1990,** *2* (4), 34–47.

Lunt, N.; Carrera, P. Medical Tourism: Assessing the Evidence on Treatment Abroad. *Maturitas* **2010,** *66* (1), 27–32.

Ma, H.; Tan, J. Key Components and Implications of Entrepreneurship: A 4-P Framework. *J. Busi. Ventur.* **2006,** *21*, 704–725.

Miquel-Romero, J. M.; Caplliure-Giner, E. M.; Adame- Sánchez, C. Relationship Marketing Management: Its Importance in Private Label Extension. *J. Busi. Res.* **2014,** *67*, 667–672.

Pardellas, X. X.; Figueroa, P. Patrimonio y turismo en la Eurorregión Galicia-Norte de Portugal= Heritage and Tourism in the Euroregion Galicia-Northern Portugal. *Polígonos. Revista de Geografía* **2017,** *29*, 143–164.

Pardellas, X.; Padín, C. The Minho Region (Galicia and Northern Portugal): A Case of Tourism Destination Planning. *Tour. Analy.* **2005,** 10(2), 97–102.

Pearce, D. *Tourist Development*; Longman Scientific & Technical: London, 1991.

Pearce, P. L.; Stringer, F. P. Toward a Symbiosis of Social Psychology and Tourism Studies. *Ann. Tour. Res.* **1984**.

Porter, M. *The Competitive Advantage of Nations*; Macmillan: New York, 1990.

Porter, M.; Millar, V. How Information Gives You Competitive Advantage. *Harv. Busi. Rev.* July–August **1985,** 149–174.

Prayag, G. Image, Satisfaction and Loyalty: The Case of Cape Town. *Anatolia Int. J. Tour. Hosp. Res.* **2008,** *19* (2), 205–224.

Quintela, J. A.; Correia, A. G.; Antunes, J. G. Service Quality in Health and Wellness Tourism–Trends in Portugal. *Int. J. Busi. Manage. Soc. Sci.* **2011,** *2* (3), 1–8.

Riel, A. C. R.; Liljander, V.; Semeijn, J.; Polsa, P. EU Deregulation and Dealer-Supplier Relations in Automotive Distribution. *J. Busi. Indust. Market.* **2011,** *26* (2), 115–131.

Rodriguez Miguel, L. El balneario: tipos de instalaciones balnearias: higiene y seguridad en los balnearios. In *María Reyes. Principios de hidroterapia y balneoterapia*; Pérez Fernández, Ed.; Madrid: Ed. McGraw-Hill Interamericana, 2005; pp 35–48.

Roget, F. M.; Dominguez, M. N. C.; Brea, J. A. F. O turismo como fonte de riqueza en Galicia: Análise dalgunhas experiencias. *Revista Galega de Economía* **2018,** *27* (3), 65–80.

Russell, R.; Faulkner, B. Entrepreneurship, Chaos and the Tourism Area Lifecycle. *Ann. Tour. Res.* **2004,** *31* (3), 556–579.

Schollhammer, H. Internal Corporate Entrepreneurship. In *Encyclopedia of Entrepreneurship*; Prentice Hall, Inc.: Englewood Cliffs, NJ, 1982.

Schumpeter, J. *The Theory of Economic Development*; Harvard University Press: Cambridge, MA, 1934.

Schumpeter, J. *Capitalism, Socialism, and Democracy*; Harper and Row: New York, 1975. Originally published in 1942.

Sharma, P.; Chrisman, J. Toward a Reconciliation of the Definitional Issues in the Field of Corporate Entrepreneurship. *Entrepren. Theor. Pract.* **1999,** *23* (3), 11–27.

Solla, X. M. S. Los turismos de interior en Galicia. Balance y perspectivas. *Polígonos. Revista de Geografía* **2012,** *23*, 213–234.

Sousa B.; Silva A.; Malheiro A. Differentiation and Market Loyalty: An Approach to Cultural Tourism in Northern Portugal. In *Advances in Tourism, Technology and Smart Systems. Smart Innovation, Systems and Technologies*; Rocha, Á., Abreu, A., de Carvalho, J., Liberato, D., González, E., Liberato, P., Eds., vol. 171; Springer: Singapore, 2020; pp 681–690. https://doi.org/10.1007/978-981-15-2024-2_58

Sousa, B.; Alves, G. The Role of Relationship Marketing in Behavioural Intentions of Medical Tourism Services and Guest Experiences. *J. Hosp. Tour. Insights* **2019,** *2* (3), 224–240. https://doi.org/10.1108/JHTI-05-2018-0032

Sousa, B.; Rocha, A. T. The Role of Attachment in Public Management and Place Marketing Contexts: A Case Study Applied to Vila de Montalegre (Portugal). *Int. J. Public Sect. Perf. Manage.* **2019,** *5* (2), 189–205. https://doi.org/10.1504/IJPSPM.2019.099094

Sousa, B. Dynamic Differentiation and the Creative Process in Tourism Management Destinations. *Revista Brasileira de Pesquisa em Turismo—RBTur* **2015,** *10* (1), 3–17. jan./abr. 2015, São Paulo, e-ISSN 1982-6125.

Sousa, B. M. O impacto do place attachment e de emoções na satisfação e lealdade em regiões turísticas transfronteiriças: uma perspetiva de marketing de nichos aplicado à Euro-região Norte de Portugal e Galiza, Tese de Doutoramento, Universidade do Minho, 2014.

Sousa, B.; Simões, C. Comportamento e perfil do consumidor de turismo de nichos: o caso de euro-regiões. *Revista Tékhne* **2010,** *VIII* (14).

Stamboulis, Y.; Skayannis, P. Innovation Strategies and Technology for Experience-Based Tourism. *Tour. Manage.* **2003,** *24*, 35–43.

Stevenson, H.; Jarillo, J. A Paradigm of Entrepreneurship: Entrepreneurial Management. *Strat. Manage. J.* **1990,** *11*, 17–27.

Tajeddini, K.; Ratten, V.; Denisa, M. Female Tourism Entrepreneurs in Bali, Indonesia. *J. Hosp. Tour. Manage.* **2017,** *31*, 52–58.

Tidd, J.; Bessant, J.; Pavitt, K. *Gestão da Inovação: Integração das Mudanças Tecnológicas, de Mercado e Organizacionais*; Monitor: Lisboa.

Ye, B. H.; Yuen, P. P.; Qiu, H. Z.; Zhang, V. H. Motivation of Medical Tourists: An Exploratory Case Study of Hong Kong Medical Tourists. Paper presented at the APTA Annual Conference, Bangkok, Thailand, 2008.

CHAPTER 2

THE CONTRIBUTION OF TOURIST EVENTS TO LOCAL DEVELOPMENT: A CASE STUDY

ANA RITA DIAS[1,*], BRUNO BARBOSA SOUSA[2], and ADRIAN LUBOWIECKI-VIKUK[3]

[1]*Polytechnic Institute of Cávado and Ave (IPCA), Portugal*

[2]*Polytechnic Institute of Cávado and Ave (IPCA), Portugal. Applied Management Research Unit (UNIAG) and CiTUR, Portugal*

[3]*SGH Warsaw School of Economics, Poland*

[*]*Corresponding author. E-mail: ritzdias2@gmail.com*

ABSTRACT

Territorial marketing is a fundamental tool for territorial management and strategic planning in order to promote and highlight a territory, place, or city. It is a fact that a city marketing policy is able to promote and transform an image, boost economic activity, and enhance all aspects of the city that present competitive advantage over other territories, so this is why associate marketing with cities makes sense. This study aims a better understanding of the city's marketing management. Nowadays, it is imperative to define and to develop a territorial strategy plan, whose accomplishment leads to the responsible development of the territory in question, creating more opportunities and attracting the target audience. In

Tourist Behavior: Past, Present, and Future. Narendra Kumar, PhD, Bruno Barbosa Sousa, PhD & Swati Sharma, PhD (Eds.)

this research is made a literature review, using scientific articles, centered on the concepts of tourism, event management, and territorial marketing. It is also presented a particular case of the city of Braga: the event "Braga Romana," considered as a case study. In order to understand and comprehend the importance of territorial marketing in promoting this event, an analysis of the results obtained through a qualitative approach is made. Three semi-structured interviews were conducted in a focus group of nine participants. As a conclusion of this project, we made an evaluation of the strategies already used, also pointing out alternative solutions to be implemented, which will contribute to the development, enhancement and prominence of "Braga Romana" event.

2.1 INTRODUCTION

Given the nature of event services, event planners must be adept at strategic planning (Ruyi and Pegg, 2007). The reason for this statement is that event tourism is a complex process that entails planning leisure, business, or civic events to attract tourists, to forecast future demand, to promote a locale's brand mage, and to maximize revenue for the event hosting community (Tsiotsou et al., 2010). Increasingly, cities are known for what they can develop and provide, or more accurately, for the virtual capabilities that are ensured through the marketing campaigns or representations that come out (Rola et al., 2018). Territorial marketing becomes an important tool, allowing the development of certain factors, both internally and externally, of a defined territory, obtaining more investments and attract more individuals to this same region. The assumption that triggered this article was built from concern and analysis of the various factors of strategic development, in the case study of the Braga Romana event, to ascertain all the strategic planning applied to date. Therefore, considering Braga as a city with a high tourist potential, but in which, however, it is still necessary to implement strategic plans in order to involve all residents, traders, possible stakeholders for a better reception of tourists, we formulate the following question: What is the impact of territorial marketing regarding the Braga Romana event? The scrutiny of this question has as its general objective the understanding of territorial marketing's value in promoting the Braga Romana event. This article is divided into four parts. First, we present the main theoretical foundations, which are the crucial point of this investigation, to understand the studied content for a better analysis of results. Here, topics focused on tourism, events and event management and territorial marketing are introduced as a form of strategic development for the city. The second part deals with the case study, where we present a brief characterization of the "Braga Romana." Afterwards, we explain the methodology applied to the investigation of this case study. Subsequently, we approached the analysis and discussion of the results obtained through the questions implemented to three respondents and also the questions asked to a focus group addressed to nine participants. The last chapter aims at the closure of this article, where the main conclusions drawn from the case study are addressed and where we refer possible future implications and limitations.

2.2 EVENTS IN TOURISM AND MARKETING MANAGEMENT

"Tourism is one of the most relevant engines and has the most significant impact on the development and growth of various economies worldwide" (Sousa et al., 2017). Is a concept in constant evolution, and as such, within the restricted scope and adapted to the theme of this study, we consider the most appropriate definition of "tourism" the one established by the World Tourism Organization (UNWTO), which considers that tourism "comprises the activities developed by people when traveling and staying in places outside their usual framework, for a consecutive period of time (not exceeding one year) for leisure, business and other purposes" (Recommendations on Tourism Statistics, 1994: 5). This definition characterizes tourism, in a general way, and serves as the basic concept of the term tourism. However, there are various types of tourism, namely, business, leisure, health, sports, events, the latter being the concept that will be approached. "Event Tourism" is defined as a set of activities performed by people, whose purpose is to participate in an event, which can be professional, sporting, religious-related among others. Generally speaking, we can say that an event is a special happening, planned and organized for a certain target audience, with a defined goal, at a particular time and location (Zitta, 2014).

According Ratten et al. (2010), event management is a fast growing professional field that incorporates such fields as tourism, leisure, hospitality, marketing, and sport (Getz, 2008). Graham et al. (2001) defined sport event management as "the organization, marketing, implementation, and evaluation of any type of event related to sport" (p. 3). Events can increase media attention and create positive outcomes among community residents (Chalip et al., 2003; Kim and Petrick, 2005). In specific, previous studies indicate that there has been acknowledgement of the social and economic benefits that events can create to the host city, region, and/or country (Ratten, 2016). According to Ribeiro and Sousa (2016), events can be classified in several categories and there is a wide range of ways to define the different existing types. These can be distinguished either by their complexity, size or range according to the conceptor goal.

- Mega events—These are considered the large-scale ones, whose impact is perceived in the communities, nations, or even continents, requiring resources of various orders and in large numbers, such as financial, human, etc. (Douglas et al., 2001; Tideswell and Faulkner, 1997). The mega events are defined by the volume of visitors, by their cost or by the psychological criteria, still emphasizing the political component, which through its approval will grant prestige. The controversy arises from the perspective of Getz (2005), who believes that if we relate the term mega with its size, then a significant example of this type of events will be the Olympic Games and World Fairs. However, if the association is made having in consideration the mega impacts inherent to events, a mega event can reffer to a small music festival due to its positive impacts in a small town, whether economic, image, or tourist. So, for this author, "Mega events, the size of its meaning, are those that yield extremely high levels of tourism, media coverage, prestige, or economic impact for the host community to the venue or the organization same" (Getz, 2005: 407).
- Hallmark events—Such events are usually of a recurrent nature and are kept deliberately with the intention of stimulating tourism of the community that welcomes them, attracting, bringing renowned, good image, and city brand recognition in order to equip the community competitive advantages. The increase in tourism is a natural consequence of this type of recurring events, to the extent that over time, tend to become synonyms of the cities that host them, as well as the spirit and philosophy of the event that spend also characterize the place and its inhabitants, as for example with the Carnival in Rio de Janeiro (Douglas et al., 2001).

- The Community-based event—These are the smallest in the scale and are also the most common. These type of events can range from exhibitions to concerts, through fund-raisers. Although they are of smaller dimension, that does not mean that they do not have a very active role in the dynamics and sense of identity of a community, because, in fact, they may have. (Douglas et al., 2001).

Events hold great importance in the life of cities and therefore also a wide range of impacts on the cities hosting them (Miragaia et al., 2018). It becomes crucial to understand the concept of perception so that we can classify an event as beneficial or not to the community that receives it, since it is through the perceptions that residents have and how exposed the influence of impacts on their lives (Getz, 2005). In this context, tourism is a phenomenon that moves millions of people around the world, taking as a major driver of the global economy. Every year, much due to the frequent changes in the tourism environment, fosters competition between and within tourist destinations (Sousa and Simões, 2010). The literature review conducted for this study allowed us to identify a set of factors that will influence more markedly the support of residents in relation to the Carnival of Ovar. These factors can be divided into intrinsic and extrinsic. Thus, we find the existence of a set of facts that influence both the perception of residents of the impacts of Braga Romana, as their attitudes toward development of this event. To be able to analyze the influence of these factors on the perception of residents and their attitudes, we developed the following model, where there are three factors that may influence the perceptions of residents (sociodemographic characteristics, connection to the event and the level of satisfaction and loyalty) which together with the perceptions themselves may influence attitudes toward the development of the Braga Romana. Based on the research model to propose and through the literature review given above, the questionnaire developed includes the main objective of realizing the attitudes of Braga residents regarding the completion of the Braga Romana event, as well as the necessary information on the factors that may influence these same attitudes.

According to Isidoro (2013), the event aims to innovate, create, and introduce new trends in the market. To ensure the success of an event, it is necessary to ally originality and creativity to many aspects related to the planning of the event, innovating, creating and making a difference to achieve a better market position. On a narrower level, every event to be held must follow a well-defined strategy to be successful. For this, regardless of their typology, it is necessary to overcome three important phases: Pre-Event, Event and Post-Event, defined by Pedro et al. (2012). The first phase translates into the idealization and planning of the event to be held. The second phase is defined as the phase of realization and monitoring of all previously established planning. The third and final phase, concerns the evaluation and analysis of event results. As the main focus of this article, we orientate this study around the concept of marketing that reflects not only in terms of advertising and sales but also in the ability to develop possible strategies that meet, in this way, the needs of consumers. Thus, companies or cities that want to create development strategies must target on creating value for consumers and building a relationship with them, acquiring value in return. For a more complex perspective of this concept, based on the ideas of Kotler and Armstrong (2011), the authors present a five-step model: (1) to understand the markets, and also the needs and wants of consumers, (2) to define a consumer-based marketing strategy, (3) to create an integrated marketing program with value offered, (4) to build profitable relationships and generate customer satisfaction, and (5) to make profits through customer value.

In a more restricted analysis of the concept of marketing, it includes several categories and marketing types, in the course of this study, we turn to "city marketing." This term is still an underdeveloped subject, and through the analysis of several authors, we can say that it is a relatively recent and contradictory concept. "City marketing" is characterized as one of the most prominent and leading areas of marketing in terms of the development of cities (Sousa, 2017). Hence, the city needs the support of residents, leaders and institutions to become welcoming and receptive to new challenges, and thus, attract new companies and investments (Kavaratzis, 2009). In such a way is possible to create a differentiation approach encompassing the elaboration of a marketing strategy according to the characteristics of each location. In general, marketing and its tools make it possible to adapt more quickly to external constraints, in order to equate the differentiation of the city compared with competing cities (Sousa and Rocha, 2019).

In this sense, and in order to avoid the homogeneity and standardization of marketing used in promotional campaigns, defined by various companies, it is important to implement a marketing strategy related to the cultural identity of the place and increase creative strategies according to nowadays (Martinez, 2011). Thus, we believe that the main challenge of the regions is to define a strategy of identity and project their image, developing modern points of attraction based on past culture and implementing enriched intellectual and spiritual strategies open to new concepts (such as music, sport, and cooking). The term "city marketing" includes several other terms, namely, marketing of place, destination marketing, territorial marketing, geographic marketing, public, urban and local marketing, and also place branding (e.g., Anholt, 2007; Avraham and Ketter, 2008; Hospers, 2004; Kavaratzis, 2004; Kotler et al., 1993). The meaning of this concept is related to the exchange process applied and established in a particular region, city, country, or even local, between the residents, visitors, organizations and investors associated with it (Fig. 2.1). Thus, "territorial marketing" covers several different points and can be directed according to many activities, which are created for greater attractiveness and competitiveness of the place in foreign market level.

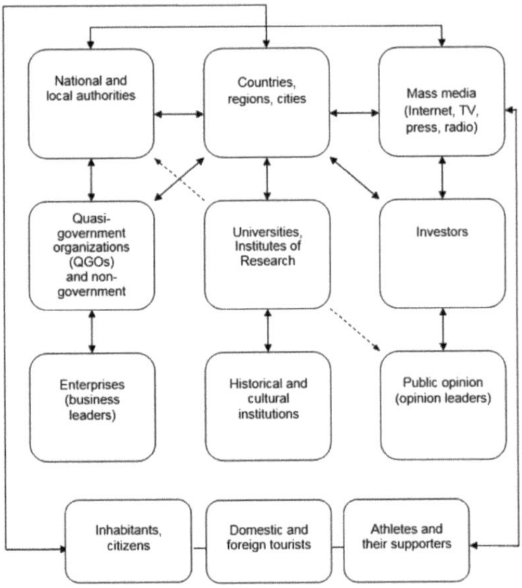

FIGURE 2.1 Entities involved in the process of building the territorial marketing.
Source: Lubowiecki-Vikuk and Basińska-Zych (2011: 37).The competitive advantage of cities becomes problematic, and based on Kanter (1995), prove to be centers of excellence and three c's: concept, competencies, and connections. Being the first characterized as advanced ideas and projections for new products and services, in order to create value for consumers. The second is distinguished as the ability to translate innovative ideas and apply them to the market. Finally, the third translates into business alliances in order to promote core competencies, more value, and enable access to new horizons through the globalization of economies (Sousa and Silva, 2019).

In a more restricted way, it is possible to verify a multiplicity of concepts related to "city marketing," which is analyzed according to different approaches due to its multidisciplinary nature. Referring to Kanter (1996), we consider important to mention that the source of competitive advantage of cities, and regions, is characterized by three scenarios:

- Can become venues for promoting new strategies by getting in touch with new ways of thinking and finding support to turn ideas into viable business.
- Differentiate themselves when enhancing productive competence while maintaining manufacturing standards and highly skilled labor.
- These could be the regions that build connections in global networks to find resources and partners that can bridge other markets.

These are referred to as forms of a global link to the local, with someone always thinking, someone implementing, and someone playing the middle part.

When applied to territorial marketing, the concept of brand and brand image show to be relevant for the theme. The first translates the differentiation of the location and the second conveys the identity of the destination (Morgan et al., 2011; Pike, 2015; Tasci and Kozak, 2006). As Keller (1993) indicates, the brand image is classified with the set of associations (present in the consumer's memory regarding the brand), the level of the attributes (personal value that the consumer attaches to the product and services) and finally, on the attitudes (consequence of the consumer's evaluation of the brand). It is confirmed in one of the researches by Chen and Phou (2013) in which the authors support brand relationship theory, indicating that tourists form emotional relationships with destinations. Thus, the brand provides differentiation and imposes value to the product/service. From the perspective of territorial marketing, the brand must be built in such a way that the tourism product is identified by a sustainable advantage that makes a difference and stands from the competition.

2.3 METHODOLOGY AND CONTEXT

In order to answer the research question "What is the impact of territorial marketing regarding "Braga Romana,?" we consider the qualitative paradigm to be the best research strategy. Throughout this method, the problem, which arises in the course of the investigation, is described more generally (Coutinho, 2011). Regarding the data collection technique, we opted for the case study as it consists of "detailed observation of a context, or individual, from a single source of documents or specific event," having a delimited area of research (Bogdan and Biklen, 1994). For this, we used two research tools: semi-structured interview and focus group. The first one allows us to draw a better analysis of the perspectives and experiences of the interviewees (Wahyuni, 2012). The focus group, our second tool, consists of listening to a group of people talking, in an open discussion, about the subjects of interest, offering a deep insight and an enlarged perception. "Braga Romana" is a cultural event that occurs on the last days of May and portrays the city of Braga 2000 years ago, when the city integrated the Roman Empire. Here, the city's commonplace as capital of Galicia (a Roman province) is remembered and recreated. This event, which was first held in 2003, counts already with 16 editions. The latest of them happened between 22nd and 26th May 2019. "Braga Romana" will be the subject of this article, and therefore, the spotlight of the practical part (semi-structured interviews and focus group). The interviews were conducted at the beginning of June 2019, the first lasted 30 min, the second about 15 min and the third about 50 min.

2.4 RESULTS AND DISCUSSION

From a general perspective, our study aims to discuss the impact of implementing a strategic plan tailored to the needs of an event of this magnitude, like the case of "Braga Romana." In a slender context, we consider the best strategy for this study to be the qualitative paradigm, as we believe, like Guba and Lincoln (1988), that the reality of promoting an event is extremely complex and that only a holistic view of it can ensure broad understanding of this process. At the first stage of the project (interviews), an analysis will be made of the results of the interviews conducted with interviewed 1 (owner of an establishment in the city of Braga), interviewed 2 (participant in the market of "Braga Romana" event, resident in Braga), and interviewed 3 (councilwoman). Subsequently, we proceed with the demonstration of the results of the covered issues, in the form of topics that resulted from the established dialogue. In this context, the city marketing, or marketing of cities, is one of the areas of marketing where there has been a higher growth and more relevant for the development and growth of cities. While the need to differentiate and assert their individuality in various aspects, whether economic forum, social or political, is already a concern that dates back almost to the beginning of the governments and the cities of marketing have been used since the 19th century was only in recent decades that cities began to employ more intensively, techniques derived from the marketing, since it was at this time that the competition for investment, tourism revenues, and concern in attracting new residents intensified and they became an object of study in how cities should be managed.

The growing business opportunity in the services sector, where cities are incorporated, and reinforces the similarities that this sector has with the business activity and the way some tools adapted to this new reality can be fundamental knowledge of how will the cities affected by any changes. Even with regard to this issue, the concepts of social marketing, the nonprofit marketing, and the notion that the image to which individuals associate a product with poorly contoured outlined, as the cities seen as a product are they were also major drivers, since it showed that the marketing did not enjoy just to make a profit, but also to creating initiatives that define the main objective of the common product. There is a consensus among many authors to classify the urgency to meet the needs of various types of consumers as essential, whether businessmen, tourists or residents, both current and potential, through social processes and management guidance policy (Ribeiro and Sousa, 2016). This study had several limitations. First, it focuses on the opinions of a small number of people. This group should be expanded to include tourists, mass media representatives, investors, and other important stakeholders. Secondly, historical events attract a specific audience. Therefore, identification of the city's image attributes is necessary.

2.5 CONCLUSIONS AND NEXT STEPS

The present study highlights the importance of strategic planning at the territorial marketing level as a way of developing a particular destination, and in this case, as a way of developing the Braga Romana event. It was found that in addition to having a good development strategy and targeting a large number of tourists, it is still essential to think about various strategies so that both residents, tourists and visitors feel interconnected. Through this study, it was possible to find that residents like and seek to be part of this event and somehow they are proud to be able to convey this passage of Roman traditions that can be passed on the various days of the event, however, there are some concerns that have such as communication, transport, and even traffic infrastructure. During the focus group interview, the importance of a development strategy plan for the event was evident and several suggestions were mentioned by the interviewees. Regarding the issue of the importance of a strategy plan, the focus group members considered that it is important to define and prepare a strategic plan to improve the dissemination and communication of the event. The interviewed elements also consider that it is important to have a translation of the programs for foreigners, since the event already has a very large tourist dimension and mention that this implementation does not have a paper-based indicator, mainly because, nowadays, technologies are already an important and central means of publicizing a particular product. Thus, the interviewees conclude that one of the strategies to be applied for better dissemination of the event would be the creation of an application related to Braga Romana. They also suggest that it does not necessarily need to be only used during the event, but also for anyone visiting the city at other times of the year, as it is a city with great Roman impact. At the second stage (focus group) of the project, we analyze the data results obtained through the focus group. Initially, a general characterization is made, followed by an individual characterization of each of the participants involved in the data collection. The four subpoints analyzed constitute the main themes of analysis and result from content analysis and data interpretation.The focus group consists of nine elements, that is, six female elements and three male elements, with the participation of three young people aged between 11 and 26 and 6 elements aged between 44 and 63 years. Regarding the academic and professional level of the participants, two are students and seven are in the labor market, distinguishing themselves in quite different areas of work. However, the common feature among the interviewees is the fact that they attended and experienced the Braga Romana event. As a future approach, we believe that it is important to implement a case study of a quantitative nature to know the opinion of a larger number of residents and even to perceive the opinion of those who visited the event, in order to understand if they were well received and how they were aware of the event. Also as a way of improving and recognizing the city of *Bracara*—Roman city, it would be important to carry out a project to develop a digital application, as a way of spreading the various local monuments and even the customs and traditions of the Roman past that once lived on the city of Braga. This work shows how event planners in Braga developed a means to evaluate value equity among business and leisure travelers. We intend, in a second stage, to present the results of group comparisons in structural equation modeling, to demonstrate that event planners need to consider the classificatory status of their target customers,that is, as business or leisure travelers—because both groups assess aspects of a venue's value equity actions similarly and differently. Future research should reveal several potentials of individual destination attributes to cause satisfaction and/or dissatisfaction, and it shed light on the most determinant and critical attributes in explaining the overall tourist experience, in specific, the case of events and local development.

KEYWORDS

- **Braga Romana**
- **strategic plan**
- **territorial marketing**
- **tourism marketing**

REFERENCES

Almeida, C. O Marketing das cidades. Instituto Universitário de Desenvolvimento e Promoção Social, Pólo de Viseu do Centro Regional das Beiras da Universidade Católica Portuguesa, 2004.

Anholt, S. *Competitive Identity: The New Brand Management for Nations, Cities and Regions*; Palgrave Macmillian: New York, 2007.

Avraham, E.; Ketter, E. *Media Strategies for Marketing Places in Crisis*; Butterworth-Heinemann: Oxford, 2008.

Bogdan, R.; Biklen, S.; *Investigação Qualitativa em Educação—uma introdução à teoria e aos métodos*; Porto Editora: Porto, 1994.

Chalip, L.; Green, B. C.; Hill, B. Effects of Event Media on Destination Image and Intention to Visit. J. Sport Manage. **2003,** *17*, 214–234.

Chen, C. F.; Phou, S. A Closer Look at Destination: Image, Personality, Relationship and Loyalty. *Tour. Manage.* **2013,** *36*, 269–278.

Douglas, N.; Douglas, N.; Derrett, R. *Special Interest Tourism: Context and Cases*; John Wiley & Sons: Australia.

Getz, D. Residents Attitudes towards Tourism, a Longitudinal Study in Spey Valley, Scotland. *Tour. Manage.* **1994,** *15* (4), 247–258.

Getz, D. *Event Management and Event Tourism*, 2nd ed.; Cognizant Communication Corporation: New York, 2005.

Getz, D. *Event Tourism: Definition, Evolution, and Research. Tour. Manage. 2008*, 29, 403–428.

Graham, S.; Neirotti, L. D.; Goldblatt, J. J. *The Ultimate Guide to Sport Marketing*, 2nd ed., McGraw-Hill: New York, 2001.

Guba, E.; Lincoln, Y. Naturalistic and Rationalistic Enquiry. In *Education Research, Methodology, and Measurement: An International Handbook*, 1998; pp 81–85.

Hospers, G. J. Place Marketing in Europe: The Branding of the Øresund Region. *Interecon. Rev. Eur. Econ. Pol.* **2004,** *39* (5), 271–279.

Isidoro, A. M. *Manual de Organização e Gestão de Eventos*. Edições Silabo, 2013.

Kanter, R. M. *World Class-Thriving Locally in the Global Economy*; Prentice-Hall, 1995.

Kanter, R. M. *A Vantagem Competitiva das Cidades, Executive Digest*; Janeiro, 1996; pp 52–59.

Kavaratzis, M. From City Marketing to City Branding: Towards a Theoretical Framework for Developing City Brands. *Place Brand.* **2004,** *1* (1), 58–73.

Kavaratzis, M. Cities and Their Brands: Lessons from Corporate Branding. *Place Brand. Public Diplomacy* **2009,** *5* (1), 26–37.

Keller, K. Conceptualizing, Measuring, and Managing Customer-Based Brand Equity. *J. Market.* **1993,** *57* (1), 1–22.

Kim, S. S.; Petrick, J. F. *Residents' Perceptions on Impacts of the FIFA 2002 World Cup: The Case of Seoul as a Host City. Tour. Manage. 2005*, *26*, 25–38.

Kotler, P.; Armstrong, G. *Principles of Marketing*, 14th ed.; Prentice Hall: New Jersey, 2011.

Kotler, P.; Haider, D.; Rein, I. *Marketing Places. Attracting Investment, Industry, and Tourism to Cities, States, and Nations*; Free Press: Nova Iorque, 1993.

Lambin, J. J. *Marketing Estratégico*, 4ª Edição; McGraw-Hill: Lisboa, 2000.

Lubowiecki-Vikuk, A.; Basińska-Zych, A. Sport and Tourism as Elements of Place Branding. A Case Study on Poland. *J. Tour. Chall. Trends* **2011,** *IV* (2), 33–52.

Martinez, N. M. City Markting and Place Branding: A Critical Review of Practice and Academic Research. *J. Town City Manage.* **2011,** 2 (4), 369–394.

Miragaia, D. A. M.; da Costa, C. D.; Ratten, V. Sport Events at the Community Level: A Pedagogical Tool to Improve Skills for Students and Teachers. *Educ. Train.* **2018,** 60 (5), 431–442.

Morgan, N.; Pritchard, A.; Pride, R. *Destination Brands: Managing Place Reputation*; Butterworth-Heinemann: Oxford, 2011.

Pedro, F.; Caetano, J.; Christiani, K.; Rasquilha, L. *Gestão de Eventos*; Escolar Editora: Lisboa, 2012.

Pike, S. D. Destination Brand Performance Measurement over Time: Tracking Consumer Perceptions of a Competitive Set of Destinations over a 10 Year Period. *Acta Turistica* **2015,** *27* (2), 135–164.

Ratten, V. Sport Innovation Management: Towards a Research Agenda. *Innovation* **2016,** *18* (3), 238–250.

Ratten, V.; Tsiotsou, R.; Ko, Y. J.; Kim, Y. K.; Kim, M. K.; Lee, J. H. The Role of Involvement and Identification on Event Quality Perceptions and Satisfaction. *Asia Pacific J. Market. Logistics* **2010**.

Recommendations on Tourism Statistics. United Nations and World Tourism Organization: New York, 1994.

Ribeiro, I.; Sousa, B. City Marketing and Tourist Events: An Exploratory Study of Residents Regarding the Carnival of Ovar. *Int. J. Sci. Technol. Res.* Oct **2016,** *5* (10), 18–21. ISSN 2277-8616.

Rola, M.; Malheiro, A.; Sousa, B. O papel da imagem dos eventos culturais na construção da marca de um destino turístico: O caso do evento "Braga Romana". *Dos Algarves: A Multidisciplinary e-Journal* **2018,** *33*, 1–21. doi: 10.18089/DAMeJ.2018.33.1 ISBN 2182-5580

Ruyi, X.; Pegg, S. Exploring the Motivations for Participant Engagement in the Australian University Games. *Int. J. Event Manage. Res.* **2007,** *3* (2), 39–46.

Sousa, B.; Ribeiro, I. City Marketing e os eventos: Um estudo de caso aplicado ao Carnaval de Ovar. *Eur. J. Appl. Busi. Manage.* **2018,** Special Issue, 73–84.

Sousa, B.; Silva, M. Creative Tourism and Destination Marketing as Asafeguard of the Cultural Heritage of Regions: The Case of Sabugueiro Village. *Revista Brasileira de Gestão e Desenvolvimento Regional, G&DR.* **2019,** *15* (5), 78–92, set-dez/2019. Taubaté, SP, Brasil. ISSN: 1809-239X.

Sousa, B.; Simões, C. Comportamento e perfil do consumidor de turismo de nichos. *Tékhne-Revista de Estudos Politécnicos* **2010,** *14*, 137–146. ISSN: 1645-9911.

Sousa, B. M.; Rocha, A. T. A afetividade na gestão dos territórios e na fixação da população: uma abordagem ao contexto transmontano português. *Revista Brasileira de Gestão e Desenvolvimento Regional* **2019,** *15* (1), 2–18.

Tasci, A. D. A.; Kozak, M. Destination Brands vs. Destination Images: Do We Know What We Mean? *J. Vacation Market.* **2006,** *12*, 299–317.

Tideswell, C.; Faulkner, B. A Framework for Monitoring Community Impacts of Tourism. *J. Sustain. Tour.* **1997,** *5* (1), 3–28.

Tsiotsou, R. H.; Ratten, V.; Rosenbaum, M. S.; Wong, I. A. Value Equity in Event Planning: A Case Study of Macau. *Market. Intell. Plan.*, **2010**.

Wahyuni, D. The Research Design Maze: Understanding Paradigms, Cases, Methods and Methodologies. *J. Appl. Manage. Acc.* **2012,** *10* (1), 69–80.

Zitta, C. Organização de eventos: da ideia à realidade, 5. ed.; Senac: Brasília, DF, 2014.

CHAPTER 3

MUSEUM VISITOR EXPERIENCE IN SRI LANKA (SPECIAL REFERENCE TO THE DEPARTMENT OF NATIONAL MUSEUMS)

J. A. R. C. SANDARUWANI[1,*] and ATHULA GNANAPALA[2]

[1]*Department of Tourism Management, Sabaragamuwa University of Sri Lanka, Belihuloya, Sri Lanka*

[2]*Department of Tourism Management, Sabaragamuwa University of Sri Lanka, Belihuloya, Sri Lanka*

[*]*Corresponding author. E-mail: rudmi@mgt.sab.ac.lk*

ABSTRACT

Tourists are eager to experience and learn about new cultures in their visit. Museums are the ideal arena in which to explore the cultural and historical timeline of the destination visiting, which housed for the repository of cultural and historical monuments and artifacts. It is required to understand visitor expectations and experience in order to offer them a better service. This chapter looks into the international visitor experience at museum encounters; reception/ticket office, museum galleries and exhibits, service staff, museum stores/souvenir selling area, surrounding of the museum, and tourist convenience and comfort especially referring to the eight museums administered by the Department of National Museums in Sri Lanka. The data were collected in between January and July 2019

Tourist Behavior: Past, Present, and Future. Narendra Kumar, PhD, Bruno Barbosa Sousa, PhD & Swati Sharma, PhD (Eds.)

through a questionnaire survey where it ended up with 356 usable and completed questionnaires using the stratified random sampling technique. The data were analyzed using SPSS 22.0 for the mean ranking analysis and bivariate analysis. The mean ranking analysis of survey data shows that the visitors to the museums of the Department of National Museums have immensely satisfied with the museum galleries/exhibits and the service staff and dissatisfied with the tourist convenience and comfort facilities within the museums. The correlation analysis between the visitor experience variables and revisit intention shows that "museum galleries/ exhibits" ($r = 0.491$), "museum staff" ($r = 0.491$) have a moderate positive relationship with museum revisit intension and "museum souvenir selling area" ($r = 0.190$) represents a poor positive relationship with museum revisit intension. The correlation analysis to measure the relationship between the museum variables and the tourists' recommendation, results in the highest positive relationship between "surrounding of the Museum" and the "Recommendation" ($r = 0.619$, $P = 0.000$) which derives the meaning that the majority of the museum visitors have recommended for potential visitors as a result of the better surrounding of the museums. The lowest relationship shows between "Museum souvenir selling area" and "Recommendation" ($r = 0.163$, $P = 0.046$). Other variables represent moderate positive relationships with "recommendation." These results finally conclude that the Department of National Museums should more focus on enhancing visitor convenience and comfort facilities to attract more foreign visitation. That basic would then create space for tourism suppliers to promote museum tourism in Sri Lanka.

3.1 INTRODUCTION

In the tourism evolutionary journey around the world over the past couple of decades, we have been continually impressed that cultural tourism has the greatest demand even though it is the oldest of the alternative tourism phenomena. Cultural tourism has recently been reaffirmed by the UNWTO as a major element of international tourism consumption, accounting for over 39% of tourism arrivals (UNWTO, 2018).

Travel involves a cultural element by its very nature as traveling itself removes tourists from their home culture and places them temporarily in a different cultural setting. But cultural tourism is seen as offering something more or different both to the tourist and the community that hosts the tourist. Tourism researchers realized "cultural tourists" as some people traveled specifically to gain a deeper understanding of the culture or heritage of a destination (Tighe, 1986).

In understanding the concept of "cultural tourism," McKercher and Du Cros (2002) identified four different types of cultural tourism definitions by reviewing the number of definitions given by different scholars. They are (1) tourism derived definitions, (2) motivational definitions, (3) experiential/conceptual definitions, and (4) operational definitions. Tourism-derived definitions place cultural tourism within a broader framework of tourism and tourism management theory. For example, cultural tourism recognized as a form of special interest tourism (McIntosh and Goeldner, 1990; Zeppel, 1992; Ap, 1999); interrelationships between people, places, and cultural heritage (Zeppel and Hall, 1991); temporary movement of people (Richards, 1996c); and as well as a business perspective (Goodrich, 1997). Motivational definitions emphasize that motivation must be considered an important element when defining cultural tourism (Richards, 1996b; WTO, 1985; Silberberg, 1995). Experiential or aspirational definitions focus on cultural tourism as it involves experiencing or having the contact of differing intensity with the unique social fabric, heritage, and special character of places (TC, 1991; Blackwell, 1997; Schweitzer, 1999) where tourists become educated as well as entertained (VICNET, 1996), learn about the community (IOCCA, 1997), or significance of a place and its associations with the local community, its heritage, and a cultural or natural landscape (AHC and TCA, 1999). Finally, operational definitions highlight cultural tourism as participation in any one of almost limitless arrays of activities or experiences. The tourism literature identifies the range of cultural tourism activities as including the use of such cultural heritage assets as archaeological sites, museums, castles, palaces, historical buildings, famous buildings, ruins, art, sculpture, crafts, galleries, festivals, events, music and dance, folk arts, theatre, "primitive cultures" subcultures, ethnic communities, churches, cathedrals and other things that represent people and their cultures (Richards, 1996a; Goodrich, 1997; Miller, 1997; Jamieson, 1995).

Within the cultural sector, museums play an important role in the transmission of knowledge, learning, and understanding of cultural identities locally and internationally. Researchers conducted in France, the UK, and the US point out "museums" are a constituent part of the cultural tourism supply being the second most popular attractions after "historic monuments and sites" (EIU, 1993). McManus (2016) stated that over 50% of cultural tourists have visited a museum at some stage during their stay in a destination. Therefore, it's obvious "museums" kept their classic role serving for "mass" or "cultural" tourists for many years.

The International Council of Museums (ICOM) is an international organization of museums comprised of 44,686 museum professionals in over 138 countries who voice on the international stage and raises public cultural awareness through global networks and co-operation programmes. The ICOM definition is recognized as the backbone of ICOM undergoing continuous revisions. The latest version is from 2007: defined the museum as a non-profitable, permanent institution in the service of society and opened to the public, which acquires, preserves, studies, exhibits, and disseminates the tangible and intangible heritage of mankind for study, education, and recreation purposes (ICOM, 2019).

In its traditional nature, museums focus on their collection, preservation, research, and educational aspects. But currently, museums are undergoing several changes as a repercussion of their histories and professional developments, and new social and cultural contexts (Trotter, 1998). In recent years, museums compete with other museums and also other leisure and educational institutions (Bergadaà, 2005; Lehn, 2006). Therefore, museums now focus more on attracting visitors. This situation creates a shift from traditional "curators' dictation" to "visitor orientation."

There are several studies in the literature about museum visitor experiences and satisfaction but the researchers could observe the limited literature done relevant to Sri Lankan museums. This literature gap influenced the researchers to conduct this study. Sri Lanka has more than hundreds of museums administered by different state and private institutions. To narrow down this study, the researchers focused only on museums administered by the Department of National Museums (DNM) as it represents the pioneer state organization responsible for maintaining the National Museums in Sri Lanka. The museums administered by the Department of National Museums are scattered around the country dedicated to different themes; National Museums, Natural History Museum, Folk Museum, Marine Museum, Dutch Museum, Memorial Museum, Heritage Museum, Ancient Technology Museum, where they exhibit natural, ethnical, cultural, and heritage and religious diversity compacted in the nation. DNM museums attract foreign travelers to a limited extent and their role in tourism cannot be overemphasized because DNM museums were and still remain in attracting 1.7–3.8% of foreign tourists from the total arrivals to the country (See Fig. 3.1).

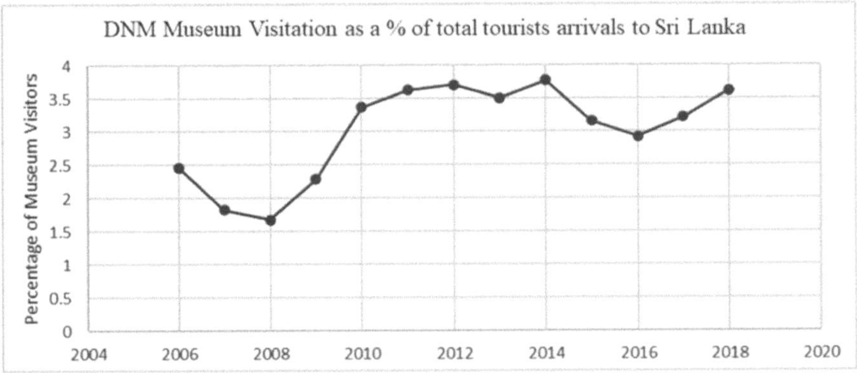

FIGURE 3.1 Museum visitation as a percentage of the total arrivals in Sri Lanka.
Source: Calculated based on SLTDA annual reports (2006–2018).

The less visitation is not a surprise as Weerasingha (2015) identified that a considerable amount of travel agencies have not included museums in their tour packages and mostly 90% of agencies promote Colombo National Museum (administered by the DNM) and it is only 20% of agencies have included the museum as a symbol of promoting their tour packages.

Based on the preliminary literature analysis, the researchers were directed toward the research question of "What is the level of the foreign visitor experiences in Sri Lankan national museums and how does visitor experience affects tourists satisfaction, WOM recommendation and revisit intention with special reference to the museums administered by the Department of National Museums?"

3.2 LITERATURE REVIEW

3.2.1 MUSEUM VISITOR EXPERIENCE

Museums compete for visitors with other recreational and educational facilities (Bergadaà, 2005; Lehn, 2006) and research into the museum visitor experience has exploded in popularity in recent years (Goulding, 2000; Kelly, 2004).

According to studies on museum visitor activity, museum visits are no longer just "informative cultural visits," but rather "experiences" (Falk and Dierking, 1992; Prentice et al., 1997; Kawashima, 1998; Goulding, 1999; Thyne, 2001; Bollo, 2004; Hume et al., 2006; Rojas and Camarrero, 2006; Slater, 2007). "Experiences" are fun, entertaining, and unforgettable moments from the consumer's point of view (Oh et al., 2007). Museum marketing strategies form the messages regarding the artifacts and experiences offered by museums within the marketing sector (Caldwell, 2000). The visitor's experience becomes a crucial factor in measuring satisfaction and museum visitors seek "total experience" that includes recreation, entertainment, education, and social interaction (Rojas and Camarrero, 2006).

The interactive experience model introduced by Falk and Dierking (1992) shows that the 'experience' is an interplay of three contexts: personal, social and physical. The personal context includes motivations, expectations, interests, beliefs, prior knowledge, and experiences. The social context includes social interactions within groups and between visitors and staff, and the physical context includes institutional rules, architecture, layout, activities, and facilities. Each museum visitor's experience is unique since they create their own interpretations within museums. The museum experience is subjective due to the visitors' different expectations (Chang, 2006), past museum encounters (Chang, 2006; McCarthy and Ciolfi, 2008), level of perceptual skills (Chang, 2006), lifestyles (Tucker, 1991), and cultures (McCarthy and Ciolfi, 2008).

Various researchers have come up with reasons that motivate visitors to visit museums. Dragicevic et al. (2012) have made many attempts to explain factors that affect visitors to museums and art galleries, including the museum's appearance, curiosity, education, entertainment, the value of its collections, price, and self-reflection and experiencing something new and also Kelly (2009); Poria et al. (2004) and Falk (1998), illustrated what tourists want from museum and art galleries experience. Museums are one-of-a-kind learning environments, also referred to as "free-choice" learning environments for visitors. Kelly (2009) revealed that factors such as an exciting location, search for knowledge and education, and cultures encourage tourists to visit museums. According to Kamal and Pramanik (2015), there are seven (7) reasons that strongly attract visitors to Bangladeshi museums, for example, motivation, service, advertising, learning, aesthetic, archeological, and cultural values. Museums succeeded in attracting visitors irrespective of their ages through innovative and high-quality exhibits, fascinating collections, and year-round family programs, and special events (NMDC, 2014). Visiting or not-visiting a museum is determined by two psychological elements: cognitive elements and effective elements. Benediktsson (2004) examined how socio-demographic factors affect museum visits, as well as people's attitudes toward museums. In summary, previous research studies have shown that visitor behavior is guided by motivational factors (Brida et al., 2013; Mengich, 2011; Niemela and Titta, 2010; Williams, 2009; McIntyre, 2007); psychological factor (McIntyre, 2007; Packer and Ballantyne, 2002); physiological factor, and learning factors (Williams, 2009; McIntyre, 2007).

3.2.2 MUSEUMS VISITOR SATISFACTION

Customer satisfaction is defined as the customer's assessment of a product or service in terms of whether it fulfilled the customer's needs and expectations. Consumer dissatisfaction is expected if customer's needs and expectations are not met (Zeithaml et al., 2009). Visitor satisfaction is influenced not only by the technical aspect of the product, but also by the emotional, cognitive, and social values associated with it (Caldwell, 2002; Martín-Ruiz et al., 2010; Rowley, 1999). Affective and emotional aspects of service experience and image (being with visitors, experiencing human interactions, etc.) have a greater impact on a visitor's image of a museum (quality, general appearance, museum shop, etc.) (Gil and Ritchie, 2009). Jeong and Lee (2006) demonstrated that a visitor's emotional affect is determined by the museum's environmental qualities, which serve as a mediating variable influencing their satisfaction. According to McCall and Weber (2006), it is critical for the museum and its administration to keep track of institutional characteristics such as the architectural environment, programming, employee relationships, and a variety of other factors that affect the visitor's experience. According to their observations, the biggest influence on museum occupancy patterns was the parking fee and museum access.

According to Mattila and Wirtz (2001), museum's physical context has a significant effect on visitor experiences. Further, they elaborated that certain museum contexts (specially music, scent, lighting, and customer service) played a role in promoting either a positive or negative perception (Mattila and Wirtz, 2001). Jeong and Lee (2006) examine three types of environmental characteristics: exhibition environment, ambient environment, and museum size. The emotional effect is positively affected by exhibition environment (exhibition techniques, museum content, lighting, and resting areas) and negatively influenced by the ambient environment (density of visitors, noise, complexity of circulation, and thermal comfort). The exhibition environment in a museum setting, the labels, and the design have been shown to significantly impact the visitor experience at a museum. In the other hand, size of the museum (both the overall size of the display space and the overall time spent there) has the greatest impact on fatigue (Jeong and Lee, 2006). In aligned with this, Milman (2009) found that different services, as well as comfort factors (i.e., safety, food service attributes), staff behavior, prices, and several other factors were critical in establishing the correlation of visitor experiences to museum attributes. Milman (2009) highlight that the design of exhibit labels should be easy to read and understand, fulfill the visitor's learning objectives, and also enhance visitor satisfaction via added visitor experience. The same research has explored how the arrangement of artifacts, as well as the layout used in exhibition rooms or galleries, affect the visitors' experience in a museum (Milman, 2009). According to Packer (2008), directional aids in the museum setting, facilitate easy navigation within the museum. Museums are well-positioned to meet a variety of customer demands due to their distinctive features, rich connotations, convenient opening hours, numerous exhibition facilities, shops, and restaurants (Stephen, 2001). Even though the main function of a museum is to offer learning and educational opportunities, it is not always the only motivator for the visitors. Some visitors come to take a break from their routine life. The visitor experience is also enhanced through other features, that is, entertainment (Milman, 2009). Nowadays, visitors want to experience along with having fun, socializing with other people, and exploring every aspect of their fascination (Packer, 2008). Kotler and Kotler (2000) also highlighted that the museum entertainment encourages repeat visitation, while at the same time attract customers to their exhibits (Kotler and Kotler, 2000).

3.3 METHODOLOGY

The Department of National Museums reflects its existence to protect the country's culture and natural heritage of resources scientifically and systematically and their presentation for the evaluation, education, and entertainment of the public. DNM administer altogether eleven museums including Colombo National Museum, Colombo Natural History Museum, Kandy National Museum, Ratnapura National Museum, Galle National Museum, Galle Maritime Museum, Dutch Museum at Fort, Anuradhapura Folk Museum, Independence Memorial Museum at Colombo, Ruhunu Heritage Museum at Magampura and Ancient Technology Museum at Polonnaruwa. These museums are to show the cultural and natural history, heritage features that are specific to those areas.

The researchers conducted a questionnaire survey to understand the foreign visitors' mindset toward visitor experience on different museum visitor encounters; reception/ticket office, educational entertainment, responsiveness and the quality of the service staff, museum stores/souvenir selling area, surrounding of the museum, and tourist convenience and comfort. The questionnaire was developed based on the selected literature and was personally administered in English.

The targeted population was all foreign visitors to the museums administered by the DNM during the study period of January to July 2019. The researchers had to exclude the National Museum of Rathnapura and Colombo Dutch Museum as they are closed for renovations from 2017 up to date. The forecast represents 76,597 foreign visitors in total and the calculation shows it is required to collect data from the sample of 382 foreign museum visitors for this study (Krejcie and Morgan, 1970).

The choosing of sample size is based on Krejcie and Morgan (1970) in their article "Determining Sample Size for Research Activities" (Educational and Psychological Measurement, #30, pp. 607–610). It is explained further with the following equation;

$$n = \frac{X^2 \times N \times P(1-P)}{(\mathrm{ME}^2(N-1)) + (X^2 \times P \times (1-P))}$$

Where:

n = sample size

X^2 = Chi-square for the specified confidence level at 1 degree of freedom

N = population size

P = population proportion (0.50 in this table)
ME = desired margin of error (expressed as a proportion)

The total sample of 382 was divided among the currently operating museums of the DNM in aligned with proportionate stratified random sampling technique and the sample for each museum shows in the following Table 3.1.

TABLE 3.1 Sample Calculation of Museum Visitors.

Museums	Forecasted arrivals in 2018 (Population)	Percentage	Sample
Colombo National Museum	61,532	80.3	307
National Museum of Natural History	6139	8.0	31
Kandy National Museum	1955	2.6	10
Ratnapura National Museum	(Closed for Renovations)		
Galle National Museum	2557	3.3	13
Galle Maritime Museum	2942	3.8	15
Anuradhapura Folk Museum	192	0.3	1
Dutch Museum	(Closed for Renovations)		
Independence Memorial Museum	1246	1.6	6
Hambanthota National Museum	34	0.0	0
Total	76,597	100	382

Source: Developed by the researchers (2019).

But the study has to be limited up to 356 usable and complete questionnaires as some of the visitors have not properly completed the whole questionnaire.

Quantitative data collected through the satisfaction survey were analyzed through manually editing and computer processing. The effective questionnaires were given unique data codes, filed, and encoded into the computer database and computer processing was done through the use of SPSS 22.0 statistical software as the data analytical tool. Afterward, the analysis was done with a mean value analysis to identify the tourists' level of satisfaction on museum encounters, that is, museum reception/ticket office, educational entertainment, responsiveness and the quality of the service staff, museum stores/souvenir selling area, surrounding of the museum, tourist convenience, and comfort facilities. Further, the study was done with a mean ranking analysis to identify the individual attributes that would give tourists the highest level of experience and the lowest level of experience. The researchers run a correlation analysis to understand how each encounter variable is related to overall museum visitor satisfaction, revisit intention, and WOM recommendation (Fig. 3.2 and Table 3.2).

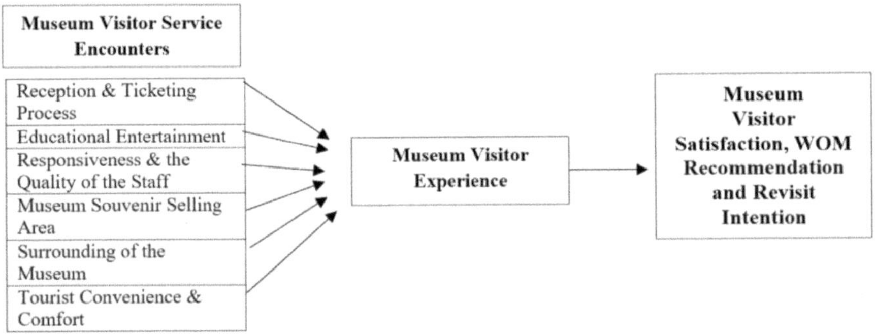

FIGURE 3.2 Conceptualization of museum visitor satisfaction.
Source: Developed based on literature (2019).

TABLE 3.2 Operationalization.

Dimension	Indicator	Q. Num	Measurement
Museum visitor experience			
Reception/ ticketing process	The ticketing process is simple	14.1.1	Likert scale
	Friendly and sensitive personnel	14.1.2	
	Ticketing office personnel provides informative documents such as brochures, catalogues, and display maps at the entrance of the museum	14.1.3	
	Computerized system of tourist information	14.1.4	
	Warnings of possible problems and potential danger	14.1.5	
	Accessibility of information about forbidden or limited behavior	14.1.6	
Educational entertainment	The displayed products are educational	14.2.1	Likert scale
	The content of the museum is attractive	14.2.2	

TABLE 3.2 *(Continued)*

Dimension	Indicator	Q. Num	Measurement
	The mode of display can enhance the understanding of the displayed products	14.2.3	
	New things can be learnt from this museum visit	14.2.4	
	The information appealed to my emotions	14.2.5	
Responsiveness and the quality of the service staff	The service staff is friendly and sensitive	14.3.1	Likert scale
	The service staff did not ignore the needs of tourists although they are busy	14.3.2	
	The service staff has good attitude in treating problems for tourists	14.3.3	
	The representative officer has rich professional knowledge	14.3.4	
	The mode of narration by the representative officer is easy to understand	14.3.5	
	The personal appearance of the staff is good	14.3.6	
	The service desks provide good consulting services for tourists	14.3.7	
Museum stores/ souvenir selling area	The products sold from the museum stores are interesting and educational	14.4.1	Likert scale
	The products sold from the museum stores are quality products	14.4.2	
	Availability of a wide range of choice	14.4.3	
	Available items are reasonably priced	14.4.4	
Surrounding of the museum	The overall landscape design is spacious and pretty	14.5.1	Likert scale
	The overall environment is clean and comfortable	14.5.2	
	The parking space is sufficient	14.5.3	
Tourist convenience and comfort	The opening time meets the needs of the tourists	14.6.1	Likert scale
	The exhibition area has sufficient chairs around for a rest	14.6.2	

TABLE 3.2 *(Continued)*

Dimension	Indicator	Q. Num	Measurement
	The lighting for the museum area is appropriate	14.6.3	
	The air conditioning is comfortable	14.6.4	
	Signs facilitative orientation and movement	14.6.5	
	Have enough sanitation facilities	14.6.6	
Satisfaction	My overall satisfaction with the museum experience is	14.7	Likert scale
Revisit intension	I plan to visit this museum again in the future	14.8.1	Likert scale
	I hope to visit the other National Museums in Sri Lanka as well	14.8.2	
W-O-M recommendation	I recommend this museum to others	14.9.1	Likert scale
	I encourage my friend or relatives to come to this museum	14.9.2	
	I take pride in telling other people about my experiences in the museum	14.9.3	

According to Sekaran (2010), reliability indicates the stability and consistency with which the instrument measures the concept. For testing the internal consistency of the instruments, Cronbach's coefficient alpha was used. All the alpha values were above the rule of thumb of 0.7 for a reliable scale; Hence, Table 3.3 shows the study of all the dimensions in independent and dependent reliability in an acceptable level.

TABLE 3.3 Reliability Statistics.

Cronbach's alpha	No. of items
0.847	8

Source: Survey Data (2019).

Sekeran (2010) says that the researchers should test the construct and content test as the least requirement.

The convergent test testifies whether there is a high correlation between when two similar measures reveal the same results. The convergent validity test accepted when the factor loadings are equal to or larger than 0.5 will be statistically significant (Hair et al., 1998). The study convergent validity was confirmed. The results are shown in Table 3.4. KMO is a test that measures the adequacy of the index in which examines the factor appropriateness. The value of 0.856 reveals the generalizability of the sample to the population (Table 3.5).

TABLE 3.4 KMO and Bartlett's Test.

Kaiser–Meyer–Olkin measure of sampling adequacy		0.856
Bartlett's test of sphericity	Approx. Chi-square	552.614
	Df	28
	Sig.	0.000

Source: Survey Data (2019).

TABLE 3.5 Analysis of Factors for the Actual Museum Experience.

Item	Factor loading extraction	Feature value
RTO1	0.854	
RTO2	0.859	
RTO3	0.768	
RTO4	0.842	0.634
RTO5	0.807	
RTO6	0.863	
EE1	0.790	
EE2	0.784	
EE3	0.771	0.873
EE4	0.725	
EE5	0.642	
RQSS1	0.717	
RQSS2	0.844	
RQSS3	0.843	
RQSS4	0.856	0.873
RQSS5	0.858	
RQSS6	0.776	
RQSS7	0.701	

TABLE 3.5 *(Continued)*

Item	Factor loading extraction	Feature value
MS1	0.875	
MS2	0.902	0.820
MS3	0.821	
MS4	0.894	
SM1	0.756	
SM2	0.737	0.478
SM3	0.702	
TCC1	0.591	
TCC2	0.685	
TCC3	0.819	0.666
TCC4	0.696	
TCC5	0.812	
TCC6	0.667	

Source: Survey Data (2019).

3.4 DATA ANALYSIS

3.4.1 DEMOGRAPHIC CHARACTERISTICS OF RESPONDENTS

This section of the study gives an overview of the respondents' profile of those who have visited the museums of the Department of National Museums and volunteered in participating in the satisfaction survey. The demographic characteristics of the respondents consist of visitor origin, profession, gender, age, marital status, travel party, source of information, and the influencing factor for museum visitation as all indicated in the details of Table 3.6.

TABLE 3.6 Demographic Profile.

		N	%
Visitor origin	British	47	13.2
	Chinese	44	12.4
	Indian	43	12.1
	Australian	36	10.1

TABLE 3.6 *(Continued)*

		N	%
	American	22	6.2
	German	20	5.6
	Dutch	18	5.1
	French	15	4.2
	Italian	12	3.4
	New Zealander	9	2.5
	Other	90	25
Profession	Student	45	12.6
	Teacher	39	11.0
	Retired	25	7.0
	Doctor	24	6.7
	Engineer	11	3.1
	Nurse	11	3.1
	Researcher	11	3.1
	Banker	8	2.2
	Businessman	7	2.0
	Consultant	7	2.0
	Other	168	47.2
Gender	Male	185	52.0
	Female	171	48.0
Age	Less than 20 years	11	3.1
	20–29	96	27.0
	30–39	86	24.2
	40–49	66	18.5
	50–59	53	14.9
	60 and over	44	12.4
Marital status	Married	160	44.9
	Single, yet not married	165	46.3
	Single, widowed	12	3.4
	Single, divorced	19	5.3

TABLE 3.6 *(Continued)*

		N	%
Travel party	A couple	107	30.1
	Alone	96	27.0
	Friends and relatives	70	19.7
	Family members	60	16.9
	Organized group	23	6.5
Sources of information	Internet	125	35.1
	Travel guide books	101	28.4
	Friends/relatives	57	16
	Travel agent	33	9.3
	Previous visit experience	14	3.9
	Travel brochures	5	1.4
	Trade fairs	2	0.6
	Tour operator	2	0.6
	Airline	2	0.6
	SL tourist board	2	0.6
	Other	13	3.7
Main influencing factor for museum visitation	Learning and discovery	231	64.9
	Passive enjoyment	66	18.5
	Social interaction	16	4.5
	Self-fulfillment	29	8.1
	Test of differences within sites	10	2.8
	Restoration	3	0.8

3.4.2 MEAN RANKING ANALYSIS OF FOREIGN VISITORS' MUSEUM EXPERIENCE

This study measured the level of foreign museum visitors' experience at different museum service encounters and Table 3.2 summarizes the mean ranking analysis done to identify how different service encounters fall within the range of the highest to the lowest level of experience.

TABLE 3.7 Mean Ranking Analysis.

Item		Mean	Std. deviation	Ranking
14.1 Reception/ticketing				
14.1.1 The ticketing process is simple	RTO1	4.29	0.851	6
14.1.2 Friendly and sensitive personnel	RTO2	4.35	0.808	3
14.1.3 Provides informative documents	RTO3	3.42	1.060	23
14.1.4 Computerized system of tourist information	RTO4	2.88	0.888	30
14.1.5 Warnings of possible problems and potential danger	RTO5	3.47	1.178	22
14.1.6 Accessibility of information about forbidden or limited behavior	RTO6	3.86	1.016	14
14.2 Educational entertainment				
14.2.1 The displayed products are educational	EE1	4.32	0.972	4
14.2.2 The content of the museum is attractive	EE2	4.17	0.973	9
14.2.3 Mode of display can enhance the understanding	EE3	4.02	0.953	11
14.2.4 New things can be learnt from this museum visit	EE4	4.36	0.742	1
14.2.5 The information appealed to my emotions	EE5	3.73	0.930	16
14.3 Responsiveness and the quality of the service staff				
14.3.1 The service staff is friendly and sensitive	RQSS1	4.35	0.752	2
14.3.2 The service staff did not ignore the needs of tourists although they are busy	RQSS2	3.92	0.922	13
14.3.3 The service staff has good attitude in treating problems for tourists	RQSS3	3.96	0.924	12
14.3.4 Representative officers has rich professional knowledge	RQSS4	3.75	1.011	15
14.3.5 The mode of narration by the representative officer is easy to understand	RQSS5	3.49	1.178	21
14.3.6 The personal appearance of the staff is good	RQSS6	4.32	0.801	5

TABLE 3.7 *(Continued)*

Item		Mean	Std. deviation	Ranking
14.3.7 The service desks provide good consulting services for tourists	RQSS7	3.62	1.070	20
14.4 Museum stores/souvenir selling area				
14.4.1 The products sold from the museum stores are interesting and educational	MS1	3.34	1.036	25
14.4.2 The products sold from the museum stores are quality products	MS2	3.23	0.945	26
14.4.3 Availability of a wide range of choice	MS3	2.99	0.857	29
14.4.4 Available items are reasonably priced	MS4	3.22	0.829	27
14.5 Surrounding of the museum				
14.5.1 The overall landscape design is spacious and pretty	SM1	4.26	0.887	7
14.5.2 The overall environment is clean and comfortable	SM2	4.11	0.974	10
14.5.3 The parking space is sufficient	SM3	3.66	0.928	19
14.6 Tourist convenience and comfort				
14.6.1 The opening time meets the needs of the tourists	TCC1	4.25	0.856	8
14.5.2 The exhibition area has sufficient chairs around for a rest	TCC2	3.41	1.038	24
14.5.3 The lighting for the museum area is appropriate	TCC3	3.73	1.095	17
14.5.4 The air conditioning is comfortable	TCC4	2.82	1.224	31
14.5.5 Signs facilitative orientation and movement	TCC5	3.73	1.038	18
14.5.6 Have enough sanitation facilities	TCC6	3.20	1.048	28

Source: Survey Data (2019).

Table 3.7 represents the mean ranking analysis of the museum visitor experience. The item "New things can be learned from the museum" holds the overall highest rank with the mean value of 4.36. The lowest rank with the least mean value (2.82) holds with the item of "the air conditioning is comfortable."

In considering the reception and ticketing process, the lowest mean value holds for "computerized system with tourist information" (2.88) and comparably, the highest mean for "friendly and sensitive staff" (4.35) in the front office.

When considering the educational entertainment aspect of the museum, most visitors are happy (4.36) about the content of the museum, as they could gain new knowledge from the visit. At the same time, visitors are moderately satisfied (3.73) with content delivery as it shows the lowest mean value for "the information appealed to my emotions."

In considering the responsiveness and the quality of the staff, the visitors are highly satisfied with their experience with friendly and sensitive staff (4.35) but comparatively low mean value for "the clear and understandable mode of narration by the interpreters" (3.49).

Overall visitor experience with the museum store or the souvenir selling area is moderate as the mean analysis shows moderate mean values less than 3.45. In the analysis of the visitor experience with the surrounding of the museum, the majority appreciate (4.26) the "landscape design as spacious and pretty." Controversy, the visitors have comparatively low mean value (3.66) for "spacious parking."

Finally, from the visitor convenience and comfort factors, the visitors are positive with the museum opening times (4.25) but comparatively low mean value (2.82) for ventilation inside (Table 3.8).

TABLE 3.8 Mean Analysis.

Variables	N	Minimum	Maximum	Mean	Std. deviation
Reception and ticketing	356	1.83	5.00	3.7984	0.65987
Educational entertainment	356	2.00	5.00	4.0209	0.69080
Responsiveness and the quality of staff	356	2.00	5.00	4.0209	0.69080
Museum souvenir selling area	151	1.00	5.00	3.1937	0.88137
Surrounding of the museum	355	1.00	5.00	4.2023	0.66994
Tourist convenience and comfort	356	1.33	5.00	3.7007	0.85673

Source: Survey Data (2019).

3.4.3 BIVARIATE ANALYSIS

The projected results shown in Table 3.9 indicate the direction, strength, and significance of the bivariate relationship among independent and dependent variables.

3.4.3.1 CORRELATION BETWEEN MUSEUM EXPERIENCE AND VISITOR SATISFACTION

Table 3.9 indicates the correlation between the visitor experience variables and overall visitor satisfaction. It explains a positive relationship between museum experience variables with overall visitor satisfaction.

TABLE 3.9 Correlation Analysis-Museum Experience vs. Visitor Satisfaction.

		mean. RTO	mean. EE	mean. RQSS	mean. MS	mean. SM	mean. TCC	OS
mean. RTO	Pearson Correlation	1						
	Sig. (two-tailed)							
	N	356						
mean. EE	Pearson Correlation	0.635**	1					
	Sig. (two-tailed)	0.000						
	N	356	356					
mean. RQSS	Pearson Correlation	0.635**	1.000**	1				
	Sig. (two-tailed)	0.000	0.000					
	N	356	356	356				
mean. MS	Pearson Correlation	0.222**	0.157	0.157	1			
	Sig. (two-tailed)	0.006	0.055	0.055				
	N	151	151	151	151			

TABLE 3.9 *(Continued)*

		mean. RTO	mean. EE	mean. RQSS	mean. MS	mean. SM	mean. TCC	OS
mean. SM	Pearson Correlation	0.432**	0.468**	0.468**	0.028	1		
	Sig. (two-tailed)	0.000	0.000	0.000	0.732			
	N	355	355	355	150	355		
mean. TCC	Pearson Correlation	0.591**	0.450**	0.450**	0.376**	0.449**	1	
	Sig. (two-tailed)	0.000	0.000	0.000	0.000	0.000		
	N	356	356	356	151	355	356	
OS	Pearson Correlation	0.607**	0.587**	0.587**	0.414**	0.416**	0.628**	1
	Sig. (two-tailed)	0.000	0.000	0.000	0.000	0.000	0.000	
	N	356	356	356	151	355	356	356

**Correlation is significant at the 0.01 level (two-tailed).

Source: Survey Data (2019).

The highest positive relationship is between "Tourist Convenience & Comfort" and the "Overall Satisfaction on Museum Visitation" which shows $r = 0.628$, $P = 0.000$, significance at 95% level. The other variables of "Reception & Ticketing," "Responsiveness & the Quality of Staff," and "Surrounding of the Museum" represent a moderate positive relationship with the overall satisfaction of museum visitation. On the other hand, "Museum Souvenir" ($r = 0.414$, $P = 0.000$) shows the least positive relationship with overall museum visitor satisfaction.

3.4.3.2 CORRELATION BETWEEN MUSEUM EXPERIENCE AND REVISIT INTENTION

Table 3.10 statistically proves the correlation between the visitor experience variables and revisit intention.

TABLE 3.10 Correlation Analysis (Museum Experience vs. Revisit Intention).

		mean. RTO	mean. EE	mean. RQSS	mean. MS	mean. SM	mean. TCC	mean. RI
mean. RTO	Pearson Correlation	1						
	Sig. (two-tailed)							
	N	356						
mean. EE	Pearson Correlation	0.635**	1					
	Sig. (two-tailed)	0.000						
	N	356	356					
mean. RQSS	Pearson Correlation	0.635**	1.000**	1				
	Sig. (two-tailed)	0.000	0.000					
	N	356	356	356				
mean. MS	Pearson Correlation	0.222**	0.157	0.157	1			
	Sig. (two-tailed)	0.006	0.055	0.055				
	N	151	151	151	151			
mean. SM	Pearson Correlation	0.432**	0.468**	0.468**	0.028	1		
	Sig. (two-tailed)	0.000	0.000	0.000	0.732			
	N	355	355	355	150	355		
mean. TCC	Pearson Correlation	0.591**	0.450**	0.450**	0.376**	0.449**	1	
	Sig. (two-tailed)	0.000	0.000	0.000	0.000	0.000		
	N	356	356	356	151	355	356	
mean. RI	Pearson Correlation	0.401**	0.491**	0.491**	0.190*	0.481**	0.431**	1
	Sig. (two-tailed)	0.000	0.000	0.000	0.020	0.000	0.000	
	N	355	355	355	150	355	355	355

*Correlation is significant at the 0.05 level (two-tailed).

**Correlation is significant at the 0.01 level (two-tailed).

Table 3.10 represents the relationship between the museum experience and the revisit intension of foreign museum visitors. The highest positive relationship is between the "Educational Entertainment" (r = 0.491, P = 0.000) and "Responsiveness of the Staff" (r = 0.491, P = 0.000). Further, it highlighted that "Reception & Ticketing," "Educational Entertainment," "Responsiveness & the Quality of the Staff," "Surrounding of the Museum," and "Tourist Comfort and Convenience" have a moderate positive relationship with revisit intension. Controversy, "Museum Souvenir Selling" represents a poor positive relationship with revisit intention.

3.4.3.3 CORRELATION BETWEEN MUSEUM EXPERIENCE AND RECOMMENDATION

Table 3.11 indicates the correlation between the visitor experience variables and recommendation.

TABLE 3.11 Correlation Analysis (Museum Experience vs. Recommendation).

		mean. RTO	mean. EE	mean. RQSS	mean. MS	mean. SM	mean. TCC	mean. WOMR
mean. RTO	Pearson Correlation	1						
	Sig. (two-tailed)							
	N	356						
mean. EE	Pearson Correlation	0.635**	1					
	Sig. (two-tailed)	0.000						
	N	356	356					
mean. RQSS	Pearson Correlation	0.635**	1.000**	1				
	Sig. (two-tailed)	0.000	0.000					
	N	356	356	356				
mean. MS	Pearson Correlation	0.222**	0.157	0.157	1			
	Sig. (two-tailed)	0.006	0.055	0.055				
	N	151	151	151	151			
mean. SM	Pearson Correlation	0.432**	0.468**	0.468**	0.028	1		
	Sig. (two-tailed)	0.000	0.000	0.000	0.732			
	N	355	355	355	150	355		

TABLE 3.11 *(Continued)*

		mean. RTO	mean. EE	mean. RQSS	mean. MS	mean. SM	mean. TCC	mean. WOMR
mean. TCC	Pearson Correlation	0.591**	0.450**	0.450**	0.376**	0.449**	1	
	Sig. (two-tailed)	0.000	0.000	0.000	0.000	0.000		
	N	356	356	356	151	355	356	
mean. WOMR	Pearson Correlation	0.503**	0.558**	0.558**	0.163*	0.619**	0.498**	1
	Sig. (two-tailed)	0.000	0.000	0.000	0.046	0.000	0.000	
	N	355	355	355	150	355	355	355

*Correlation is significant at the 0.05 level (two-tailed).

**Correlation is significant at the 0.01 level (two-tailed).

Table 3.11 shows the relationship between the museum experience and the recommendation. The highest positive relationship is between "Surrounding of the Museum" and the "Recommendation" ($r = 0.619$, $P = 0.000$) which derives the meaning that the majority of the museum visitors have recommended as a result of the better surrounding of the museums. The lowest relationship shows between "Museum souvenir selling area" and "Recommendation" ($r = 0.163$, $P = 0.046$). Other variables represent moderate positive relationships with recommendation.

3.5 DISCUSSION

The researchers here evaluated the level of visitor experience encountered during the museum visit based on six variables; Reception/Ticketing, Educational Entertainment, Responsiveness & the quality of the service staff, Museum Stores/Souvenir Selling area, Surrounding of the museum, Tourist Convenience & Comfort facilities.

The mean ranking analysis of survey data shows that the visitors to the museums of DNM are highly experienced (satisfied) with the knowledge they could gain through visiting the museum (mean = 4.36) and friendly and sensitive staff with good looking appearance (mean = 4.35). On the other hand, lowly experienced (dissatisfaction) with prevailing ventilation and air circulation inside the museum as it got a mean value of 2.82.

In the variable analysis, the visitors are highly satisfied with their experience with the surrounding of the museum. The especially high mean value gained for the overall landscape and the clean and spacious environment. The next highest variable mean value goes to educational entertainment and responsiveness & the quality of the staff that carries equal mean values (4.0209). The lowest variable mean is with the museum souvenir selling area, where most of the respondents skip answering as they couldn't find and not visible for the visitors' attention. The next lowest mean value is with the variable of visitor convenience and comfort facilities within the premises.

The correlation analysis results in identifying the highest positive relationship between "Tourist Convenience & Comfort" and the "Overall Satisfaction on Museum Visitation" which shows $r = 0.628$, $P = 0.000$, significance at 95% level. Controversy, "Museum Souvenir" ($r = 0.414$, $P = 0.000$) shows the least positive relationship with overall museum visitor satisfaction. Results highlighted that "Reception & Ticketing," "Educational Entertainment," "Responsiveness & the Quality of the Staff," "Surrounding of the Museum," and "Tourist Comfort and Convenience" have a moderate positive relationship with museum revisit intension and "Museum Souvenir Selling" represents a poor positive relationship with museum revisit intension. When comparing these results with the tourists' responses toward the number of repeat visitations, the majority answered as it's their first-time visit. It reveals that museum visitors are rarely returning after their first visit.

The correlation analysis to measure the relationship between the museum variables and the tourists' recommendation, results in the highest positive relationship between "Surrounding of the Museum" and the "Recommendation" ($r = 0.619$, $P = 0.000$) which derives the meaning that the majority of the museum visitors have recommended for potential visitors as a result of the better surrounding of the museums. The lowest relationship shows between "Museum souvenir selling area" and "Recommendation" ($r = 0.163$, $P = 0.046$). Other variables represent moderate positive relationships with recommendation.

These results finally conclude that the Department of National Museums should more focus in enhancing visitor convenience and comfort facilities to attract more foreign visitation. That basic would then create space for tourism suppliers to promote museum tourism in Sri Lanka.

3.6 CONCLUSION AND RECOMMENDATIONS

Museums, being a part of the lucrative cultural tourism segment increasingly attract millions of tourists. Even though Sri Lanka is blessed with a massive number of museums covering all parts of the island dedicated to different themes, the international visitors' interest in visiting museums is likely at a slow pace. Therefore, the overall idea of the research has developed based on the objective of identifying the level of the foreign museum visitor experience and its impact on tourists' satisfaction, WOM recommendation, and revisit intention.

Other countries in the globe that is richly endowed with extensive cultural resources are capable of creating demand, creating tourist traffic, and satisfying the travel needs of tourism to create wealth, reducing unemployment, and poverty at the destination. For Sri Lankan museums, in order to be well placed in the attention of the world tourists market and build substantial tourist traffic to the country, the immediate need is there to move from traditional asset-focus to visitor-focus. There, Sri Lankan museums should facilitate required visitor services, interactive learning opportunities, and effective marketing in a way that does not ever damage its originality. Museum visitor convenience and comfort facilities need to be developed to support their museum visitation and inter-cultural learning. The traditional role of the museums, that is, gathering artifacts, storage, and research have to be refocused toward improving its service quality in order to achieve maximum visitor satisfaction. The museum should tap on the visitors' curiosity on relevant subjects and convert them into activities allowing the tourists to take part in learning with fun. Through these activities, the tourists will receive both education and entertainment which will increase the interest of others who may want to visit the museum.

KEYWORDS

- **museums**
- **visitor experience**
- **Department of National Museums**
- **Sri Lanka**

REFERENCES

AHC and TCA. *Draft Heritage Tourism Guidelines*; Australian Heritage Commission and Tourism Council of Australia: Canberra, 1999.

Ap, J. Arts/Cultural Tourism. Position Paper Prepared for the Arts/Cultural Tourism Working Group. Hong Kong Tourist Association and Hong Kong Arts Development Council, Hong Kong Polytechnic University, 1999; 6 pp.

Bergadaà, M. Le Musée Olympique: La Structuration D'un Objet Culturel À Caractère Universel À L'aide D'enquêtes Qualitatives. In *8th International Conference on Arts and Cultural Management (3–6 July)*. Montréal: Aimac, 2005.

Blackwell, C. Tourism and Cultural Tourism: Some Basic Facts. *Preservation* **1997,** *7* (3).

Bollo, A. Indagine Sul Pubblico Dei Musei Lombardi. Fondazione Fitzcarraldo, 2004. http://www.Fitzcarraldo.İt/Ricerca/Pdf/Musei_Lombardia.Pdf

Caldwell, N. G. The Emergence of Museum Brands. *Int. J. Arts Manage.* **2000,** *2* (3), 28–34.

Chang, E. Interactive Experiences and Contextual Learning in Museums. *Studies in Art Education* **2006,** *47* (2), 170–186.

Chou, H. J. The Effect of Experiential and Relationship Marketing on Customer Value: A Case Study of International American Casual Dining Chains in Taiwan. *Soc. Behav. Person.* **2009,** *37* (7), 993–1008.

Davey, G. What Is Museum Fatigue? *Visit. Stud. Today* **2005,** *8*, 17–21.

Department of National Museums. *Annual Performance Report*; Department of National Museums: Colombo 07, 2018.

Falk, J.; Dierking, L. *Learning from Museums: Visitor Experiences and the Making of Meaning*; AltaMira Press: Walnut Creek, 2000.

Falk, J. H.; Dierking, L. D. *The Museum Experience*; Howells House: Washington, DC, 1992.

Gil, M. S.; Ritchie J. R. B. Understanding the Museum Image Formation Process: A Comparison of Residents and Tourists. *J. Travel Res.* **2009,** *47* (4), 480–493.

Goodrich, J. N. Cultural Tourism in Europe. *J. Travel Res.* 1997, *35* (3), 91.

Goulding, C. Contemporary Museum Culture and Consumer Behaviour. *J. Market. Manage.* **1999,** *15*, 647–671.

Goulding, C. The Museum Environment and the Visitor Experience. *Eur. J. Market.* **2000,** *34* (3/4), 261–278.

Hume, M.; Mort, G. S.; Liesch, P. W.; Winzar, H. Understanding Service Experience in Non-Profit Performing Arts: Implications for Operations and Service Management. *J. Oper. Manage.* **2006,** *24*, 304–324.

ICOM. Retrieved from International Council of Museums, 2019. www.icom.museum

Jamieson, W. The Use of Indicators in Monitoring: The Economic Impact of Cultural Tourism Initiatives. *COMOS Canada* 1995, *4* (3). http://www.icolnos.org/canada/bulletin/vol4_no3jalnieson_e.html

Jeong, J. H.; Lee, K. H. The Physical Environment in Museums and Its Effects on Visitors' Satisfaction. *Build. Environ.* **2006,** *41* (7), 963–969. doi: 10.1016/j.buildenv.2005.04.004

Kamal, M.; Pramanik, S. A. Identifying Factors Influencing Visitors to Visit Museums in Bangladesh and Setting Marketing Strategies for Museums. *IOSR J. Busi. Manage. (IOSR–JBM)* **2015,** *17* (10), 85–92.
Kawashima, N. Knowing the Public. A Review of Museum Marketing Literature and Research. *Museum Manage. Curatorship* **1998,** *17* (1), 21–39.
Kelly, L. Evaluation, Research and Communities of Practice: Program Evaluation in Museums. *Arch. Sci.* **2004,** *4*, 45–69.
Kotler , N.; Kotler, P. Can Museums Be All Things to All People? Missions, Goals, and Marketing Role. *Museum Manage. Curatorship* **2000,** *18* (3), 271–287.
Lehn, D. V. Embodying Experience. *Eur. J. Market.* **2006,** *40* (11/12), 1340–1359.
Mattila, A. S.; Wirtz, J. Congruency of Scent and Music as a Driver of In-store Evaluations and Behaviour. *J. Retail.* **2001,** *77* (2), 273–289.
McCall, J.; Weber, J. Predicting Attendance At Southeastern US Amuseument Parks Using Accessibility-and Amenity Based Models. *Geo. Bull.* **2006,** *47*, 45–59.
McCarthy, J.; Ciolfi, L. Place as Dialogue: Understanding and Supporting the Museum Experience. *Int. J. Herit. Stud.* **2008,** *14* (3), 247–267.
Mclntosh, R. W.; Goeldner, C. R. *Tourism Principles, Practices, Philosophies*, 6th ed.; John Wiley and Son: New York, 1990.
McKercher, B.; du Cros, H. *Cultural Tourism: The Partnership Between Tourism and Cultural Heritage Management.* The Haworth Press, Inc., 2002.
McManus, P. *Archiological Displays and the Public*; Routledge: London, 2016.
Miller, J. Cultural Tourism Worthy of Note. *Hotel Motel Manage.* **1997,** *2* (15), 7.
Milman, A. Evaluating the Guest Experience at Theme Parks: An Emperical Investigation of Key Attributes. *Int. J. Tour. Res.* **2009,** *11* (4), 373–387.
NMDC. *Museums and Tourism*, 2014. Retrieved from National Museum Director's Council. https://www.nationalmuseums.org.uk/media/documents/what_we_do_documents/nmdc_museums_and_tourism_briefing_2013.pdf
Oh, H.; Fiore, A. M.; Jeoung, M. Measuring Experience Economy Concepts: Tourism Applications. *J. Travel Res.* **2007,** *46*, 119–132.
Packer, J.; Ballantyne, R. Motivational Factors and the Visitor Experience: A Comparison of Three Sites. *Curator: Museum J.* **2002,** *45* (3), 183–198.
Packer, J. Beyond Learning: Exploring Visitors' Perception of the Value and Benefits of Museum Experiences. *Curator: Museum J.* **2008,** *51* (1), 33–54.
Poria, Y.; Butler, R.; Airey, D. How Tourists Decide Which Heritage Site to Visit? *Tour. Rev.* **2004,** *59* (2), 12–16.
Prentice, R.; Davies, A.; Beeho, A. Seeking Generic Motivations for Visiting and Not Visiting Museums and Like Cultural Attractions. *Museum Manage. Curatorship* **1997,** *16* (1), 45–70.
Richards, G.; Gratton, C. *Cultural Tourism in Europe*; CABI: Wallingford, 1996.
Rojas, M. C.; Camarrero, M. C. Experience and Satisfaction of Visitors to Museums and Cultural Exhibitions. *Int. Rev. Public Non-Profit Market.* **2006,** *3* (1), 49–65
Rowley, J. Measuring Total Customer Experience in Museums. *Int. J. Contemp. Hosp. Manage.* **1999,** *11* (6), 303–308.
Schweitzer, C. The Hot Ticket to Cool Meetings. *Association Management* **1999,** *51* (8), 121–130.

Silberberg, T. Cultural Tourism and Business Opportunities for Museums and Heritage Sites. *Tour. Manage.* **1995,** *16* (5), 361–365.

Slater, A. Escaping to the Gallery: Understanding the Motivations of Visitors to Galleries. *Int. J. Nonprofit Voluntary Sect. Market.* **2007,** *12*, 149–162.

SLTDA. *Annual Statistical Report 2017*; Sri Lanka Tourism Development Authority: Colombo, 2017.

Stephen, A. The Contemporary Museum and Leisure: Recreation as a Museum Function. *Museum Manage. Curatorship* **2001,** *19* (3), 297–308.

Thyne, M. The Importance of Values Research for Nonprofit Organizations: The Motivation-Based Values of Museum Visitors. *Int. J. Nonprofit Volunt. Sect. Market.* **2001,** *6* (2), 116–130.

Tighe, A. J. The Arts/Tourism Partnership. *J. Travel Res.* **1986,** *2*, 2–5.

Trotter, R. The Changing Face and Function of Museums. *Media Int. Aust.* **1998,** *89* (1), 47–61. https://doi.org/10.1177/1329878X9808900108

Tucker, R. B. Ten Driving Forces of Dynamic Change. *Exec. Excell.* **1991,** *8* (3), 16.

UNWTO. *UNWTO Annual Report 2017* United Nations World Tourism Organization VICNET **1996:** Madrid, 2018.

Weerasingha, I. Involvement of Travel Agencies in Sri Lanka to Promote Museums in Tourism Industry through the Medium of Tour Packages. *3rd Biennial Conference of the International Association for Asian Heritage*, 27th–28th December 2015; Centre for Asian Studies, University of Kelaniya & International Association for Asian Heritage (IAAH), **2015;** p 69.

Zeithaml, V. A.; Bitner, M. J.; Gremler, D. D. *Services Marketing: Integrating Customer Focus across the Firm*, 5th International ed.; McGraw Hill: Singapore, 2009.

Zeppel, H. *Cultural Tourism in Australia: A Growing Travel Trend. Material Culture Unit, Department of Tourism*; Jatnes Cook University: Townsville, Australia, 1992.

Zeppel, H.; Hall, C. M. Selling Art and History: Cultural Heritage and Tourism. *J. Tour. Stud.* **1991,** *2* (1), 29–45.

CHAPTER 4

THE ATTITUDES OF LOCAL RESIDENTS TOWARD TOURISTS IN THE DESTINATION OF MARMARIS, TURKEY

GIZEM ÖZGÜREL[1,*] and CEVDET AVCIKURT[2]

[1]*Muğla Sıtkı Koçman University, Marmaris Tourism Vocational School, Muğla, Turkey*

[2]*Balıkesir University, Faculty of Tourism, Balikesir, Turkey*

[*]*Corresponding author. E-mail: gizemozgurel@mu.edu.tr*

ABSTRACT

Being one of the known tourism destinations in the world, Marmaris, a coastal city, is preferred by many local and foreign tourists. In 2019, about 1,350,000 foreign tourists and 419,000 local tourists stayed overnight in Marmaris, while local residents, consisting of about 95,000 people, hosted tourists. This study aims to determine the settled attitudes of local residents in Marmaris, which is located in the southwestern region of Turkey and which has transformed from a small fishing town into a preferred tourism destination, toward tourists. In the chapter, the studies of Var, Kendel, and Tarakcioglu (1985) examining Marmaris constituted the starting point. The researchers tried to identify how the local community of this small town that was declared as a tourism development area by the Turkish government in the 1960s perceives the social impacts of tourism. In their

Tourist Behavior: Past, Present, and Future. Narendra Kumar, PhD, Bruno Barbosa Sousa, PhD & Swati Sharma, PhD (Eds.)

study, they stated that the Marmaris community was not studied from this perspective before and they chose the Marmaris community as the target population of the study as they thought this community symbolized a series of cultural values. The authors conducted a survey for the Marmaris community in the peak season of 1980–1981 and analyzed valid data of 98 participants. Among the results of their study, the following points become prominent: local residents perceive the positive and negative effects of tourism; they find meeting with tourists valuable as an educational experience; and they least prefer to see in Marmaris the tourists who are Turkish locals, who come from the Middle East, and who are Turkish workers living abroad. In this context, it was aimed to determine whether there was any change in the local residents' attitude toward tourism and tourist in Marmaris since the 1980s, and to determine its direction and level, if any. Besides, the issues such as the dissolution of the USSR in 1989, Russian tourists visiting Marmaris with an increasing number of each year, and residents' attitudes toward Russian tourists will be examined for the first time in the scope of the study. In the relevant literature section of the study; the titles such as the effects of tourism, attitudes towards tourism, stereotypes of Turkish people for different nations, and Marmaris destination were mentioned.

In the study, a quantitative research method was adopted, and a questionnaire consisting of 3 parts was applied to collect data from the field, and the data were collected from the population formed by the residents of Marmaris through an online survey. The first part of the questionnaire included questions aiming to determine demographic characteristics, while the second part covered local residents' attitudes towards tourism and its effects, and the third and the final part included questions aimed at determining the stereotypes of local residents for tourists of different nationalities. The data obtained were analyzed in the SPSS 22.0 program, the findings were interpreted, the results and suggestions were made, and the study was concluded. The assumptions and limitations of the research were: the reliability and validity of the data collected in the research were limited by the features of the data collection methods; it was assumed that the participants answered the research questions correctly; the questionnaires planned to be made face-to-face were carried out online due to the Covid-19 and the measures taken by the Turkish government and by social responsibility; finally, insufficient time. The literature of the study has been listed according to the effects of tourism in destinations, perceptions and attitudes towards tourism, Turkish stereotypes of different nations and the people in Marmaris.

4.1 INTRODUCTION

Tourism, which has become an integral part of modern life, directly and indirectly causes change in both the sending and receiving societies. Perceptions, attitudes, and behaviors of local people play a key role in ensuring the development and continuity of tourism in places that develop and turn into tourism destinations with the start of tourist visits and become centers of tourism activities. Marmaris district, which was a small fishing town in the 1950s, was declared a tourism development area with the decision of the Turkish government in the 1960s and the central government took the first steps for the development of the tourism sector. Tourism activities in Marmaris, which has hosted more than 2 million local and foreign tourists together with daily visitors in recent years, create various effects with social, cultural, economic, and environmental dimensions as in all other tourism destinations.

In tourism literature, it is known that touristic activities create both positive and negative effects, creating change, and transformation in tourism destinations, and many researchers try to reveal the extent and direction of change by measuring the effects of tourism. It is known that the development of tourism in Marmaris, besides economic effects, creates changes in the natural and physical environment and social life. However, it is possible to trace the social and cultural effects of tourism, with the lifestyle, participation of women in the workforce, family relations, level of commitment to traditions and customs, leisure activities, changes in beliefs, attitudes, and values in Marmaris.

The study tried to reveal how the local people perceive the effects of tourism in the Marmaris and how their attitudes toward tourists are. It has been tried to determine whether there has been a change in the perceptions and attitudes toward the effects of tourism in the local people of Marmaris, stereotypes regarding tourists, and, if there has been a change, its direction and level. The literature of the study has been listed according to the effects of tourism in destinations, perceptions, and attitudes toward tourism, Turkish stereotypes of different nations, and the people in Marmaris. Quantitative research method was used in the study. With the scale consisting of three parts, data was collected from 393 local people online between 07 and 31 July 2020. The obtained data were analyzed in the SPSS 22.0 program. The study was structured in the form of relevant literature, findings, comments, results, and recommendations.

4.2 EFFECTS OF TOURISM

The tourism sector, whose importance is increasing on a global scale and attracting attention with its effects on societies, has reached a large scale today. Tourism creates various effects in the social, cultural, economic, and natural environments of the societies. Countries compete intensely to be within the targets of international tourist flows and to benefit more from the positive economic effects of tourism. There are various reasons for the efforts made to gain a greater share of tourism. In countries that are successful in attracting tourists by transforming their tourism potential into tourism supply and by creating artificial attractions, the positive effects of tourism in economic, social, cultural, and environmental dimensions turn into gains. In contrast, the intensification of tourist flows to tourism destinations, unplanned tourism development, exceeding the economic, social, ecological, and physical transportation capacities, creates various negative effects and threatens tourism sustainability.

Tourism is an economic activity that directly or indirectly affects many people around the world in terms of production and consumption (Bahar and Kozak, 2015). The positive economic effects of tourism—monetary character and real economic effects—are analyzed with qualitative measures. Among the effects of monetary character—the effects of tourism on foreign payments, national income, domestic prices, investment, and government revenues—among the real effects are the effects on employment, other sectors (agriculture, service, industry, and transportation sectors) and infrastructure (Olalı and Timur, 1988; Çeken, 2014). The economic effects of tourism differ in developed and developing countries, and statistics show that the countries that receive and send tourists the most are the developed countries.

Especially for underdeveloped and developing countries, tourism activities are a driving force in their economies, and tourism is supported and encouraged by government policies due to its positive effects on economic growth and development. Tourism contributes to the national economy, the current account balance is positively affected, foreign exchange and income increase are created, large employment areas are opened, and new business opportunities arise. In places which have potential to become tourism destinations, infrastructure is completed, existing infrastructure is strengthened, new investments are attracted to the region, and added value are created for the local economy. The tourism sector can be strengthened with the existing sectors by reviving the local economy with tourism. Various adverse economic effects can also arise, such as the opportunity cost of economic choices made in tourism destinations, inflationary pressure, the danger of over-dependence on tourism, low rate of return on investments, rising land values, increased import tendency, seasonal production, and external costs caused by crowding. Opportunities that can be missed in other sectors as a result of transferring the resources owned in destinations to tourism development can create a negative impact on the economy as a cost factor. However, the dependence of the economy on tourism activities, the decrease in demand, income, employment losses, bankruptcies, and economic damages that are difficult to correct can be caused by the negativities experienced in the fragile tourism sector. The pressure created by the touristic demand causes an increase in the prices of the products and services in the destinations, and the resulting inflationary situation is observed as a negative economic effect. In the tourism sector, where investment costs are relatively high, especially under the influence of seasonality, tourism enterprises cannot operate at full capacity throughout the year, which prolongs the investment return process and negatively affects the investment decisions of entrepreneurs in the sector. In the tourism destinations that have reached the growth and development stage, the presence of both national and international capital with investments leads to high increases in land prices, and local people's land ownership and local control can change hands and cause many negativities. In tourism enterprises, mainly beverages, furnishings, and so on, the supply of many products by exporting creates the money that is desired to remain within the country's economy to go abroad. Tourist influx may have a negative impact on the economic climate of the destinations, especially in the high season, due to the crowd and various economic negativities caused by this. The economic effects created by tourism activities continue to be studied with many studies in the literature: Tourism and its economic effects (Comerio and Strozzi, 2019; Tohmo, 2018; Hepaktan and Çınar, 2010); tourism and economic growth (Brida et al., 2020; Kanca,

2015); productivity and economic growth in tourism (Liu and Wu, 2019); tourism demand and economic growth (Santamaria and Filis, 2019; Esen and Özata, 2017; Holzner, 2011); tourism and economic development (Sokhanvar et al., 2018); tourism and local development (Özgürel, 2012); and tourism and regional development (Mesci and Dönmez, 2016; Çeken, 2008) are some of the studies on tourism and economic recovery (Dogru and Bulut, 2018; Jucan and Jucan, 2013).

Tourism has become one of the most important social events at the present times and has attracted attention to the changes and transformations it has created in social life. It has the potential to affect both at the macro level in the societies that send and receive tourists, in the tourists who are the subjects of tourism, and in the individuals, who have contact and relations with the tourists at the micro level. Although globalization, international migration, developments in transportation, and communication technologies affect individuals and societies, the transformation of tourism into a lifestyle provides an environment for social effects and interaction. However, it should be emphasized that it is almost impossible to explain and measure how much of social change and transformation is due to conjunctural developments and how much of tourism. Tourism has the potential to create positive effects in establishing individual social relations, increasing the quality of life of the local people with the social environments it creates, developing the understanding of tourism, strengthening the sense of community in the destination, being proud of the place, the formation or development of social organizations, the increase of social capital, and the socialization of local people. However, different personality traits and expectations can also create conflict environments for tourist—local people, tourism workers, and tourist in tourism destinations. In other words, the effects of tourism on the social structure are explained by considering urbanization, social layers, and changes in political structure and family structure (Doğan, 2004). As the tourists begin to come to the places of attraction, change, transformation, and a city view begin to emerge with the mediation effect of tourism, especially in underdeveloped and developing regions. A small town or a settlement that continues its existence traditionally with its own resources, after being opened to tourism, turns into a presentation to tourists with scenes prepared according to the tourist, and the local people can experience the urban development and transformation of the place where they live their daily life. Tourism affects the family structure in the society, the roles of family members, especially the family institution, which is the smallest unit of the society, with the participation of women in working life. The work migration created by those who come to the destination for working purposes can create changes in the demographic structure, the increase in second residences, the settlement of lifestyle immigrants, and quantitative and qualitative changes in the social structure can be experienced, especially in coastal destinations. Tourism destinations provide social interaction environments for local people and tourists, and social interaction with mutual social relations can affect tourism, attitudes, and behaviors toward the tourist and the visited society. Positive social interaction is likely to have an impact on the contribution of social peace and tourism to world peace. The tourist can play a role in reducing the cultural distance and stereotypes and prejudices between societies by visiting the destination again, recommending the destination to others, experiencing the life style, and values of the society they visit, and transferring them to their surroundings. The arrival and destination of the tourist do not remain the same, social norms, behavioral styles, communication styles, socialization styles, consumption styles, social institutions, and habits may differ. The local community can act under the "show-off effect" by taking care of tourists, class difference,s and differences in status between tourist and local people can create an environment of conflict, drug use, excessive alcohol consumption, gambling habits, prostitution, conflicts of interest, infectious diseases, and tourism crimes increase can occur as effects. There is an increasing number of academic studies in recent years on the social effects of tourism (Özgürel, 2020; Matarrita-Cascante and Suess, 2020; Çavuş and Baki, 2020; Kang and Lee, 2018; Scholtz and Slabbert, 2016; Deery et al., 2012; Tayfun and Kılıçlar, 2004; Berber, 2003; Gürkan, 1996).

In some studies, in the literature, the cultural effects of tourism are examined as a dimension in itself, and in other studies, the socio-cultural effects dimension is examined by considering the social and cultural effects from an integrated perspective. The social effects of tourism have been examined separately earlier, and in this section, the cultural effects dimension of tourism has been handled and explained as a separate dimension. Effects of tourism on culture is the transformation and transformation of religious beliefs and behaviors, moral values, lifestyle, language, art, tourism as a cultural institution, and as a part of the culture. Tourism is one of the important sectors that mediate cultural change and spread. Promotion of Turkish culture in terms of Turkey's tourism, customs and traditions, eating and drinking behavior, hospitality, religious practices, lifestyle is the synthesis of Eastern and Western cultures, many values of the traditional architecture of the crafts are hosting a huge potential for cultural dissemination (Avcıkurt, 2015). The encounters and contacts of different cultures through tourism can cause local people to become aware of their own culture, to take care of their values and to be proud of their cultureIn contrast, local people, who have the opportunity to get to know different cultures, have the opportunity to increase their knowledge. Cultural communication between the tourist–local people broadens the horizons of the parties (Richards, 2006), and the direction of cultural interaction is determinant in the increase of the tolerance level of different cultures. In addition to the positive effects of tourism on culture, negative effects such as the adaptation of foreign words, the adoption of the lifestyle of tourists, the degeneration of religious and moral values, the transformation of cultural values into commercial commodities, and the increase in cultural distance between the parties can be observed in tourism destinations. There are many studies in the literature on the cultural effects of tourism (Wei et al., 2020; Jingyi and Chung-Shing, 2018; Göktaş and Türkeri, 2016; Uslu and Kiper, 2006; Ünlüönen and Tayfun, 2003).

In terms of its structure, tourism is a whole with the natural and cultural environment that contains many attractions. Tourism provides positive environmental impacts in terms of protecting natural areas, increasing environmental awareness, improving the quality of the environment, improving infrastructure, protecting archaeological and historical sites, increasing the value of the environment, and ensuring the transfer of local architecture to future generations (Avcıkurt, 2015). However, as a result of exceeding the ecological and physical carrying capacity in tourism destinations, the negative environmental effects of tourism emerge and endanger the sustainability of tourism. The unplanned development of tourism activities predominantly creates negative effects on the natural and cultural environment that are difficult to reverse. Among these negative effects, disturbance of ecological balance, destruction of historical man-made artefacts, loss of characteristics of geographical attractions, change of water, air and soil quality, pollution, loss of biodiversity, and noise effect pollution are some of them. There are many studies focused on the environmental effects of tourism (Chen, 2020; Sun et al., 2020; Robinson et al., 2019; Wang and Zhou, 2019; Özsavaş Uluçay, 2018; Şafaklı, 2017; Tuna, 2007; Kahraman and Türkay, 2004; Demir, 2002).

4.3 PERCEPTION AND ATTITUDE TOWARD TOURISM

The issue of the effects of tourism in the tourism literature attracts a great deal of attention, as stated in the earlier studies. In this context, it should be noted that with the academic studies conducted in recent years, the perceptions and attitudes of the local people toward tourism have remained important, and the literature on theoretical and practical studies continues to develop. The perceptions and attitudes of local people toward tourism depend on the development level of tourism destinations, the demographic structure of the local community, dependency on tourism, the status of being a local or an immigrant, the duration of residence, etc. It can differ according to many variables. However, it cannot be said that there is a full consensus on the variables in the relevant literature. Local people's perceptions of the impact of tourism can be positive or negative in different situations. In some studies, a positive relationship was found between demographic variables and local people's perception of tourism effects, while others may have the opposite situation (Akova, 2006; Yılmaz and Filiz, 2017).

In tourism destinations, local people are faced with many positive and negative effects of tourism, their view on tourism and how they perceive tourism, it seems worth examining because there cannot be a tourism development without local support and tourism cannot be sustained. The fact that the local people are among the main actors in tourism stakeholders and the support of the well-known local people in the success of tourism planning make it important to know the perceptions of the local people toward tourism and tourists.

Perception means "by directing attention to something, to become conscious, to realize that thing" (Türk Dil Kurumu, 2020). When the concept of perception is evaluated in terms of tourism and tourist, it explains how local people see, understand, interpret, and create awareness about tourism and tourist effects. Local people perceive tourism effects in terms of social, cultural, economic, and environmental aspects, and tourist effects as gains and losses (negativities) predominantly on the basis of social and cultural effects.

Işık and Turan (2018) investigated the perceptions of local people on the effects of tourism in Izmir province. They concluded that the general average of local people's perception was high. The researchers stated that the positive economic, environmental, and socio-cultural effects of tourism, such as creating new business opportunities, supporting the local economy, ensuring the protection and development of the natural environment, improving the environmental quality, ensuring the development of cultural activities, and contributing to the cultural development of the local people are perceived at a high level by the local people. In the study, tourism, socio-cultural negative effects such as increasing the prices of products and services in the region and causing the cost of living, creating environmental pollution, negatively affecting the nature, negatively affecting the attitudes and behaviors of the local people, and causing the destruction of cultural values are also perceived at a moderate level by the local people. According to the results they obtained, the perception of the local people toward the effects of tourism differs according to demographic characteristics. They found that female participants had higher perceptions of socio-cultural negative effects than male participants, married participants perceived negative socio-cultural effects more than single participants, and that there was no significant difference between their perceptions of working status in tourism and the effects of tourism. The researchers also found that the higher the education level of the local population, the higher their perception of economic positive effects.

Yılmaz and Filiz (2017) investigated the tourism perceptions and attitudes of local people two destinations (Mazı and Türkbükü) with different development levels, with Butler's "Tourism Life Cycle Model" and Doxey's "Tolerance Model." The researchers concluded that the local people in Mazı village have insufficient tourism knowledge and negative perception of tourism. In Türkbükü, they reached the conclusion that the general tourism perception of the local people is positive, the decrease in the number of tourists and the reflection of the political events negatively affect the tourism perception of the local people, but they are in a cautiously optimistic approach. It was stated that the participants from the local population of Mazı village perceive the tourists as a financial resource, while the participants in Türkbükü are perceived as people who are associated with cultural differences in the perception of tourists.

In the study of Özdemir and Kervankıran (2011), they concluded that 52.4% of the local people see tourism as an activity that improves the economy, 17.4% enables the increase of culture, and 15.7% preserves nature, historical, and cultural values. In line with the findings of the researchers, they came to the conclusion that local people in Afyonkarahisar expect the economy to develop, natural, historical, and cultural values to be recognized, protected, and cultural diversity will be caused by tourism. In addition, it has been stated that it is not regarded as an activity that destroys or regresses natural, cultural, economic, and social values.

The concept of attitude that has been defined many times from different perspectives (Albu and Nicolau, 2010); it explained as the tendency of individuals to react positively or negatively to any object, opinion, or environment. It is generally stated that attitudes are a part of individuals' socialization process or a kind of social knowledge (Daruwalla and Darcy, 2005). The body of reactions that assume a guiding mission toward something that an individual, who is a social entity, associates with himself in the socialization process, and daily life can be defined as attitude. Although it is not possible to observe the attitudes of individuals in society, attitudes are important due to the approach that there are tendencies that reveal the behaviors that can be observed and examined. Although the concept of attitude is generally used together or interchangeably with the concepts of belief, value, and opinion, it can be said that it creates some gaps in its explanation. The concept of attitude in the literature to close this gap is tried to be explained by taking emotional, cognitive, and behavioral dimensions, and it requires a comprehensive examination because of the effect of each dimension on the other dimension. Cognitive dimension is shaped by an individual's thoughts, knowledge, and beliefs about anything. Emotional dimension includes the individual's emotions and consists of emotional reactions. Behavioral dimension, in contrast, is the behavioral tendency toward what is the subject of an individual's attitude and it is operational (Odabaşı and Barış, 2016). To achieve the desired attitude development in individuals, there should be consistency between attitude dimensions.

Attitudes toward tourism is the tendency of individuals to react positively or negatively to the phenomenon of tourism. Differences between individuals' attitudes toward tourism can arise from any of the cognitive, emotional, or behavioral dimensions, or the characteristics of the combination of all dimensions. Among these features; the strength of the attitude toward tourism, the level of learning about tourism experiences and tourism through social relations, the state of changing attitude, the formation of attitude with a holistic structure, the complexity of attitude dimensions, and the consistency or inconsistency of the attitudes. Attitudes toward tourism undertake various functions in reaching the needs or wishes of the local people. These functions include creating benefit, providing value, protecting self-image, and gaining knowledge. For the attitudes toward tourism to turn into a behavior in the local community, some conditions must be formed. Among these conditions, behaviors can occur as a result of the realization of the attitude, the high belief in the attitude, the fact that the attitude is fully known, situational factors such as economic facts are not effective, and knowledge and awareness about the attitude toward tourism or can be formed immediately (Odabaşı and Barış, 2016). There are various sources of information that have a high impact on the formation of attitudes toward tourism. Among the main sources, personality, experiences, personal influence, digital platforms, and mass media may be listed. Introverted or extroverted personality structure, family, peer groups, past experiences, and, today, Facebook, Twitter, Instagram, etc. social media channels, blogs etc., play a key role. Television, newspapers and magazines also play key roles in attitude formation and change as channels for information transfer, sharing, and access to information. It is stated that especially persuasive communication is effective in attitude change (Albu and Nicolau, 2010).

Since the 1970s, the number of studies focusing on the attitudes of local people toward tourism and the effects of tourism on the local community has been increasing. In the studies conducted, it has been determined that tourism can create positive and negative effects, they review the positive and negative effects while determining the attitudes of local people toward tourism, and determine whether local people's attitudes toward tourism will support tourism (O'Leary, 1976; Pizam, 1978; Var et al., 1985; Andereck and Vogt, 2000; Berber, 2003; Pennington-Gray, 2005; Akova, 2006; Gursoy et al., 2010; Deery et al., 2012; Scholtz and Slabbert, 2016; Bal et al., 2016; Kang and Lee, 2018; Tohmo, 2018; Robinson et al., 2019; Batmas et al., 2020; Baykal and Ataberk, 2020; Brida et al., 2020; Çakır and Aydın, 2020; Eryılmaz, 2020; İmre, 2020; Koçoğlu et al., 2020; Koday and Akbaş, 2020; Ng and Feng, 2020; Uğur, 2020; Vatan and Bildin, 2020). The attitude of the local people toward tourism and tourists, depends on the destination on tourism, therefore, whether the local people earn income from tourism, the excessive tourism that occurs with the increase in the number of tourists, tourism types in the destination, tourist typologies, the duration of the tourist, the seasonality characteristic of tourism. It is likely that many factors such as the dominance of the local people in the destination, the social and cultural distance between the tourist and the local people, the level of interaction and the tourism awareness of the local people are the major factors. It is quite difficult to examine the attitudes toward tourism and the tourist and how these positive or negative attitudes affect tourism activities and the success of these activities. However, it would be appropriate to state that the attitudes of local people toward tourism and tourists require multi-dimensional research (Sharpley, 2014). In the following section, the concept of stereotype and the stereotypes of the local people (Turks) that constitute the universe of the study will be explained.

4.4 STEREOTYPES OF TURKS RELATED TO DIFFERENT NATIONS

When considered etymologically, the concept of stereotype consists of the Greek words "stereos" (solid, continuous) and "typos" (attribute, character, type), and it means stereotype or stereotypical thought in Turkish. The concept began to be used in printing in 1798 in naming a type of printing plate (Encyclopædia Britannica, 2020), the first theoretical view in social sciences took place in Walter Lippman's (1922) book "Public Opinion." With the concept of stereotype, the researcher tried to explain the ideas, ideologies, or expectations that belong to human beings and have a guiding mission in decision making (Lippmann, 1998). Stereotypes are "complex belief clusters that are sometimes held together by cultural glue or prejudices, but more often joined by theories that reflect a combination of experience and culture" (Schneider, 2004). The concept can be explained as the expression of the individual's established ideas about out-group members, and the set of beliefs and judgments toward social groups. In this context, stereotypes shape the individual's knowledge, belief, and expectations toward the other/others in the distinction between self and other, we and others (Hamilton and Sherman, 1994). Stereotypes are formed by bringing together the information heard from others, read in the form of a title, and accepted without questioning its accuracy, and it is evaluated as the effort of the individual to classify the unknown by producing meanings (Gökdayı, 2017). In this context, it is possible to define it as personal characteristics that are believed to be found in a certain group of people, learned from others and learned (Yağan, 2020). In other words; by placing others/other people in descriptive categories, groups of people, for example, the concept of stereotype can be defined as the tendency to categorize and attribute characteristics as a target group in the form of age, gender, social class, occupation, sexual orientation, political orientation, minority groups, and nationalities. Stereotypes are examined by different disciplines, in line with the approaches is tried to be explained under the headings of cultural and individual stereotypes (Blum, 2004; Carter et al., 2006):

> *Personal stereotypes are treated as individual thoughts and beliefs about members of another group as a result of experience. Cultural stereotypes are defined as beliefs shared by the whole culture about the characteristics of a group. For example, while it is a cultural stereotype that women are fond of housework and motherhood, it is a personal stereotype that they define women as unreliable as a result of experience (Stangor and Schaller, 1996, as cited in Aktan and Sakallı Uğurlu 2013).*

According to stereotypes functions by Snyder and Miene (1994), it is distinguished as cognitive, social, and self-defense function. Cognitive function is about the facilitative aspect of stereotypes of information processing. Social function—the individual's integration with the group and social environment of which he is a member; self-defense function is related to the individual's affirmation of his/her self by using stereotypes (Aktan and Sakallı Uğurlu, 2013). Although stereotypes are not formed for a specific purpose, it can serve as a mediator function in perceiving its environment, reducing uncertainties against facts, positioning itself in the society, reflecting its importance and valued things to the outside, adapting to the environment, shaping social and mental representations about reality, and a defense mechanism in dealing with negativities. The way to examine the concept with all its dimensions is through the features it carries. Some of the features of stereotypes are as follows (Karabayev and Tutkun, 2001; Apalı, 2011; Gökdayı, 2017; Yağan, 2020):

- Stereotypes can have positive or negative qualities,
- Usually based on intangible beliefs that need proof,
- Expresses a predictive and emotional value,
- As it may be wrong, it does not reflect the facts exactly,
- It stands out as a way for the individual to obtain information in a short way,
- It is spreading rapidly within the group or society,
- Based on oral culture,
- Resistance to change, and
- It contains simple information.

Stereotypes are shaped by a cognitive process and have an emotional character. It reflects the mentality of prejudice and preconceptions especially in the relations between people from different cultures and forms behavioral patterns. For example, similar prejudices and hatreds can be seen according to different countries, religions, and races. The individual can weaken or eliminate the stereotypes that emerge with thought processes and the negative attitudes that can be developed with these stereotypes, through life experience and intellectual accumulation over time (İmançer, 2004). The individual/individuals have a tendency to generalize toward large groups of people; therefore, they ignore the individual characteristics of the others due to the overgeneralization and lack of them in the structure of stereotypes, and self-fulfilling prophecies (The Pigmalion Mythos or the Pigmalion Effect) creates its own truth (Sürgevil, 2008; Myers, 2017). Stereotypes, especially regarding large groups of people, resist change and change slowly with the influence of social, political, and economic factors. Stereotypes are used more when social tensions occur between groups and may manifest themselves in the form of hostile behavior (Harlak, 2000). The literature, which began with Katz and Braly's (1933) first empirical study on the contents of nation stereotypes, made significant contributions to understanding the cognitive and affective processes underlying discriminatory behavior. Stereotype content model (SCM), which has recently contributed to and guided the stereotype related literature, is its continuation behaviors from intergroup affect and stereotypes map (BIAS Map), the social-structural sources of stereotypes. created a theoretical framework that integrates contents, affective, and behavioral consequences (Aktan et al., 2019).

Serving as a bridge between East and West, economic and cultural aspects of the transitional location on the itinerary in Turkey, Turkish society, its values, and social norms are a society that has different characteristics under the influence of external factors, especially the West and globalization. In Turkey, "is tolerant and welcoming of Turkey," "the history of racism is not like in the West," "which are tolerant toward other nations in the Ottoman Empire" as there are strong stereotypes. Textbooks contain sentences supporting these judgments (Çayır, 2018, as cited in Yağan, 2020). Stereotypes that lead to incorrect cultural generalizations, such as "belligerent" Serbs, "cunning" Greeks, "lazy" Turks, "wild" Albanians, which emerged in the historical process, were reproduced according to periodic government policies (Bitirim Okmeydan, 2019). For a Greek, the "Turks are hospitable" stereotype eliminates the prejudice of "Turks are barbarians," and for a Turk, the "Greeks are hot" stereotype fails to preserve its existence in front of the prejudice "Greeks are traitors." However, the positive relations established with the social, political. and economic developments that live in time can create a moderate atmosphere even if they do not erase the negativity completely (Özsüer, 2012).

A limited number of studies have been reached on the stereotypes of Turks toward different nations, and the findings obtained from these studies have been tried to be presented by taking into consideration the tourists of different nationalities who have chosen Marmaris destination. A limited number of studies on national stereotypes in Turkey were examined:

Harlak (2000) worked with two separate sampling groups in the Kuşadası and Kiraz districts, compared the stereotypes of the villagers living in touristic villages with and without contact with foreigners, and that the contact resulted in more positive stereotypes, and that the Turks saw themselves as Europeans. He reached the conclusion that certain nations such as Israelis and Iranians were put in the category of enemies. The researcher stated that "Western" nations and stereotypes generally contain more positive adjectives, and the stereotypes related to "neighboring" nations contain negative adjectives and that the most positive adjectives are used for the Japanese and the most negative adjectives are for the Armenians.

Işık (2012) researched the stereotypes of university students regarding Germans in the axis of intercultural communication and collected data from 139 participants. Işık (2012) stated that, participants marked more positive adjectives for Germans, such as hardworking (90.6%), successful-disciplined (88.5%), self-confident (88.4%), nationalist (83.5%), smart (82%)), active / social (82%), punctual (81.3%), sportive (81.3%), patriotic (80.6%), cultured (79.1%), and regular (74.8%). In addition, those with the highest rates of adjectives that can be described as negative for Germans, he stated that they only care about money (83.5%) and racist (82%) and that the participants describe Germans with these adjectives and that the participants generally have positive stereotypes toward Germans.

Demir (2016) investigated the stereotypes of 8th grade students and collected data from 400 participants. According to the findings obtained from the study, the participants agree that Germans are "disciplined" (31.4%), Greeks are "enemies of Turks" (32.5%), Japanese "read a lot of books" (35.3%), Germans are "blond-haired, blue-eyed," Indians are "not lagging" (49%), Japanese reminds me of robots (52%), Brazilians are "good football players" (48.3%), Singaporeans are "not good at math" (35.8%), and Russian girls are "beautiful" (60.3%).

Bakan and Canöz (2017) investigated the stereotypes of Turkish university students toward Japanese and collected data from 486 participants. Researchers concluded that the Japanese had a place with positive adjectives in the minds of the majority of the participants. The strongest among Turkish university students' judgments about the qualifications of the Japanese are the judgments that the Japanese are hardworking (93.1%), successful (92%), and smart (88.7%). The rate of those who find Japanese disciplined (87.9%), cultured (83.8%), respectful (83.2%), and active (81.2%) is also quite high. Based on this, researches stated that Japanese of university students have stereotypes about being respectful, cultured, disciplined, and active. In general, according to the Turks, the British have cultural stereotypes of "noble," "cold," Arabs "treacherous," "dirty," Japanese "terrible," Germans "terribly disciplined," French "gracious," Italians "hot," Swiss "soulless" (Harlak, 2000; Akyıldız, 2017).

Tourism causes different cultures to encounter, to experience intercultural relations, and to affect different groups such as tourists, local people, and tourism workers positively or negatively from the relations experienced. Frequent contacts in tourism destinations and stereotypes in cultural relations undoubtedly move on two separate axes from the perspective of the local people, various stakeholders in the tourism destination such as tourism workers and the tourist perspective, and it is thought that the relations and interactions between them can have an effect on the level and direction. Stereotypes that exist almost everywhere can affect the tourism sector positively or negatively and support the shaping of tourism destinations. However, stereotypes can be re-condensed or restructured as a result of interactions in tourism destinations. Personal and cultural stereotypes of tourists from different nationalities visiting tourism destinations are a must-know situation for the sustainability of tourism. However, in the tourism literature, a limited number of studies have been conducted on the stereotypes of local people toward tourists in the interaction between tourist and local people and have been examined superficially (Var et al., 1985; Ward and Berno, 2011; Monterrubio, 2018).

Harlak (1991) conducted the field study of the research titled "Categories and perception repertoires for various tourists" with participants working in the tourism business in Kusadasi, where one of Turkey's most important tourism destinations. Participants' most used positive adjectives for tourists of seven different nationalities (German, American, French, British, Scandinavian, Italian and Greek) were "curious" and "smiling," Germans ranked first, British ranked second, and the least popular. He reached the conclusion that the favorite tourists are Greeks.

Köroğlu, Avcıkurt, Köroğlu, and Karaman (2007) concluded that the nationalities of the tourists that the guides most wanted to work with are Americans, Japanese, and Germans. In addition, American tourists "spend a lot of money," "tolerant," "seek comfort," British tourists are "sincere and friendly," "respectful and polite," "cold character," "looking for comfort," German tourists are "fun and cheerful," "principled and disciplined," "seeking comfort," Russian tourists are "fun and cheerful," "disorganized," "reckless," "disrespectful and rude," "respectful and polite," "cold character" of the French; they concluded that Japanese tourists are "tolerant," Israeli tourists are "complacent," "intolerant," "disrespectful and rude," and "do not like to spend money."

Güzel and Köroğlu (2015) investigated the characteristics and behavior patterns of nationalities from the perspective of the guides. They concluded that the majority of positive statements regarding the characteristics are concentrated in Northern Europe (England, Scandinavian countries, and Scotland) and Western Europe (Austria, Belgium, France, Germany, and Holland). In negative statements, they stated that Azerbaijan, the Czech Republic, Poland, Russia, and Kazakhstan came to the fore. In terms of behavior patterns, positive expression of the majority of Far-East Asia (Korea, Japan, and China) for the country, the majority are in negative statements Southern Europe (Turkish Republic of Northern Cyprus, Italy, Turkey, Spain, Israel, Greece, and Serbia) stated that for the country. The stereotypes of uncertainty and anxiety caused by lack of information, uncertainty about how to behave, and relationships with tourists from different nationalities or indigenous-subcultures who are temporarily present in tourism destinations can be seen as a guide depending on personal characteristics. Depending on the personality traits of the individuals, the way of obtaining information from scratch by getting to know the stranger and establishing relationships can be preferred. In the following section, information about the Marmaris destination where the field study of the research was carried out was given.

4.5 MARMARIS DESTINATION

The study area-Marmaris, is on the southwest of Turkey is a major port and the coastal city of Muğla province. The district of Marmaris is on 36°51'48 latitude and 28°16'30 longitude coordinates and its surface area is 963.73 km^2. Reşadiye Peninsula and Kerme Bay are located in the west of the district, Balan Mountain, Karadag, and Günlük hills are in the east, Ula in the north and the Mediterranean in the south. Adaköy, which is connected to the mainland by a thin isthmus to the west of the bay, is located in front of the islands of Bedir, Keçi ve Güvercin. There are regular bus services from Istanbul, Ankara, Izmir, Denizli, Muğla, and Antalya to Marmaris. Marmaris is 56 km from Muğla city center, 90 km from Dalaman Airport, 140 km from Bodrum-Milas Airport, 281 km from Antalya Airport and 284 km from İzmir Adnan Menderes Airport. There are boat trips between Bodrum-Çamlık and catamaran services between Marmaris and Rhodes (T.C. Muğla Büyükşehir Belediyesi, 2020; Marmaris Belediyesi, 2020).

Marmaris is under the influence of the Mediterranean climate where summers are hot and dry and winters are warm and rainy, and it has suitable climatic features for coastal tourism. About three quarters of its land is covered with forest and maquis vegetation. Its features, natural environment, vegetation, suitable climate, low humidity, about 10 km long beach, Cennet (Nimara / Yıldız), Bedir, Keçi islands, boat tours to Turunç, Kumlubük, Çiftlik bays (Blue Voyage) are important geographical features that played a leading role in the initiation and development of coastal tourism in Marmaris (Bozkurt, 2019).

Since ancient times, Marmaris has been an important transit point between the Aegean and the Mediterranean. The indented coastal structure, well-preserved bays and harbors, made it an important place among the connection points between Asia, Europe, and Africa. Marmaris, where the first settlement dates back to 3200 B.C., was named "Physkos," which means "city of nature" during the Carian Civilization. Herodotus mentions the Marmaris as "Happy people living under the most beautiful view" (Özgürel, 2018). Marmaris looks like a unique museum of civilizations displaying traces from the civilizations of Caria, Rhodes and Island, Egypt, Ionian, Doric, Persian, Macedonian, Syrian, Roman, Byzantine, Seljuk, and Ottoman (Birsin, 2017).

There are 13 districts in Muğla province including the central district, Marmaris is the 5th in terms of population (www.nufusu.com, 2020). In the district with 25 neighborhoods, a significant portion of the settlements transformed from village status to neighborhoods are located on the coastline, while a small number of other residential areas are located close to the sea or within the forest area. There are various tourism activities such as coastal tourism, rural tourism, agro tourism, camping-caravan tourism, cultural tourism, yacht tourism, bicycle tourism, and event tourism in the Marmaris. City center, Armutalan, Beldibi, İçmeler, Turunç, Selimiye, and Bozburun neighborhoods stand out, especially with coastal tourism.

Agriculture, animal husbandry, sponge fishing, forestry, and fishing were important economic activities until the 1960s, while tourism activities that started with house boarding have taken their place as the most important sector today.

The population of the district of Marmaris, which was 12,780 in 1965, reached 94,749 people in 2019 according to the Address Based Population Registration System (ADNKS) and the annual population growth rate of the district is 5.3 per thousand compared to the previous year (Türkiye İstatistik Kurumu, 2019). In parallel with the development of tourism in Marmaris, it has become a target destination for those who migrate from neighboring provinces and abroad for various purposes, especially after the 1990s. Migrations from neighboring and eastern provinces for purpose of working in tourism sector increased the population. Furthermore, foreign brides settle in Marmaris due to international marriage, immigrant resident aliens for retirement, or a better life has been effective in increasing population.

In the district, where there were 10 certified hotels and 1 holiday village in the 1960s, the accommodation needs of the tourists were tried to be met by house boarding. Depending on the development of tourism, there has been a significant increase in the number of facilities and bed capacity in the destination. Services are provided with a total of 935 facilities with tourism operation and municipality certificates, 39,419 rooms, and 80,095 beds in the Marmaris. Due to COVID-19, a total of 89 accommodation facilities (Ministry Certified and Municipal Certified) participated in the Safe Tourism Certification Program, 61 facilities received certification, and the certification process of 28 facilities continues (T.C. Kültür ve Turizm Bakanlığı, 2020). There are 216 A Group travel agency in the district (Özgürel, 2020).

A total of 1,024,150 tourists, including 815,458 foreign and 208,692 local (Turkish) tourists, stayed overnight at the facilities with Tourism Management Certificate in Marmaris in 2019 (Yatırım ve İşletmeler Genel

Müdürlüğü, 2019). In Municipal Certified Facilities, a total of 744.536 tourists, including 534.101 foreigners and 210.435 local (Turkish), stayed overnight (Yatırım ve İşletmeler Genel Müdürlüğü, 2019).

In Marmaris, which is especially preferred by British, Russian, German, and Dutch tourists, local tourists visit the district for the trio of sea, sand and sun during Ramadan and Eid al-Adha, which coincides with the summer season. In recent years, it is known that tourists from Middle Eastern countries such as Israel, Iran, Jordan, and Lebanon have increasingly preferred Marmaris destination.

4.6 MATERIALS AND METHODS

4.6.1 STATISTICAL ANALYSIS OF DATA

The study of Var et al. (1985) was the source of inspiration for this study. In this study, quantitative research method was adopted and benefited from the scale used by Var et al. (1985). In the field study, a scale consisting of three parts and 51 questions was used.

In the first part, socio-demographic characteristics (9 questions), in the second part, perceptions about tourism and its effects (26 questions—5 Likert 1 = Strongly Disagree to 5 = Strongly Agree), in the third and final part, there are questions aimed at determining the stereotypes of local people about different tourist nationalities (8 Questions—5 Likert 1=Absolutely Not Preferred 5= Definitely Preferred, 4 Questions E=Yes/ H=No and 4 Questions 5 Likert 1= Absolutely Not Participating 5= Definitely Participating).

Fieldwork was conducted during 07 and 31 July 2020, the surveys were conducted online due to the COVID-19 pandemic, and data were collected from 393 local people. The data obtained in the study were analyzed using the Statistical Package for Social Sciences (SPSS) for Windows 22.0 program. Number, percentage, mean, and standard deviation were used as descriptive statistical methods in the evaluation of the data. The *t*-test was used to compare the quantitative continuous data between two independent groups, and the One-way Anova test was used to compare the quantitative continuous data between more than two independent groups. After the Anova test, the Scheffe test was used as a complementary post-hoc analysis to determine the differences.

4.6.2 DATA COLLECTION TOOL

The internal consistency coefficient "Cronbach Alpha" was calculated to calculate the reliability of 20 items in the perception scale about tourism and its effects. The overall reliability of the scale was found to be very high as alpha = 0.877. To reveal the construct validity of the scale, the explanatory factor analysis method was applied. As a result of the Barlett test ($p = 0.000 < 0.05$), it was determined that there is a relationship between the variables included in the factor analysis. As a result of the test (KMO = 0.876 > 0.60), it was determined that the sample size was sufficient for factor analysis. The varimax method was chosen in the factor analysis application and the structure of the relationship between the factors was kept the same. As a result of factor analysis, variables were grouped under two factors with a total variance of 52.156%. According to the alpha found for its reliability and the variance value explained, it was understood that the perception scale for tourism and tourism is a valid and reliable tool.

Var et al. (1985) included 6 statements from 26 statements in the perception of tourism and its effects (TA) scale:

TA4. "Tourists have led to changes in our lifestyle.,"

TA6. "I would like to see that the government attaches more importance to protecting the Marmaris environment.,"

TA9. "Tourism increases property values and housing prices.,"

TA10. "Local businesses must pay additional taxes to cover some public expenditure." ,

TA16. "I am proud to live in Marmaris.,"

TA22. "I feel there is a decline in the ecological environment.," items were excluded because co-loading and factor load were low.

The factor structure of the scale of perception toward tourism and its effects is shown in Table 4.1 below.

TABLE 4.1 Factor Structure of the Perception Scale Toward Tourism and Its Effects.

Dimensions	Factor loadings
Perception Dimension Toward Tourism Gains	
(Eigenvalue = 7.030; Explained variance = 33.015; Alpha = 0.903)	
TA20. Tourism creates jobs.	0.818
TA7. Tourism earns foreign currency.	0.795
TA12. Tourism provides income and increases revenues.	0.769

TABLE 4.1 *(Continued)*

Dimensions	Factor loadings
TA2. Tourists contribute to international peace and the development of peace understanding.	0.743
TA18. I am aware that growing tourism increases government revenues.	0.738
TA3. I'm willing to meet tourists from Turkey and many countries of the world.	0.720
TA8. Tourism encourages local people to participate in various cultural activities.	0.710
TA1. Meeting with tourists provides an educational experience.	0.708
TA24. I would like to see more government spending to protect the environment.	0.669
TA17. I feel very good when I see the tourists enjoying the views of Marmaris.	0.663
TA21. Tourism reduces unemployment.	0.640
TA23. The government can control the impact of tourism on the environment by making long-term plans.	0.571
TA5. I endure discomfort to earn/earn money from tourism and tourists.	0.403
Perception dimension of negative effects of tourism	
(Eigenvalue = 3.401; Explained variance = 19.141; Alpha = 0.845)	
TA25. Tourism increases the crime rate in Marmaris.	0.825
TA26. Crime is increasing in Muğla due to tourism.	0.805
TA13. Tourism increases traffic congestion.	0.746
TA14. Tourism increases the prices of basic goods.	0.704
TA19. I think tourists have negative effects on children and young people.	0.680
TA15. Tourism increases the crowd.	0.634
TA11. Tourism promotes a greedy business ethic.	0.613
Total variance = 52.156%; overall reliability (Alpha) = 0.877	

In the factor analysis of the dimension of tourism gains, attention was paid to handling factors with eigenvalues greater than one, the high factor loadings indicating the weight of the variables in the factor, and not being close to each other for the same variable. The fact that the reliability coefficients of the factors that make up the scale and the high variance rates explained show that the scale has a strong factor structure. It was considered as the items in the first factor, the reliability of the 13 items constituting the dimension of perception toward tourism gains was determined as alpha = 0.903, and the variance value explained was 33.015%. It was considered as the items in the second factor, and the reliability of the seven items that constitute the dimension of perception about the negative effects of tourism was determined as alpha = 0.845, and the explained variance value was found as 19.141%. While calculating the scores of the factors in the scale, the factor scores were obtained by dividing the values of the items in the factor by the number of items (arithmetic mean).

4.7 FINDINGS AND COMMENTS

In this section, the results obtained from the analysis of data collected through scales from the participants in the research are included in order to solve the research problem. Explanations and comments were made based on the findings obtained. Table 4.2 below shows the findings regarding the socio-demographic characteristics of the participants from the local people of Marmaris.

TABLE 4.2 Findings Regarding Socio-Demographic Characteristics of Participants.

Groups	Frequency (N)	Percent (%)	Groups	Frequency (N)	Percent (%)
Gender			Foreign language speaking status		
Male	217	55.2	Yes	93	23.7
Female	176	44.8	No	300	76.3
Total	393	100.0	Total	393	100.0
Age			English-speaking level		
20 years and under	34	8.7	Can't speak	93	23.7
Between 21 and 30 years	113	28.8	Little	6	1.5
Between 31 and 40 years	94	23.9	Middle	105	26.7
Between 41 and 50 years	98	24.9	Good	104	26.5

TABLE 4.2 *(Continued)*

Groups	Frequency (N)	Percent (%)	Groups	Frequency (N)	Percent (%)
51 years and over	54	13.7	Very good	85	21.6
Total	393	100.0	Total	393	100.0
Duration of residence			Level of speaking Russian		
5 years and below	71	18.1	Can't speak	241	61.3
Between 6 and 10 years	50	12.7	Little	103	26.2
Between 11 and 15 years	40	10.2	Middle	33	8.4
16 years and over	113	28.8	Good	10	2.6
Since he was born	119	30.2	Very good	6	1.5
Total	393	100.0	Total	393	100.0
Education level			Level of speaking Arabic		
Secondary school	97	24.7	Can't speak	317	80.7
Associate degree	99	25.2	Little	58	14.8
Undergraduate	143	36.4	Middle	8	2.0
Graduate	54	13.7	Good	2	0.5
			Very good	8	2.0
Total	393	100.0	Total	393	100.0
Income level			Level of speaking German		
2500 TL and below	54	13.7	Can't speak	369	93.9
Between 2501 and 5000 TL	143	36.4	Little	9	2.3
Between 5001 and 7500 TL	90	22.9	Middle	5	1.3
Between 7501 and 10,000 TL	60	15.3	Good	4	1.0
10,000 TL and over	46	11.7	Very good	6	1.5

TABLE 4.2 *(Continued)*

Groups	Frequency (N)	Percent (%)	Groups	Frequency (N)	Percent (%)
Total	393	100.0	Total	393	100.0
Sector			Level of speaking another language (Spanish, Italian, French, etc.)		
Tourism	123	31.3	Can't speak	379	96.4
Out of tourism	270	68.7	Little	5	1.3
Total	393	100.0	Middle	4	1.0
Working style in the tourism sector			Good	3	0.8
Tourism employee	83	67.5	Very good	2	0.5
Tourism employer	40	32.5			
Total	123	100.00	Total	393	100.0

55.2% of the participants (n = 217) were "men" and 44.8% (n = 176) were "women." Total population of Marmaris is 94,749 and 52.47% of the population is men and 47.53% are women. It is close to the male–female proportional distribution (45% female ratio) obtained in the field study and in the studies of Var et al. (1985). According to the "age" variable, 48.8% of the participants are in the "31–50 age range." According to the "duration of residence" variable, 30.3% (n = 119) of the participants "since their birth" and 28.8% (n = 113) have been living in Marmaris for "16 years and over." In the study of Var et al. (1985), it was stated that 64% of the respondents were residents of Marmaris since birth and only 18% lived in Marmaris for less than 5 years. According to the findings, the ratio of local people living in Marmaris for "5 years and less" is 18.1% (n = 71) similar to the results obtained from the studies of Var et al. (1985). With the development of tourism in Marmaris, there has been migration from many cities of Turkey, especially from the surrounding provinces. With the settlement of individuals coming for seasonal purposes over time, Marmaris witnesses the settlement of individuals from the different subculture. According to the "education level" variable, 75.3% (n = 296) of the local community participants have a "higher education - associate, undergraduate and graduate" degree.

Var et al. (1985) found that "8% of the local population did not receive primary education, which was mandatory for 5 years in those years." The findings obtained in this study prove that previous findings have changed over the years and that the level of education in Turkey has increased.

In the field study, it was determined that all of the participants were trained and the rate of those who completed 12-year compulsory education was 24.7% (n = 97). "Income level," according to a variable, has come from 36.4% of the participants' (n = 143) "TL 2501–5000," 22.9% (n = 90) "TL 5001–7500." About 31.3% of the participants (n = 123) are dealing with " tourism," 67.5% (n = 83) of the participants who are engaged in tourism are "tourism workers / employees," 32.5% (n = 40) is "tourism operator / employer." In their study, Var et al. (1985) found the rate of those dealing with tourism among the participants as 22%. About 76.3% of the participants (n = 300) "can speak a foreign language" and 74.8% (n = 294) "English," 12.5% (n = 49) "Russian," 4.5% (n = 18) can speak "Arabic," 3.8% (n = 15) "German" at intermediate and above level. In addition, 2.3% of the participants (n = 9) can speak other European languages (Spanish, French, Italian, etc.) at intermediate and above level. Var et al. (1985) reported that 32% of the participants were able to speak a foreign language, including participants who were not fluent. It is predicted that the high level of education of the local people of Marmaris, the unique nature of the destination, the intertwining of the tourist attractions with the living spaces of the local people, the fact that some of the local people are in the tourism sector and therefore in the communication environment with the tourist, both directly and indirectly, affect their ability to speak foreign languages.

In Table 4.3, the perception dimensions and point averages of the participants regarding tourism and tourism effects are given.

TABLE 4.3 Perception Point Average for Tourism and Tourism Effects.

Dimensions	N	Mean	SS	Min.	Max.
Perception dimension toward tourism gains	393	4.060	0.666	1.000	5.000
Perception dimension of negative effects of tourism	393	3.343	0.845	1.000	5.000

The average "dimension of perception toward tourism gains" of the participants was found to be 4.060 ± 0.666 (min = 1; max = 5). As perceived by local people participants, "Tourism creates jobs" ($\bar{x} = 4.422$), "tourism creates a foreign exchange" ($\bar{x}$ = 4.394), "tourism provides income, increases incomes" ($\bar{x}$ = 4.275) expressions very high level; "I am aware that growing with tourism increases government revenues." ($\bar{x}$ = 4.165), "tourism reduces unemployment." ($\bar{x}$ = 4.115) expressions high level; "I bear inconvenience to earn/earn money from tourism and tourists." It was determined that they agree with the expression ($\bar{x}$ = 2.924) at a moderate level. The findings show that local people perceive the positive economic effects of tourism at a very high level. The local people of Marmaris say that "tourists contribute to international peace and development of peace understanding" (x = 3.972), "Turkey and tourists from around the world to meet with a number of countries willing" ($\bar{x}$ = 3.952), "meeting tourists provides an educational experience" ($\bar{x}$ = 3.982), "I feel very good when I see tourists enjoying the sights of Marmaris" ($\bar{x}$ = 4.048) expressions highly agree. The findings show that the positive social effects of tourism are perceived at a high level by the local people. "Tourism encourages local people to participate in various cultural activities." It has been concluded that the expression ($\bar{x}$ = 4.025) is highly attended by local people, and in this context, their awareness of the positive cultural impact of tourism is high. Local people—"I would like to see more government spending to protect the environment" ($\bar{x}$ = 4.433) stated at a very high level. In contrast, they highly agree with the expression "government can control the impact of tourism on the environment by making long-term plans" ($\bar{x}$ = 4.076).

The local people who want to receive more government support for the protection of the Marmaris environment and believe that they can control the impact of tourism on the environment in line with the strategic plans of the public have a high awareness of the environmental impacts of tourism. The positive perception of the effects of tourism-social, cultural, economic, and environmental aspects—of the local people can be considered as a precursor to their support for tourism and related policies and to contribute to the development of tourism. It is thought that the positive perception of tourism by the local people of Marmaris is an important finding in terms of the support they will provide for tourism, the competitive advantage to be created with the power to be obtained from the local and the sustainability of Marmaris tourism.

The average of "dimension of perception toward the negative effects of tourism" was found to be medium 3.343 ± 0.845 (min = 1; max = 5). It was determined that local people agree with expressions—"tourism increases the crowd" ($\bar{x}$ = 4.244), at a very high level, "tourism increases traffic congestion" ($\bar{x}$ = 3.761) at high level, "tourism increases the crime rate in Marmaris" ($\bar{x}$ = 2.913) and "crime is increasing due to tourism in Muğla" ($\bar{x}$ = 2.875) at a moderate level.

The local people of Marmaris are aware of the social costs/negative effects created by tourism in the societies where it develops. Seasonality effect of tourism brings with it crowd and traffic congestion during the summer months in Marmaris, which stands out especially with coastal tourism. Intensive tourist flows, seasonal tourism workers, the presence of second home owners in Marmaris during the high season, especially on weekends, excursionists from the neighboring provinces can strain their social, and physical transport capacity are some factors. Local people have a moderate perception of tourism crimes and crime rates. It is thought that by lowering the current situation, even more, the arguments that may be subject to crime and crime can be reduced to a much lower level with public measures to be taken.

It was found that participants agreed with the expressions; "I think tourists have negative effects on children and young people" ($\bar{x}$ = 2.715) and "tourism promotes greedily governed business ethics" ($\bar{x}$ = 2.985) at a moderate level.

In this context, it can be said that the local people of Marmaris perceive the negative effects of tourism on the culture at a moderate level. It can be said that the presence of a multicultural environment in Marmaris and the encounter with different cultures with the arrival of tourists do not contain negativities that cannot be accepted by the local people. Var et al. (1985) stated that 37.5% of the participants agree that the relaxation/deterioration in commercial ethics.

It was determined that the local people of Marmaris agreed with the expression "tourism increases prices of basic goods" at a high ($\bar{x}$ = 3.906) level. The findings show that the participants perceive the negative economic impact of tourism. The negative perception of tourism effects by local people means not getting support from the local, negative perceptions have the potential to turn into negative tourism attitudes and/or behaviors. Findings obtained show that it is necessary to act with a planned and sustainable tourism understanding in order not to interrupt the desired and planned tourism development in Marmaris.

Table 4.4 shows the differentiation of the perceptions of the participants, consisting of the local people of Marmaris, toward the effects of tourism and tourism.

TABLE 4.4 Differentiation Status of Perception Toward Tourism and Its Effects According to Descriptive Features.

Demographic features	n	Perception of tourism gains	Perception of the negative effects of tourism
Gender		Mean ± SS	Mean ± SS
Male	217	4.085 ± 0.764	3.292 ± 0.915
Female	176	4.029 ± 0.521	3.406 ± 0.747
t=		0.831	−1.334
p=		0.389	0.174
Age		Mean ± SS	Mean ± SS
20 years and below	34	3.959 ± 0.726	3.189 ± 0.765
Between 21 and 30 years	113	4.193 ± 0.576	3.403 ± 0.892
Between 31 and 40 years	94	4.079 ± 0.579	3.439 ± 0.850
Between 41 and 50 years	98	3.900 ± 0.778	3.356 ± 0.859
51 years and over	54	4.104 ± 0.677	3.122 ± 0.724
F=		2.883	1.675
p=		0.022	0.155
PostHoc=		2 > 4 (p < 0.05)	
Duration of residence		Mean ± SS	Mean ± SS
5 years and below	71	4.203 ± 0.632	3.312 ± 0.881
Between 6 and 10 years	50	3.989 ± 0.574	3.494 ± 0.824
Between 11 and 15 years	40	3.619 ± 1.010	3.357 ± 0.959
16 years and over	113	4.053 ± 0.572	3.242 ± 0.828
Since born	119	4.160 ± 0.602	3.389 ± 0.807
F=		6.342	0.923
p=		0.000	0.450
PostHoc=		1 > 3, 2 > 3, 4 > 3, 5 > 3 (p < 0.05)	
Level of education		Mean ± SS	Mean ± SS
Secondary school	97	3.961 ± 0.743	3.155 ± 0.794
Associate degree	99	4.147 ± 0.597	3.420 ± 0.887
Bachelor's	143	4.063 ± 0.697	3.352 ± 0.871

TABLE 4.4 *(Continued)*

Demographic features	n	Perception of tourism gains	Perception of the negative effects of tourism
Graduate	54	4.073 ± 0.536	3.516 ± 0.733
F=		1.284	2.674
p=		0.279	0.047
PostHoc=			2 > 1, 4 > 1 (p < 0.05)
Gelir Düzeyi		Mean ± SS	Mean ± SS
2500 TL and below	54	3.906 ± 0.823	3.111 ± 0.825
Between 2501 and 5000 TL	143	4.023 ± 0.593	3.447 ± 0.808
Between 5001 and 7500 TL	90	4.015 ± 0.825	3.321 ± 0.886
Between 7501 and 10,000 TL	60	4.146 ± 0.464	3.431 ± 0.875
10,000 TL and over	46	4.334 ± 0.453	3.221 ± 0.821
F=		3.211	1.995
p=		0.013	0.095
PostHoc=		5 > 1, 5 > 2, 5 > 3 (p < 0.05)	
Foreign language speaking status		Mean ± SS	Mean ± SS
No	93	3.869 ± 0.744	3.333 ± 0.910
Yes	300	4.120 ± 0.629	3.346 ± 0.825
t=		−3.217	−0.123
p=		0.001	0.902
Sector		Mean ± SS	Mean ± SS
Tourism	123	4.151 ± 0.580	3.412 ± 0.838
Out of tourism	270	4.019 ± 0.699	3.311 ± 0.847
t=		1.822	1.102
p=		0.069	0.271
Employment type in the tourism sector		Mean ± SS	Mean ± SS
Employee	83	4.075 ± 0.617	3.492 ± 0.825
Employer	40	4.308 ± 0.463	3.246 ± 0.851
t=		−2.113	1.532
p=		0.037	0.128

According to the "gender" variable of the participants, perception points toward tourism gains and negative effects of tourism do not differ significantly ($p > 0.05$). The fact that the distribution of the participants according to the gender variable is close to each other in percentage is important for the analysis results. In some studies, it is stated that people of different genders perceive tourism differently (Milman and Pizam, 1988), while there are significant differences in other studies (Özaltın Türker and Türker, 2014).

According to the "age" variable, the perception points of the participants toward tourism gains differ significantly ($F = 2.883$; $p = 0.022 < 0.05$). The reason for the difference; the perception points of those in the "21–30 age range" toward tourism gains are higher than those in the "41–50 age range" ($p < 0.05$). It was determined that the perception scores of the participants about the negative effects of tourism according to the "age" variable did not differ significantly ($p > 0.05$). Özaltın Türker and Türker (2014) concluded that the age variable showed significant differences in the total impact score of tourism.

Participants' perception points toward tourism gains in the "durations of residence" variable ($F = 6.342$; $p = 0 < 0.05$) differ significantly. The reason for the difference; the perception points of those with a "duration of residence" 5 years and less toward tourism gains are higher than those of those with a residence time "11–15 years" ($p < 0.05$). Perception scores for tourism gains for those with "6–10 years" are higher than perception scores for tourism gains for those with "11–15 years" ($p < 0.05$). The fact that the perception points toward tourism gains of those with a "duration of residence "16 years and more" are higher than those of those with a "duration of residence of 11–15 years" ($p < 0.05$). Perception points toward tourism gains of those whose residence period "since birth" is higher than those who have a "residence period of 11–15 years" ($p < 0.05$). It can be said that the perception of tourism gains according to their score averages has the lowest perception of tourism gains among those with a "residence time of 11–15 years." The perception points of the participants about the negative effects of tourism do not differ significantly according to the "duration of residence" variable ($p > 0.05$).

The perception points of the participants about the negative effects of tourism differ significantly according to the "education level" variable ($F = 2.674$; $p = 0.047 < 0.05$). The reason for the difference; the perception scores of those whose education level is "associate degree" about the negative effects of tourism is higher than those whose education level is "secondary education" about the negative effects of tourism ($p < 0.05$). The perception scores of those whose education level is "graduate" about the negative effects of tourism is higher than those whose education level is "secondary education" about the negative effects of tourism ($p < 0.05$). The perception points of the participants toward tourism gains do not differ significantly according to the "education level" variable ($p > 0.05$). In some studies, in the literature, it is stated that the level of education of local people affects the level of perception of the effects of tourism (Caneday and Zeiger, 1991).

Participants' perception scores toward tourism gains differ significantly according to the "income level" variable ($F = 3.211$; $p = 0.013 < 0.05$). The reason for the difference; the perception points of those with an income level of "over 10,000 TL" toward tourism gains are higher than those of those with an income level of "2500 TL and below" ($p < 0.05$). The fact that the perception points of those with an income level of "over 10,000 TL" toward tourism gains are higher than those of those whose income level is between "2501–5000 TL" ($p < 0.05$). The fact that the perception points toward tourism gains of those whose income level is "over 10,000 TL" is higher than the perception points of those whose income level is between "5001–7500 TL" ($p < 0.05$). The perception points of the participants about the negative effects of tourism do not differ significantly according to the "income level" variable ($p > 0.05$). Özaltın Türker and Türker (2014) concluded that the total impact score of tourism does not differ significantly according to the income variable.

According to the variable of "speaking a foreign language"; the perception scores ($x = 3.869$) toward tourism gains of "non-foreign language speakers" were found to be lower than those of "foreign language speakers" ($x = 4.120$) ($t = -3.217$; $p = 0.001 < 0.05$). Being able to speak a foreign language enables establishing formal and informal relationships and forms an integral part of communication with tourists in social, cultural and economic relations. It is normal for local people to perceive that the gains they have achieved in this context increase and increase according to their language skills. The perception points of the participants about the negative effects of tourism do not differ significantly according to the "speaking a foreign language" variable ($p > 0.05$).

The perception points of the participants about their perception of tourism gains and negative effects do not show a significant difference according to the "sector of work" variable ($p > 0.05$). The perception points ($x = 4.075$) of those who work in the tourism sector as a "tourism worker (employee)" were found to be lower than the perception points ($x = 4.308$) of those whose employment status is "tourism operator (employer)" ($t = -2.113$) $p = 0.037 < 0.05$). Participants who are in the position of employee/operator are considered to have higher average scores due to the gains they have achieved in economic, social, and cultural aspects (most of the businesses in Marmaris are small/medium sized enterprises and the operators are in close contact with tourists) due to tourism activities. The perception points of the participants about the negative effects of tourism do not differ significantly according to the variable "employment type in the tourism sector" ($p > 0.05$).

In the following section, the findings of the perceptions and stereotypes of the participants from the local people of Marmaris about tourists are included. It has been tried to determine the level of preference of the local people of Marmaris to tourists from different countries and with different characteristics with a 5-point Likert-type scale (1 = I Definitely Do Not Prefer 5 = I Definitely Prefer).

Table 4.5 presents the distribution of the responses of the participants to the statements regarding tourist preferences.

TABLE 4.5 Distributions of Participants' Answers to the Expressions Related to Tourist Preferences.

	Absolutely I do not prefer		**I do not prefer**		**Neither I prefer nor I do not prefer**		**I prefer**		**Absolutely I prefer**			
	f	%	f	%	f	%	**f**	**%**	**f**	**%**	**Av.**	**Ss**
Tourists from Western Europe	34	8.7	26	6.6	60	15.3	158	40.2	115	29.3	3.748	1.195
Tourists from Eastern Europe	33	8.4	41	10.4	85	21.6	170	43.3	64	16.3	3.486	1.136

TABLE 4.5 *(Continued)*

	Absolutely I do not prefer		I do not prefer		Neither I prefer nor I do not prefer		I prefer		Absolutely I prefer			
Tourists from United States	44	11.2	42	10.7	80	20.4	146	37.2	81	20.6	3.453	1.245
Local Tourists	38	9.7	53	13.5	77	19.6	153	38.9	72	18.3	3.427	1.210
Turkish Tourists Living Abroad	37	9.4	57	14.5	71	18.1	166	42.2	62	15.8	3.405	1.190
Tourists from Far East	59	15.0	48	12.2	98	24.9	131	33.3	57	14.5	3.201	1.265
Hippies	88	22.4	63	16.0	123	31.3	85	21.6	34	8.7	2.781	1.255
Tourists from Middle East	112	28.5	78	19.8	101	25.7	73	18.6	29	7.4	2.565	1.278

According to the average score of the expressions in line with the answers of the participants in the study about tourist preferences; (1) ranked "Western European," (2) "East European" ranked third, "American" ranked fourth, "Native (Turkish)" ranked fourth, "Turkish Living Abroad" ranked fifth, and local people agreed with the statements at a high level. When the mean scores of the expressions of the participants are examined; "Far Eastern," "Hippies," and "Middle Eastern" tourists are in the lower ranks. Participation of "Far Easterners" and "Hippies" in the statement toward "Middle Eastern" tourists was found to be at a moderate level, and they were found to be at a low level. According to the Var et al. (1985) tourist perceptions and stereotypes of local people in Marmaris (the most preferred ones) are as follows: tourists from Western Europe, tourists from the Far East, tourists from America, tourists from Eastern European countries, tourists from Middle Eastern countries, local tourists, Turkish workers live abroad, and Hippies.

When the findings of both studies are compared, it is seen that Western European and American tourists maintain the ranking. It has been determined that Eastern European, local (Turkish) tourists, and Turkish tourists living abroad climb to the top, Far Eastern and Middle Eastern tourists fall to the bottom, and Hippies go up only one rank from the bottom row. In the works of Hippies were perceived as a separate tourist group without nationality and was included in the scale for comparison in the study of Var et al. (1985).

Table 4.6 shows the findings of the participants about the preferences of the Turkish living abroad and the domestic (Turkish) tourists living in the country. Table 4.6 shows that 77.1% of respondents '(*n* = 303) prefer "Turks living abroad to come on holiday in Turkey," but 72.0% (*n* = 283) "prefer of Turks living abroad come to Marmaris on vacation." Var et al. (1985) stated that local people strongly support "*Turkish workers abroad to come to Turkey for holiday.*" but they added that there are strong stereotypes about coming to Marmaris and that they are the least preferred tourists.

It has been determined that the stereotypes of the local people regarding the arrival of Turks living abroad to Marmaris for vacation since 1985 have changed over time. In contrast, 60.6% (*n* = 238) of the participants of the local people of Marmaris stated that "they did not learn better and new things from Turks living abroad." The findings obtained support the results found by Var et al. (1985). About 70.0% (*n* = 275) of the local people, "prefer local (Turkish) tourists to come to Marmaris on holiday." The findings obtained do not match the findings of Var et al. (1985). The researchers stated that local (Turkish) tourists are not preferred by the local people, either.

TABLE 4.6 Participants' Preferences for Turkish and Local (Turkish) Tourists Living Abroad.

Groups	Frequency (n)	Percent (%)
Would you prefer turks living abroad to come on vacation to turkey?		
Yes	303	77.1
No	90	22.9
Would you prefer Turks living abroad to come on vacation to Marmaris?		
Yes	283	72.0
No	110	28.0

TABLE 4.6 *(Continued)*

Groups	Frequency (n)	Percent (%)
Would you prefer local (Turkish) tourists to come on vacation to Marmaris?		
Yes	275	70.0
No	118	30.0
I learn better and new things from Turks living abroad		
Yes	155	39.4
No	238	60.6

In Table 4.7, expressions about how participants view tourists of different nationalities as "pleasant and thoughtful" are given by their level of participation, frequency and percentage. According to Table 4.7, the percentage of participants rated tourists "pleasant and considerate" is as follows: 57.5% (n = 226) "British tourists," 39.7% (n = 156) "Russian tourists," 39.5% (n = 155) "American tourists," and 15%, 5 (n = 61) "Arab tourists." In addition, when the answers given in the other option are examined; 30 participants from Marmaris local people "German tourists, 28 participants "Dutch tourists," 20 participants "Scandinavian tourists," 16 participants "European tourists," 15 participants "Finnish tourists," 10 participants "Japanese tourists," and 10 participants "Danish tourists" found pleasant and thoughtful. Eight participants stated that they did not find "local tourists" pleasant and considerate.

The destination of Marmaris, which has been preferred by British tourists for many years, has been hosting Russian tourists, whose number continues to increase with each year, especially after the 1990s, and the locals show a positive trend in stereotyping of British, Russian, and American tourists. The positive stereotypes of the local people of Marmaris toward European tourists are supported by their answers to the other option. Negative stereotypes of Turks about Arabs from the past were supported in the study, and 60.1% of the participants stated that they did not find Arabs pleasant and considerate. Var et al. (1985) noted that the participants expressed their perceptions of various tourists by national origin and found that 44% of the local population found American tourists pleasant and considerate.

TABLE 4.7 Distribution of Participants' Responses to the Expression of Finding Tourists of Different Nationalities as Welcoming and Considerate.

Groups	Frequency (n)	Percent (%)
British tourists' visiting Marmaris are pleasant and thoughtful		
I strongly disagree	18	4.6
I disagree	35	8.9
Neither agree nor disagree	114	29.0
I agree	152	38.7
Absolutely I agree	74	18.8
Total	**393**	**100.00**
Russian tourists' visiting Marmaris are pleasant and thoughtful		
I strongly disagree	32	8.1
I disagree	57	14.5
Neither agree nor disagree	148	37.7
I agree	114	29.0
Absolutely I agree	42	10.7
Total	**393**	**100.00**
American tourists' visiting Marmaris are pleasant and thoughtful		
I strongly disagree	24	6.1
I disagree	46	11.7
Neither agree nor disagree	168	42.7
I agree	113	28.8
Absolutely I agree	42	10.7
Total	**393**	**100.00**
Arab Tourists' visiting Marmaris are pleasant and thoughtful		
I strongly disagree	135	34.4
I disagree	101	25.7
Neither agree nor disagree	96	24.4
I agree	41	10.4
Absolutely I agree	20	5.1
Total	**393**	**100.00**

TABLE 4.7 *(Continued)*

Groups	Frequency (n)	Percent (%)
Other (nationality and participation status)		
German tourists' visiting Marmaris are pleasant and thoughtful		
I strongly disagree	1	3.0
I disagree	1	3.0
Neither agree nor disagree	1	3.0
I agree	19	57.7
Absolutely I agree	11	33.3
Total	**33**	**100.00**
Dutch tourists' visiting Marmaris are pleasant and thoughtful		
I strongly disagree	1	3.3
I disagree	1	3.3
Neither agree nor disagree	14	46.7
I agree	14	46.7
Total	**30**	**100.00**
Scandinavian tourists' visiting Marmaris are pleasant and thoughtful		
I agree	9	45.0
Absolutely I agree	11	55.0
Total	**20**	**100.00**
European tourists' visiting Marmaris are pleasant and thoughtful		
I strongly disagree	1	5.9
I agree	6	35.3
Absolutely I agree	10	58.8
Total	**17**	**100.00**
Finnish tourists' visiting Marmaris are pleasant and thoughtful		
I agree	8	53.3
Absolutely I agree	7	46.7
Total	**15**	**100.00**

TABLE 4.7 *(Continued)*

Groups	Frequency (n)	Percent (%)
Local tourists' visiting Marmaris are pleasant and thoughtful		
I strongly disagree	3	25.0
I disagree	5	41.7
Neither agree nor disagree	1	8.3
I agree	1	8.3
Absolutely I agree	2	16.7
Total	**12**	**100.00**
Japanese tourists' visiting Marmaris are pleasant and thoughtful		
Neither agree nor disagree	1	9.0
I agree	5	45.5
Absolutely I agree	5	45.5
Total	**11**	**100.00**
Danish tourists' visiting Marmaris are pleasant and thoughtful		
Neither agree nor disagree	1	9.1
I agree	4	36.4
Absolutely I agree	6	54.5
Total	**11**	100.00

4.8 CONCLUSION

In the study conducted by Var et al. (1985) in 1980–1981, it was tried to determine whether their results have changed from the past to the present. The study received various criticisms due to the lack of sampling. In this study, it was tried to be re-examined by reaching a sufficient number of samples. The following can be stated about the results obtained from the study:

About 59.11% of the respondents from the local population (16 years old and over and was born in Marmaris [$n = 232$]) lived for many years in Marmaris, 75.3% ($n = 296$) have Bachelor's, Assoc. or graduate degree, 31.3% ($n = 123$) of them are in the "tourism" sector, 67.5% ($n = 83$) of those are employees, and 32% of them ($n = 40$) are "tourism operators/ employers.

It was determined that 76.3% ($n = 300$) of the local respondents speak "foreign language" and 74.8% ($n = 294$) speak "English," 12.5% ($n = 49$) speak "Russian," 4.5% ($n = 18$) speak "Arabic," and 3.8% ($n = 15$) speak "German" at intermediate and advanced level. It is believed that the fact that most of the participants lived in Marmaris for many years makes the information valuable. The high levels of education and foreign language skills of the local people in Marmaris, which grows and develops with tourism, is promising in terms of tourism and tourist perspective.

When the answers given to the scale regarding the effects of tourism and tourism are analyzed; "Tourism gains" and "negative effects of tourism" dimensions have emerged.

The average "dimension of perception toward tourism gains" was found to be high 4.060 ± 0.666 (Min = 1; Max = 5), and the average "Dimension of perception toward the negative effects of tourism" was found to be 3.343 ± 0.845 (Min = 1; Max = 5).

According to the findings obtained, the local people agree in **very high level** with the expressions; "Tourism creates jobs." ($\bar{x} = 4.422$), "Tourism creates a foreign exchange." ($\bar{x} = 4.394$), "Tourism provides income, increases incomes." ($\bar{x} = 4.275$), "I would like to see more government spending to protect the environment." ($\bar{x} = 4.433$).

They agree in **high level** with the statements "I am aware that growing with tourism increases government revenues." ($\bar{x} = 4.165$), "Tourism reduces unemployment." ($\bar{x} = 4.115$), "Tourists contribute to international peace and the development of peace understanding." ($x = 3.972$), "Turkey and tourists from around the world to meet with a number of countries willing." ($\bar{x} = 3.952$), "Meeting tourists provides an educational experience." ($\bar{x} = 3.982$), "I feel very good when I see tourists enjoying the sights of Marmaris." ($\bar{x} = 4.048$), "Tourism encourages local people to participate in various cultural activities." ($\bar{x} = 4.025$), "The government can control the impact of tourism on the environment by making long-term plans." ($\bar{x} = 4.076$), "I would like to see more government spending to protect the environment." ($\bar{x} = 4.433$).

Their moderate agreement with the statement "I bear inconvenience to earn/earn money from tourism and tourists." ($\bar{x} = 2.924$) is important in terms of expecting positive attitudes and behaviors toward tourism and tourists. The findings, which can be considered as a sign that the tourism carrying capacity in the destination has not yet been exceeded, is important in terms of developing and making Marmaris tourism sustainable.

Regarding the negative effects of tourism; they stated that they agree with the statement "Tourism increases the crowd." ($\bar{x} = 4.244$) at **very high** level, and "Tourism increases traffic congestion." ($x = 3.761$) at **high** level, "Tourism increases the crime rate in Marmaris." ($\bar{x} = 2.913$) and "Crime is increasing due to tourism in Muğla. ($x = 2.875$) at a **medium** level. Crowd and traffic congestion must be eliminated for sustainable tourism. It is thought that by reducing the adverse effects of the seasonality of coastal tourism with different types of tourism and by providing different activities planned with a temporal and spatial distribution at the destination, crowding effect and traffic congestion can be reduced. In this context, it has been determined that the local people of Marmaris perceive the positive and negative effects of tourism, and they agree with the expressions pointing to the economic, social, cultural, and environmental aspects of tourism at different levels.

In case the participants' perceptions of tourism and its effects differ according to the descriptive characteristics:

- According to the "gender" variable, the perception points about tourism gains and negative effects of tourism do not differ significantly ($p > 0.05$),
- According to the "age" variable, the perception points of the participants toward tourism gains show a significant difference ($F = 2.883$; $p = 0.022 < 0.05$), while the perception points about the negative effects of tourism do not differ significantly ($p > 0.05$),
- According to the "residence period" variable, the perception points of the participants toward tourism gains show a significant difference ($F = 6.334$; $p = 0 < 0.05$), but the perception scores for the negative effects of tourism do not differ significantly ($p > 0.05$),
- According to the "education level" variable, the perception points of the participants toward tourism gains do not differ significantly ($p > 0.05$), and the perception scores of the negative effects of tourism differ significantly ($F = 2.674$; $p = 0.047 < 0.05$),
- According to the "income level" variable, the perception points of the participants toward tourism gains differ significantly ($F = 3.211$; $p = 0.013 < 0.05$), the perception points about the negative effects of tourism do not differ significantly ($p > 0.05$),
- According to the variable of "speaking a foreign language"; the perception scores of those who do not speak a foreign language ($x = 3.869$) toward tourism gains ($x = 3.869$) are lower than those of

those who can speak a foreign language ($x = 4.120$) ($t = -3.217$; $p = 0.001 < 0.05$), regarding the negative effects of tourism perception scores do not differ significantly ($p > 0.05$),

- The perception points of the participants about their perception toward tourism gains and their negative effects do not differ significantly according to the "sector of work" variable ($p > 0.05$).
- According to the variable "working style/employee or employer/ in the tourism sector," those who are "tourism workers (employees)" have lower perception scores ($x = 4.075$) toward tourism gains than those whose working status is "tourism operator (employer)" ($x = 4.308$) ($t = -2.113$; $p = 0.037 < 0.05$), and the perception scores of the participants about the negative effects of tourism did not differ significantly ($p > 0.05$).

The participants from the local people of Marmaris who answered the questionnaire; "Western European," "Eastern European," "American," "Domestic (Turkey) "and" Turks Living Abroad "tourists prefer a high level," Far Easterners "and" hippies "in the middle level. It can be said they do not prefer "Middle Eastern" tourists. Also, 77.1% of the respondents (*n* = 303) prefer "Turks coming on holiday to Turkey living abroad," 72.0% of them (*n* = 283) "Turks are coming on holiday to Marmaris living abroad," 70.0% (*n* = 275) preferred "local (Turkish) tourists to come to Marmaris on holiday." Only 39.4% (*n* = 155) of the participants who answered the questionnaire thought that "they learned better and new things from Turks living abroad."

The rates of being pleasant and considerate according to the nationalities are as follows: British (57.5%; *n* = 226), Russian (39.7%, *n* = 156), American (39.5%; *n* = 155) and Arab 15.5% (*n* = 61). In addition, when the answers given in the other option are examined; 30 participants from Marmaris local people consider "German tourists" are pleasant and considerate. According to the 28 participants "Dutch tourists," 20 participants " Scandinavian tourists," 16 participants "European tourists," 15 participants "Finnish tourists," 10 participants " Japanese tourists," 10 participants "Danish tourists" are pleasant and thoughtful, 8 participants stated that they did not find "local tourists" pleasant and considerate. The fact that tourists from European countries in the other option were found pleasant and thoughtful by a total of 119 participants, shows that their stereotyping toward European tourists is strong and positive. Var et al. (1985) stated that the participants expressed their perceptions of various tourists by national origin and that 44% of the local population found American tourists pleasant and considerate. The result obtained supports the findings of Var et al. (1985).

As a result, it has been presented in detail with the literature and field study results within the scope of the study that tourism creates various effects in the destinations that tourism developed. These effects are perceived by the local people, perceptions can change, although it is difficult to change stereotypes, they can change with time and experience.

The reliability and validation of the data collected, data collection techniques, participants are to assume that they give the correct answer to the research questions, carrying out the survey online due to COVID-19, and lack of time are the assumptions and limitations of the research.

KEY WORDS

- **effects of tourism**
- **perceptions and attitudes towards tourism**
- **Marmaris-Turkey**
- **Turkish stereotypes of different nations**

REFERENCES

Akova, O. Yerel Halkın Turizmin Etkilerini Algılamalarına ve Tutumlarına Yönelik Bir Araştırma. *Akademik İncelemeler Dergisi* **2006,** *2* (1), 1–34.

Aktan, T.; Sakallı Uğurlu, N. (2013). Kalıpyargı İçeriği Modeline Sosyal Bağlamsal Bir Yaklaşım: Bağlam İçinde Kalıpyargı İçerikleri. *Türk Psikoloji Yazıları* **2013,** *16* (31), 15–31.

Aktan, T.; Yalçındağ, B.; Ünal, T.; Süllü, M. E. Ulus Kalıpyargılarının İçerikleri: Sosyal-Yapısal Değişkenler Ve Kimliklenmenin Kalıpyargı İçerikleriyle İlişkisi. *Alternatif Politika* **2019,** *11* (3), 623–650.

Akyıldız, G. Uluslararası İlişkiler ve Banal Milliyetçilik: İlkokul Öğrencileri Üzerine Bir Araştırma. *Yayınlanmamış Yüksek Lisans Tezi*. TOBB Ekonomi ve Teknoloji Üniversitesi Uluslararası İlişkiler ABD, 2017, Nisan.

Albu, R. G.; Nicolau, L. C. Changing Attitudes in Tourism—A Possible Way to Win the Battle with the Current Economic Crisis. *Bull. Transilvania Univ. of Braşov Series V: Econ. Sci.* **2010,** *3* (52), 201–206.

Andereck, K. L.; Vogt, C. The Relationship between Residents' Attitudes toward Tourism and Tourism Development Options. *J. Travel Res.* **2000,** *39* (1), 27–36.

Apalı, Y. Sosyolojik Açıdan Kadınlarla İlgili Kalıp Yargılar. *Süleyman Demirel Üniversitesi İlahiyat Fakültesi Dergisi* **2011,** *26*, 49–64.

Avcıkurt, C. *Turizm Sosyolojisi: Genel ve Yapısal Yaklaşım (4 b.)*; Detay Yayıncılık: Ankara, 2015.

Bahar, O.; Kozak, M. *Turizm Ekonomisi* (7 b.); Detay Yayıncılık: Ankara, 2015.

Bakan, Ö.; Canöz, K. Kültürlerarası İletişim Bağlamında Türk Üniversite Öğrencilerinin Japonlara Yönelik Kalıp Yargıları. *Selçuk Üniversitesi Sosyal Bilimler Enstitüsü Dergisi* **2017,** *37*, 91–107.

Bal, H.; Akça, E. E.; Bayraktar, M. The Contribution of Tourism to Economic Growth. *Akademik Yaklaşımlar Dergisi* **2016,** *7* (1), 1–20.

Batmas, H. K.; Ulusoy Yıldırım, H.; Köroğlu, Ö. Yerel Halkın Kırsal Turizme İlişkin Tutumlarının İncelenmesi: Balıkesir Örneği. *J. Recreat. Tour. Res.* **2020,** *7* (2), 190–208.

Baykal, F.; Ataberk, E. Kozbeyli'de (Foça/İzmir) Kırsal Turizmin Geliştirilmesi Hakkında Yerel Halkın Görüşleri. *Ege Coğrafya Dergisi* **2020,** *29* (1), 1–17.

Berber, Ş. Sosyal Değişme Katalizörü Olarak Turizm ve Etkileri. *Selçuk Üniversitesi Sosyal Bilimler Enstitüsü Dergisi* **2003,** *9*, 205–221.

Birsin, H. Marmaris Tarihi, Jan 1, 2017. July 24, 2020 tarihinde www.muglamarmaris.com: www.muglamarmaris.com/marmaris-in-tarihi adresinden alındı.

Bitirim Okmeydan, S. Kültürlerarası İletişim ve Uluslararası Halkla İlişkiler. In *Kültürlerarası İletişim: İletişim Odaklı Güncel Yaklaşımlar*; Bitirim Okmeydan, S., Saran içinde, M., Eds.; Konya: Eğitim Yayınevi, 2019; pp. 93–124.

Blum, L. Stereotypes and Stereotyping: A Moral Analysis. *Philosophical Papers* **2004,** *33* (3), 251–289.

Bozkurt, G. Yerel ve Mekansal Dinamiklerin Turizm Temelli Girişimcilik Üzerine Etkileri: Marmaris Yerleşmesindeki Girişimcilik Yapısı ve Girişimciliği Etkileyen Faktörlerin Çözümlemesi. *Yayınlanmamış Doktora Tezi*; Gazi Üniversitesi Fen Bilimleri Enstitüsü: Ankara, 2019.

Brida, J. G.; Gomez, D. V.; Segarra, V. On the Empirical Relationship between Tourism and Economic Growth. *Tour. Manage.* Aug10, 2020 tarihinde. https://reader.elsevier.com/reader/sd/pii/S0261517720300571?token=7575AE9FAE5B24ECCFA3A6A8DF72B8F91482208CC7C0082B0661D42D7192C419C77235EA421C80C8077D89A0AD025434 adresinden alındı

Çakır, M.; Aydın, F. Yerel Halkın Termal Turizme Yönelik Görüşleri: Haymana İlçesi Örneği. *Doğu Coğrafya Dergisi* **2020,** *25* (43), 93–100.

Caneday, L.; Zeiger, J. The Social, Economic, and Environmental Costs of Tourism to a Gaming Community as Perceived By Its Residents. *J. Tour. Res.* **1991,** *30* (2), 45–50.

Carter, J. D.; Hall, J. A.; Carney, D. R.; Rosip, J. C. Individual Differences in the Acceptance of Stereotyping. *J. Res. Personal.* **2006,** *40* (6), 1103–1118.

Çavuş, A.; Baki, A. Yerel Halk Üzerindeki Sosyal Etkileri Bakımından Akçaabat Müzik ve Halk Oyunları Festivali. *Atatürk Üniversitesi Sosyal Bilimler Enstitüsü Dergisi* **2020,** *24* (1), 279–289.

Çeken, H. Turizmin Bölgesel Kalkınmaya Etkisi Üzerine Teorik Bir İnceleme. Afyon Kocatepe Üniversitesi İ.İ.B.F. *Dergisi* **2008,** *X* (II), 293–306.

Çeken, H. *Turizm Ekonomisi*; Detay Yayıncılık: Ankara, 2014.

Chen, H. Complementing Conventional Environmental Impact Assessments of Tourism with Ecosystem Service Valuation: A Case Study of The Wulingyuan Scenic Area,

China. *Ecosystem Services* **2020,** *43*. https://www.sciencedirect.com/science/article/pii/S2212041620300425 adresinden alındı.
Comerio, N.; Strozzi, F. Tourism and Its Economic Impact: A Literature Review Using Bibliometric Tools. *Tour. Econ.* **2019,** *25*, 109–131.
Daruwalla, P.; Darcy, S. Public Sympathy: Private Antipathy: Personal and Societal Attitudes Towards. *Ann. Tour. Res.* **2005,** 32(3), 549–570.
Deery, M.; Jago, L.; Fredline, L. Rethinking Social Impacts of Tourism Research: A New Research Agenda. *Tour. Manage.* **2012,** *33*, 64–73.
Demir, C. Turizm Ve Rekreasyon Faaliyetlerinin Olumsuz Çevresel Etkileri: Türkiye'deki Milli Parklara Yönelik Bir Uygulama. *Dokuz Eylül Üniversitesi İ.İ.B.F. Dergisi* **2002,** *17* (2), 93–117.
Demir, M. A. 8. Sınıf Öğrencilerinde Rastlanan Kalıpyargılar ve Kalıpyargıların Kaynakları: Aydın İli Örneği. *Yayınlanmamış Yüksek Lisans Tezi*. Muğla: Muğla Sıtkı Koçman Üniversitesi Eğitim Bilimleri Enstitüsü, 2016, Şubat.
Doğan, H. Z. *Turizmin Sosyo-Kültürel Temelleri* (2. b.); Detay Yayıncılık: Ankara, 2004.
Dogru, T.; Bulut, U. Is Tourism an Engine for Economic Recovery? Theory and Empirical Evidence. *Tour. Manage.* **2018,** *67*, 425–434.
Encyclopædia Britannica. *Stereotype*, Jan 1, 2020. July 29, 2020 tarihinde www.britannica.com: https://www.britannica.com/technology/stereotype-printing adresinden alındı
Eryılmaz, B. Yerel Halkın Turiste ve Turizme Bakışı: Amasya Örneği. *J. Tour. Gastron. Stud.* **2020,** *8* (2), 1381–1395.
Esen, E.; Özata, E. Turizmin Ekonomik Büyümeye Etkisi: Turizme Dayalı Büyüme Hipotezinin Türkiye İçin Geçerliğinin ARDL Modeli ile Analizi. *Anadolu Üniversitesi Sosyal Bilimler Dergisi* **2017,** *17* (1), 43–58.
Gökdayı, H. Bir Kalıp Yargı Olarak Türk Lokumu. *TÜRÜK Uluslararası Dil, Edebiyat ve Halkbilimi Araştırmaları Dergisi* **2017,** *5* (9), 59–70.
Göktaş, L. S.; Türkeri, İ. Turizmin Kültürel ve Ahlaki Etkisi: Şanlıurfa İlinde Uygulamalı Bir Araştırma. *Selçuk Üniversitesi Sosyal ve Teknik Araştırmalar Dergisi* **2016,** *12*, 104–117.
Gürkan, M. İ. Turizmin Sosyal Yapı Üzerindeki Etkileri. *Anatolia* **1996,** 112–119.
Gursoy, D.; Chi, C. G.; Dyer, P. Locals' Attitudes toward Mass and Alternative Tourism: the Case of Sunshine Coast, Australia. *J. Travel Res.* **2010,** *49* (3), 381–394.
Güzel, Ö. F.; Köroğlu, Ö. Turlarda Turistlerin Sergiledikleri Davranış Kalıplarının Belirlenmesi: Turist Rehberlerine Yönelik Bir Araştırma. *Uluslararası Hakemli Beşeri ve Akademik Bilimler Dergisi* **2015,** *4* (11), 1–22.
Hamilton, D.; Sherman, J. Stereotypes. In *Handbook of Social Cognition: Volume 1 Basic Processes*; Wyer, R., Srull içinde, T. K., Eds., 2 b; Psychology Press: New York, 1994; s. 1–68.
Harlak, H. *Çeşitli Turistlere İlişkin Kategoriler ve Algı Repertuvarları*; Türkiye Kalkınma Bankası Yayını: Turizm Yıllığı, Ankara, 1991; pp 237–244.
Harlak, H. *Önyargılar. Psikososyal Bir İnceleme*; Sistem Yayıncılık: İstanbul, 2000.
Hepaktan, C. E.; Çınar, S. Turizm Sektörünün Türkiye Ekonomisi Üzerindeki Etkileri. *Celal Bayar Üniversitesi Sosyal Bilimler Enstitüsü Dergisi* **2010,** *8* (2), 135–154.
Holzner, M. Tourism and Economic Development: The Beach Disease. *Tour. Manage.* **2011,** *32* (4), 922–933.

İmançer, D. Sosyal Psikolojik Açıdan Stereotip Kavramının Dil ve Metin Analizinde Kullanımı. *Selçuk İletişim* **2004,** *3* (3), 128–142.

İmre, N. Yerel Halkın Turizme Bakış Açısının Belirlenmesi: Kemaliye Örneği. *Erzincan Üniversitesi Sosyal Bilimler Enstitüsü Dergisi* **2020,** *13* (1), 1–16.

Işık, C.; Turan, B. Yerel Halkın Turizmin Etkilerine Yönelik Algısı. Güncel Turizm Araştırmaları Dergisi **2018,** *2* (2), 101–133.

Işık, Ş. Kültürlerarası İletişim Bağlamında Cumhuriyet Üniversitesindeki Gençlerin Almanlara Yönelik Kalıp Yargıları. *Cumhuriyet Üniversitesi Edebiyat Fakültesi Sosyal Bilimler Dergisi* **2012,** *36* (2), 159–191.

Jingyi, L.; Chung-Shing, C. Local Cultural Vicissitudes in Regional Tourism Development: A Case of Zhuhai. *Tour. Manage. Perspect.* **2018,** *25*, 80–92.

Jucan, C. N.; Jucan, M. S. Travel and Tourism as a Driver of Economic Recovery. *Procedia Economics and Finance* **2013,** *6*, 81–88.

Kahraman, N.; Türkay, O. *Turizm ve Çevre*; Detay Yayıncılık: Ankara, 2004.

Kanca, O. C. Turizm Gelirleri ve Ekonomik Büyüme. *Marmara Sosyal Araştırmalar Dergisi* **2015,** *8*, 1–14.

Kang, S. K.; Lee, J. Support of Marijuana Tourism in Colorado: A Residents' Perspective Using Social Exchange Theory. *J. Destinat. Market. Manage.* **2018,** *9*, 310–319.

Karabayev, B.; Tutkun, Ö. F. (2001). Türk Cumhuriyetlerinde Gelen Öğrencilerin Türkiye Cumhuriyeti Vatandaşlarına Karşı Taşıdıkları Kalıp Yargılar. *Bilig* **2001,** *18*, 1–26.

Koçoğlu, C. M.; Saraç, Ö.; Batman, O. Safranbolu'da Sürdürülebilir Turizm Gelişmesine Yönelik Yerel Halkın Tutumları. *Türk Turizm Araştırmaları Dergisi* **2020,** *4* (1), 202–219.

Koday, S.; Akbaş, F. Kırsal Turizmin Gelişimi ile İlgili Olarak Yerel Halkın Algıları: Urla (İzmir) Örneği. *Atatürk Üniversitesi Sosyal Bilimler Enstitüsü Dergisi* **2020,** *24* (2), 689–703.

Köroğlu, Ö.; Avcıkurt, C.; Köroğlu, A.; Karaman, S. The Effects of Professional Tourist Guides' Perceptions and Attitudes towards Tourist Groups on Job Behaviour Forms. *Int. Tour. Biennial (April 30–May 05)*; Canakkale Onsekiz Mart University, School of Tourism and Hotel Management: Canakkale, 2007; s. 651–659.

Lippmann, W. *Public Opinion* (2 b.); Transaction Publishers: London, 1998.

Liu, A.; Wu, C. Tourism Productivity and Economic Growth. *Ann. Tour. Res.* **2019,** *76*, 253–265.

Marmaris Belediyesi. *Marmaris Rehberi Ulaşım*, Jan 1, 2020. July 19, 2020 tarihinde www.marmaris.bel.tr: http://www.marmaris.bel.tr/icerik/24/29/ulasim.aspx adresinden alındı.

Matarrita-Cascante, D.; Suess, C. Natural Amenities-Driven Migration and Tourism Entrepreneurship: Within Business Social Dynamics Conducive to Positive Social Change. *Tourism Management*, 2020. https://reader.elsevier.com/reader/sd/pii/S0261517720300662?token=B1CBE984E824022810569351 3F6E88F3543524BDFDB1EA4D0E16A87271EA5CDD3BA392032FB35C7A4A9805B03D55D97F adresinden alındı

Mesci, M.; Dönmez, G. Turizmin Bölgesel Kalkınma ve Yerel Halk Üzerindeki Etkisi: Mudurnu Örneği. *AİBÜ Sosyal Bilimler Enstitüsü Dergisi* **2016,** *16* (16), 21–50.

Milman, A.; Pizam, A. Social Impact of Tourism on Central Florida. *Ann. Tour. Res.* **1988,** *15* (2), 191–204.

Monterrubio, C. Tourist Stereotypes and Servers' Attitudes: A Combined Theoretical Approach. *J. Tour. Cult. Change* **2018,** *16* (1), 57–74.

Myers, D. *Sosyal Psikoloji* (S. Akfırat, Çev.); Nobel Yayıncılık: İstanbul, 2017.

Ng, S. L.; Feng, X. Residents' Sense of Place, Involvement, Attitude, and Support for Tourism: A Case Study of Daming Palace, A Cultural World Heritage Site. *Asian Geo.* **2020,** *37* (2), 189–207.

Odabaşı, Y.; Barış, G. *Tüketici Davranışı* (16. b.); MediaCat Akademi: İstanbul, 2016.

Olalı, H.; Timur, A. *Turizm Ekonomisi*; Ofis Tic. Mat. Şti: İzmir, 1988.

O'Leary, J. T. Land Use Redefinition and The Rural Community: Disruption of Community Leisure Space. *J. Leisure Res.* **1976,** *8*, 268–274.

Özaltın Türker, G.; Türker, A. Yerel Halkın Turizm Etkilerini Algılama Düzeyi Turizm Desteğini Nasıl Etkiler: Dalyan Örneği. *Electr. J. Vocation. Coll.* **2014,** 81–98.

Özdemir, M. A.; Kervankıran, İ. Turizm ve Turizm Etkileri Konusunda Yerel Halkın Yaklaşımlarının Belirlenmesi: Afyonkarahisar Örneği. *Marmara Coğrafya Dergisi* **2011,** *24*, 1–25.

Özgürel, G. *Tunceli İlinin Kalkınmasında Turizm: Yerel Kurum Yöneticileri Boyutuyla Bir Araştırma. Yayınlanmamış Yüksek Lisans Tezi*; Balıkesir Üniversitesi Sosyal Bilimler Enstitüsü: Balıkesir, 2012.

Özgürel, G. Kültürel Mirasa Bir Bakış: Marmaris Örneği. *J. Soc. Hum. Sci. Res. (JSHSR)* **2018,** *5* (31), 4894–4906.

Özgürel, G. *Turizm Destinasyonlarında Yerleşik Yabancı-Yerel Halk Etkileşimi Üzerine Bir İnceleme. Yayınlanmamış Doktora Tezi.* Balıkesir Üniversitesi SBE: Balıkesir, 2020.

Özsavaş Uluçay, N. Turizmin Geleneksel Bodrum Konut Yapısı Ve Kent Sakinleri Üzerindeki Etkisi. *J. Acad. Soc. Sci. Stud.* **2018,** *72*, 299–310.

Özsüer, E. Türk-Yunan İlişkilerinde "Biz" ve "Öteki" Önyargıların Dinamikleri. *Avrasya İncelemeleri Dergisi (AVİD)* **2012,** *1* (2), 269–309.

Pennington-Gray, L. Resident Attitudes towards Tourism in A Destination in The Stagnation Stage of the Tourism Life Cycle. *WIT Trans. Ecol. Environ.* **2005,** 84, 1411–1418.

Pizam, A. Tourism's Impacts: The Social Costs to the Destination Community as Perceived by the Residents. *J. Travel Res.* **1978,** *16* (4), 8–12.

Richards, G. *The Impact of Travel Experiences on the Cultural, Personal and Social Development of Young People.* Summary Report of A Research Study Undertaken for the International Student Travel Confederation 2nd draft, TRAM (Tourism Research and.), 2006.

Robinson, D.; Newman, S. P.; Stead, S. M. Community Perceptions Link Environmental Decline to Reduced Support for Tourism Development in Small Island States: A Case Study in The Turks And Caicos Islands. *Marine Policy* **2019,** *108*. https://reader.elsevier.com/reader/sd/pii/S0308597X18300149?token=9A7220B51D7E58560799008940FB3DC81FF35F0682A872EE91BAC2D928A213255FA433760A0A051F6115D15FD979E7EB adresinden alındı.

Şafaklı, O. V. Üniversite Öğrencilerinin Turizmin Olumsuz Çevre Etkilerini Algılama Düzeyi: Kuzey Kıbrıs Örneği. In *International Conference on "Scientific Cooperation for the Future in Economics and Administrative Sciences"*; Usak University: Thessaloniki, 2017; s. 117–124.

Santamaria, D.; Filis, G. Tourism Demand and Economic Growth in Spain: New Insights Based on the Yield Curve. *Tour. Manage.* **2019,** 75, 447–459.

Schneider, D. J. *The Psychology of Stereotyping*; THe Guilford Press: New York, 2004.

Scholtz, M.; Slabbert, E. The Relevance of the Tangible and Intangible Social Impacts of Tourism on Selected South African. *J. Tour. Cult. Change* **2016,** *14* (2), 107–128.

Sharpley, R. Host Perceptions of Tourism: A Review of the Research. *Tourism Management* **2014,** *42*, 37–49.

Şıvgın, N. Cinsiyet Rolleri Eğitim Etkinliklerinin Anasınıfına Devam Eden 60–72 Aylık Çocukların Toplumsal Cinsiyet Kalıp Yargılarına Etkisinin İncelenmesi. *Yayınlanmamış Doktora Tezi*; Gazi Üniversitesi Eğitim Bilimleri Enstitüsü: Ankara, 2015.

Snyder, M.; Miene, P. K. (1994). Stereotyping of the Elderly: A Functional Approach. *Br. J. Soc. Psychol.* **1994,** *33* (1), 63–82.

Sokhanvar, A.; Çiftçioğlu, S.; Javid, E. Another Look at Tourism- Economic Development Nexus. *Tour. Manage. Perspect.* **2018,** *26*, 97–106.

Sun, Y.-Y.; Cadarso, M. A.; Driml, S. Tourism Carbon Footprint Inventories: A Review of the Environmentally Extended Input-Output Approach. *Ann. Tour. Res.*, 2020. https://reader.elsevier.com/reader/sd/pii/S0160738320300724?token=1FA951D12B02F80AA8A92E935390A93E669A085A565C27CE53265AEB15903A3F01226A531114DB4B1FA158FD01CB50F9 adresinden alındı

Sürgevil, O. Farklılık Kavramına ve Farklılıkların Yönetimine Temel Oluşturan Sosyopsikolojik Kuramlar ve Yaklaşımlar. *Balıkesir Üniversitesi Sosyal Bilimler Enstitüsü Dergisi* **2008,** *11* (20), 111–124.

T. C. Kültür ve Turizm Bakanlığı. *Güvenli Turizm Sertifikasyon Programı*, July 15, 2020. July 26, 2020 tarihinde T. C. Kültür ve Turizm Bakanlığı: https://www.ktb.gov.tr/ adresinden alındı

T. C. Muğla Büyükşehir Belediyesi. *Marmaris hakkında* Jan 1, 2020. July 19, 2020 tarihinde www.mugla.bel.tr: https://www.mugla.bel.tr/sayfa/marmaris-hakkinda adresinden alındı

Tayfun, A.; Kılıçlar, A. Turizmin Sosyal Etkileri ve Yerli Halkın Turiste Bakışı. *Gazi Üniversitesi Ticaret ve Turizm Fakültesi Dergisi* **2004,** *1*, 1–17.

Tohmo, T. The Economic Impact of Tourism in Central Finland: A Regional Input–Output Study. *Tour. Rev.* **2018,** *73* (4), 521–547.

Tuna, M. *Turizm, Çevre ve Toplum* (2. b.); Detay Yayıncılık: Ankara, 2007.

Türk Dil Kurumu. *Türk Dil Kurumu Sözlükleri*, 2020. July 10, 2020 tarihinde https://sozluk.gov.tr/ adresinden alındı

Türkiye İstatistik Kurumu. Merkezi Dağıtım Sistemi, Dec 31, 2019. TÜİK: https://biruni.tuik.gov.tr/medas/?kn=95&locale=tr adresinden alındı.

Uğur, İ. Yerel Halkın Turizmin Çevresel Etkileri Konusundaki Beklenti ve Algılarının Belirlenmesi. *Anasay* **2020,** *4* (12), 165–182.

Ünlüönen, K.; Tayfun, A. Turistlerin Yerli Halkın Tüketim Davranışlarına Etkileri Üzerine Ampirik Bir Araştırma. *Muğla Üniversitesi Sosyal Bilimler Enstitüsü Dergisi* **2003,** *10*. https://dergipark.org.tr/en/download/article-file/217151 adresinden alındı.

Uslu, A.; Kiper, T. (2006). Turizmin Kültürel Miras Üzerine Etkileri: Beypazarı/Ankara Örneğinde Yerel Halkın Farkındalığı. *Tekirdağ Ziraat Fakültesi Dergisi* **2006,** *3* (3), 305–314.

Var, T.; Kendall, K.; Tarakcioglu, E. Resident Attitudes towards Tourists in a Turkish Resort Town. *Ann. Tour. Res.* **1985,** *12* (4), 652–658.

Vatan, A.; Bildin, E. A Research Evaluating the Attitudes of Local People towards the Effects of Tourism in Bozcaada. *J. Econ. Cult. Soc.* **2020**. https://dergipark.org.tr/en/download/article-file/972031 adresinden alındı.

Wang, S.-J.; Zhou, L.-Y. Integrated Impacts of Climate Change on Glacier Tourism. *Adv. Clim. Change Res.* **2019,** *10* (2), 71–79.

Ward, C.; Berno, T. (2011). Beyond Social Exchange Theory. Attitudes towards Tourists. *Ann. Tour. Res.* **2011,** *38* (4), 1556–1569.

Wei, C.; Dai, S.; Xu, H.; Wang, H. (2020). Cultural Worldview and Cultural Experience in Natural Tourism Sites. *Journal of Hospitality and Tourism Management* **2020,** *43*, 241–249.

www.nufusu.com. *Muğla İlçeleri Nüfusları*, Jan 1, 2020. July 19, 2020 tarihinde www.nufusu.com: https://www.nufusu.com/ilceleri/mugla-ilceleri-nufusu adresinden alındı.

Yağan, E. Çakışan Öyküler-Çatışan Beklentiler: Türk ve Yabancı Öğrencilerin Ortak Deneyimleri ve Gelecek Beklentilerinin İncelenmesi (İstanbul-Pendik'te Yapılan Karma Çalışma, 2019). Yayınlanmamış Doktora Tezi. T. C. Maltepe Üniversitesi Lisansüstü Eğitim Enstitüsü: İstanbul, 2020, Şubat.

Yatırım ve İşletmeler Genel Müdürlüğü. İşletme (Bakanlık) Belgeli ve Belediye Belgeli Konaklama İstatistikleri Yıllık Bültenleri. T. C. Kültür ve Turizm Bakanlığı, 2019, Aralık 31. https://yigm.ktb.gov.tr/TR-201121/isletme-bakanlik-belgeli-tesis-konaklama-istatistikleri.html adresinden alındı.

Yılmaz, Ö. D.; Filiz, A. Turizm Gelişim Yaklaşımları Kapsamında Yerel Halkın Turizme Yönelik Algı ve Tutumları: Mazı Köyü-Türkbükü Karşılaştırmalı Analizi. *Hitit Üniversitesi Sosyal Bilimler Enstitüsü Dergisi* **2017,** *10* (2), 1799–1824.

CHAPTER 5

LUXURY BRANDS: CREATION AND SUSTAINABILITY

ARUP KUMAR BAKSI*

Department of Management and Business Administration, Alliah University, West Bengal, India

**E-mail: baksi.arup@gmail.com*

ABSTRACT

Luxury branding, in the context of tangible luxury consumables, has received academic attention. But the notion remained inconclusive in the context of consumption of luxury intangibles. The travel setting provides an excellent backdrop to explore the complex cognitive process of assigning meaning to the relationship between the traveler and luxury travel brands. The shifting image of luxury consumption from elitism to mass aspirational, too, needs to be studied for its transformative implications. Further, the growing concern about environmental decay and ethno-cultural fragility is re-shaping the perceived luxury. Sustainability issues may evolve as critical brand elements. The study focused in developing a construct framework, namely, SusTraveLux, which may be used to assess the neo-luxury perception in the context of travel, tourism, and hospitality sector. The study used hermeneutic phenomenology to assess the experiential nature of luxury travel branding and the growing consciousness regarding environmental and ethnocultural conservation. The study could identify a five-factor model capturing the essence of immersive experience, ethnocultural rootedness, socioenvironmental consciousness,

Tourist Behavior: Past, Present, and Future. Narendra Kumar, PhD, Bruno Barbosa Sousa, PhD & Swati Sharma, PhD (Eds.)

self-projection, and passion and excitement. The study reinforced and expanded the existing literature governing luxury brand concept from sustainability point of view. The study could implicate a theoretical shift in branding luxury from a price-based exclusivity to a transformative cultural experience. Further extrapolations of the study could be made by incorporating subtle behavioral patterns of travelers in perceiving sustainable luxury and subsequent evocation and predisposition toward decision-making.

5.1 INTRODUCTION

In a recent study, the Tourism Economics data measuring outbound flights found that the growth in luxury travel exceeded compared with the overall travel from 2011 to 2015, with a Compound Annual Growth Rate (CAGR) of 4.5% (4.2% for overall travel). The demand for travel has remained constant despite testing economic times, and the luxury market has remained resilient. Over the next 10 years, the growth rate in outbound luxury trips is projected at 6.2%, almost a third greater than the overall travel (4.8%). An important facet of the luxury travel experience is how well travelers feel that their needs and preferences are understood by their travel providers. This includes striking the right balance between respecting their desire for independence and lack of intrusive service, which may shift at different points along the trip cycle. Enhancing their trip by checking in to offer additional elements or assistance might be appreciated. The sweet spot is the ultimate representation of modern luxury. Marketers need to make a balance between the "high-touch" (human interaction oriented) and "low-touch" (prefer technology intervention). While designing and branding luxury travel services, the marketers need to explore the emerging traveler segments, popularly acronymed as "luxury traveler tribes". The luxury service positioning and branding (with justifiable differentiators, attributes and associationship) hovers over the six identified "luxury traveler tribes" or segments, namely, ethical travelers, cultural purists, social capital seekers, simplicity searchers, obligation meeters, and reward hunters. The creation of luxury brands is theoretically grounded on the hierarchy of luxury travel needs, which is an offshoot to Maslow's need hierarchy. The needs spanned from the "trusted travel guardian" (lowest order need) to the VIP privacy and security level (highest order need). The frequency with which travelers experience luxury travel not only influences their perceptions but also presents different levels of difficulty for offering an end to end luxury experience. One can explore this by comparing the "always luxury" and "special occasion luxury" traveler tribes or segments.

Dubois and Duquesne (1993) opined that luxury goods were acquired for what they symbolize, which are argued to be consistent with the hedonic consumption and extended self-personality models. Atwal and Williams (2007) argued that this notion reflected a shift in cognitive alignment from a transactional relationship to a holistic experience based on perceived luxury value proposition. Dumoulin (2007) posited postmodern luxury brand creation on personal creativity, fluidity of imaginations, expressiveness, and intelligence. Conceptualizing luxury brands has always been essentially embedded in the principle of hedonic consumption. The two postmodern discourses of hedonic consumption in the context of luxury travelling include (1) hyperreality and (2) image. Hyperreality refers to the blurring of distinction between the real and the unreal, in which the prefix "hyper" signifies more real than real. While constructing a luxury brand concept based on luxury travel needs, the hedonic consumption motives collapse the real product sense into an image or illusion allowing the consumers (travelers) to experience a simulated sense of luxury. For example, the "Disneyfication" of Bollywood movies to create an immersed brand concept takes the audience from a sense of reality to hyperreality. Travelers are made to believe unreal offers through simulation and simulacra, an essence of postmodern luxury brands. For example, the visitors to the Kempinski Hotel at the Mall of the Emirates enjoy an Alpine experience that features the world's third-largest indoor ski resort and the largest covered snow park. Miller and Real (1998) argued that images and symbolizers had replaced the core concepts of products and services which influenced marketers to induce experiential marketing. Experiential marketers used the new Marketing 3.0 paradigm introduced by Kotler et al. (2010) with modified marketing mix elements that replaced the traditional 4Ps of marketing. The new mix elements, cognito-affective in nature, namely, emotion, exclusivity, engagement and experience captured the essence of experiential branding, which formed the basic framework for creating luxury brands. Based on bipolar constructs developed by Pine and Gilmore (1998, 1999), Atwal and Williams (2008) identified four experiential zones, namely, entertainment, education, escapist and aesthetic to propagate luxury brand perception. These experiential zones used consumer involvement and intensity as interacting dimensions. The term "involvement" refers to the level of interactivity between the service provider and the consumers. Augmented levels of involvement essentially brings the service providers and the consumers into an interactively co-creation of experience. The term "intensity" referred to the perception of strength of feeling towards the interaction. The four experiential zones were not intended to be mutually exclusive. The richness of an experience is, however, a function of the degree to which all four zones are incorporated. For example, the launch of the Royal Tented Taj Spa (Taj Hotels Resorts and Palaces) at the Rambagh Palace in Jaipur (India) recreated the mobile palaces used by the Mughal emperors of the 16th and 17th centuries, with chandeliers, royal pennants, and Indian love swings. This effort of service designing used the "escapist" experiential zone having high degree of consumer involvement and intensity and modulated a luxury consumption. Cognitive projection based on luxury consumption motives lies fundamental to the creation of luxury brands. An orchestrated initiative would include an experience audit, creation of a brand platform, designing brand experience, brand communication, brand assimilation and equity realization. However, the present body of academic research remains inconclusive with regard to a comprehensive framework to capture the essence of luxury brand creation.

Luxury and sustainability, for long, were understood as opposing notions. The essence of luxury brands necessarily integrates customer value proposition. Wiedmann et al. (2007) proposed a tripartite model that distinguished functional, individual, and social value creation. Functional value implied that luxury products/services must meet performance standards communicated as brand commitments. Individual value referred to values governing self-identity, hedonic benefits, and an affinity to materialism. Social values were understood to be fulfilled by luxury conspicuousness. However, with the changes in perception inducted from the rather perilous ecological impacts of materialism, consumers' perception regarding luxury consumption has also changed. The anti-law of marketing which says that "price does not define luxury, but luxury determines the price" (Kapferer and Bastien, 2012) has assumed significance in unraveling the sustainable dimensions of luxury brands. Responsible tourism has reoriented travelers and their luxury travel preferences toward sustainability. Luxury brands no longer competes on "quality" but on a latent social hierarchy of good taste and symbolic capital. Hence, the composition of luxury brands cannot ignore the cultural dimension. Sustainability of luxury travel brands would require a cultural bond between the luxury brand and the travelers. However, existing academic inputs remain scarce to conceptualize a comprehensive framework of sustainable luxury travel brand.

5.2 REVIEW OF LITERATURE

Bespoke designs have continued to lure consumers as they connote subtle evocation of luxury (Novotna and Kunc, 2019). Over the years, the researchers (Amaldoss and Jain, 2005; Beverland, 2005; Yeoman, 2011; Kapferer and Tabatoni, 2012; Godey, 2013; De Barnier et al., 2013; Kapferer, 2015; Adams et al., 2018; Atwal et al., 2019; Kunz et al., 2020, etc.) have tried to decode the structural scaffold of luxury brands and converged on seven factors with assorted discriminatory impact: (1) elitism, (2) creativity, (3) uniqueness, (4) distinction, (5) refinement, (6) quality, and (7) power. The ambivalence in the semantic roots of the term is omnipresent across the geospatial existence and through the cultures and their hybrids. Conventionally, the term epitomized "wealth and status." Over the last two decades, the emerging globalized and digitized market forces (Okonkwo, 2007) ensured that luxury products and services reach emerging economies and the consumption is guided by symbolic and nonutilitarian entities (Sreejesh et al., 2016; Kapferer and Laurent, 2016). These changing paradigms induced challenges for luxury brand builders to embed, in the luxury brand architecture, the "aspirational quotient" in the context of mass consumption (Nobre and Simoes, 2019; Sreejesh et al., 2016) without compromising the value-metaphor of premium pricing associated with luxury consumption (Kapferer et al., 2016). For example, the products of Louis Vuitton are presently mass consumed as neo-luxury item compared with Rolls Royce, grounded on conventional luxury platform, reveals this luxury paradox (Cristini et al., 2017). Conventional marketing instruments to foster branded posits for luxury consumption, therefore, also became redundant as they fail to capture the emerging brand essence of co-creation (Merz et al., 2018; Kim and Ko, 2012) to resonate with consumer psyche. With aristocracy being replaced by meritocracy, the role of luxury has changed. Luxury brands no longer compete on "quality" but on a latent social hierarchy to accrue symbolic capital. Therefore, luxury needs brands for propagation.

Theoretically, the concept of luxury branding went through a process of evolution. The "phenomenological" nature of value co-creation of the luxury brands was emphasized by Tynan et al. (2010), while Roper et al. (2013) highlighted the "interpretivist" perspective to explain the construction and internalization of luxury value creation. The contemporary theory governing luxury branding rejected the age-old managerial view of a brand which looks to "translate the brand identity into a perceivable brand image" (Schroeder, 2009) and adopted a polysemic approach that focused on contextuality and established a brand on strong cultural footing (Seo and Buchanan-Oliver, 2015; Bengtsson et al., 2010). Theories propagating frameworks for luxury branding have considered a number of inputs over time, namely, luxury disposition behavior (Lee, 2013), ethnocultural infusion (Liao and Wang, 2009), utilitarian benefits and symbolic meanings (Sung et al., 2015), value co-creation (Tynan et al., 2010), culture-driven brand personification metaphors/motifs (Aguirre-Rodriguez, 2014), aesthetics and ideologies (Townsend and Sood, 2012, Chitturi et al., 2010)), emotions of exclusivity and superiority (Dion and Arnould, 2011) etc. Becker et al. (2018) proposed the BECKER model of luxury branding which essentially focused on synchronising luxury product qualities with consumer psyche and forwarded the concept of proxy-quality. Proxy-quality, associated with perceived value-driven luxury, is likely to promote equity judgments, socioemotional justice and help in calibrating the relationship between a consumer and luxury brands. Theories revealing consumers' attitude toward luxury brands were also studied and three major clusters were identified, namely, the elitist, the democratic and the distant (Dubois et al., 2005) with varying degrees of consumption motives.

5.2.1 BRANDING LUXURY TRAVEL AND HOSPITALITY

Research work focusing niche luxury branding manifested in identifying the psychic roots and the psychosomatic drivers that activate participation in luxury travel and stimulate to comprehend brand experience on parameters of perceived luxury (Novotna and Kunc, 2019; Han and Hyun, 2018). According to Husic and Muris (2009), the orientation of luxury branding in the context of travel, tourism and hospitality has been shifted from a price-driven elitism to an experience-driven hedonism. Tutek et al. (2018) posited the neo-luxury travel as "transformational and soul-searching experience" and identified a set of five perspectives for transformational impact, namely, purpose, personalization, experiential, balanced disconnection and seamless journey that may be embodied as focal brand elements to connect with the cultural layout. The enactment of "traveling" has to interact intensely with culture, ethnicity, artifacts, heritage and prevailing art forms embedded in the host communities. The structural proximity of luxury brands with the world of art offers a metaphorical bridge to relate consumers (with luxury consumption mindset) with luxury brands (Joy et al., 2012). That art and culture can be a joinder in consumer–luxury brands dyad and have also been studied extensively by Koronaki et al. (2017), whereby they found that art-based initiatives as a relationship-strengthening tool transcends geographical barriers. Artification of travel brands, therefore, embarked on philanthropic and experimental collaboration (Baumgarth et al., 2014; Kapferer, 2014) for luxury embeddedness in culture. Luxury travel brands, therefore, were viewed to possess ethnocultural meanings (Wilcox et al., 2009) targeted to gratify the inherent social appetite of consumers (Bian and Forsythe, 2012; Kim and Ko, 2012) and luxury consumption, conceptually shifted from the notion of "having-to-being" to "owning-to-experiencing" (Cristini et al., 2017). The traveler–brand relationship, in the context of luxury travel consumption, was therefore explained with brand experiences (Chun et al., 2017; Payne et al., 2009) evoked by a bundle of (travel) brand stimuli (Brakus et al., 2009) responsible for the construction of the brand image. These relational constructs framed a brand architecture for luxury travel, which essentially infused cultural consumption with the fundamental brand offer (Baksi, 2017, 2016). Prahalad and Ramaswamy (2004) observed that value co-creation process was also an evolving dimension in experiential traveling which assigned meanings to the luxury travel brands (Vargo and Lusch, 2004). The notion of value co-creation as luxury branded content in travel context reflected in the observations of Lee and Hwang (2011) as they postulated a triangulation of functionalism, experiential symbolism, and metaphorical interaction to understand the derived sensorial pleasure and gratification of actualization needs.

As services shifted from transactional mode to relationship-driven operations, the consumer-based brand-equity (CBBE) was used to explain the traveler–brand affinity (Liu et al., 2017; Yang and Lau, 2016; Huang and Cai; 2015) in the context of luxury consumption. One of the major issues in branding luxury travel is the range of consumptions (from high-tangible to high-intangible) associated with the travel itself and hence a pooled-brand concept might bridge the abstract and asymmetric nature of the relationship between the traveler and the luxury–travel brand. Sumbly and Siraj (2019) talked about brand clues that could take care of this asymmetry and propagate a homogeneous perception of the brand. Earlier, Vigneron and Johnson (2004) adopted a bifurcated proposition in their luxury brand framework and posited nonpersonal perceptions (conspicuousness, uniqueness, and quality) and personal perceptions (hedonic and extended-self) as drivers of luxury consumption. It was reinforced in the studies made by Yang and Mattila (2016) and Jain et al. (2015). The inward view of luxury consumption explained the dynamism of relationship between the brand and the consumer as self-indulgence in luxury consumption manifested the aspirational instinct of actualization and esteem seekers (Liang et al., 2017; Canziani et al., 2016). While developing an instrument to capture the nature of psychosomatic luxury alignment of a traveler, it is only but imperative to understand these relational realms that trigger luxury embracement (Escalas, 2013). This would provide insights into the gradual shift from nuclear luxury travel (constrained to elitism) (Kapferer et al., 2014) to peripheral luxury travel (unconstrained aspirational affinity) (Bizarris et al., 2017; Seo and Buchanan-Oliver, 2015) and would serve as building blocks to construct a framework for luxury travel brands.

The luxury concept in travel sector emerges from luxury-specific experiences, which are likely to form the scaffold of luxury–travel brands. However, these experiences in shaping traveler–luxury brand relationship remain largely unexplored. Literature also remained inconclusive about the process by which travelers ascribe meaning to luxury brands which demand construction of invigorating brand narratives in communication and marketing (Beverland, 2004; Kapferer, 2006). One of the major shortcomings of the emerging theories of branding luxury–travel services, which includes a bouquet of personalized services, namely, beauty salons, spas, private travel services (e.g., private jets, cruises, yatch, etc.), tour-operatives, event designing, etc., is its predominant focus on either corporate or consumer-based brand equity. Luxury brands are not limited to value-added assets; they also possess sociocultural, ideological, and political identities that infuse consumption with meaning (Holt et al., 2004; Schroeder, 2009). An understanding of the luxury travel brands requires insights into the culture and ideologies that encapsulate their consumption, in addition to conventional branding concepts, such as strategy, equity, and value (Schroeder, 2009). Additionally, the brand–consumer relationship, in the context of luxury–travel consumption, is expected to assume critical role in transformational experience. This study, therefore, focused on designing an instrument to capture the relational notion of luxury travel brand and test the same to explain perceived value of luxury–travel brands as opined by travelers in social media.

The neo-luxury consumption posited travelers as value co-creators (Tynan et al., 2010) and in the process, they relived the luxury experience (Merz et al., 2018). The absorptive and immersive experience, travelers derived from the value co-creation process, embedded sense of exclusivity, excellence, novelty and creativity in the configuration of luxury brand (Cristini et al., 2017). Self-brand connection (Brandão et al., 2019) was found to explain the complexity of interactions, perceived image, and aspirational experiences governing consumer–luxury brand relationship (Kaufman et al., 2016). Social identification theory was also used to explain the consumers' inherent desire to construct their social identity based on commonality of luxury brand attributes (Berrozpe et al., 2017), and the hedonic nature of luxury consumption captured the complexity of intimacy of the relationship (Nobre and Simoes, 2019).

5.2.2 SUSTAINABILITY ISSUES IN THE MAKING AND CONSUMPTION OF LUXURY BRANDS

With the growing environmental concern, a gradual shift in responsible brands has been observed in the context of neo-luxury brand consumption (Anthony and Jules, 2006). Being associated with a luxury product or service is no longer constrained in manifesting status and stylistic independence. Acquaintance with the ethnocultural heritage, craftsmanship, and product provenance governing the built-up of a luxury notion represent a deeper interest of consumers in socioethical and environmental issues underpinning the desire to be acknowledged as responsible consumer in luxury consumption and endorsement. This sustainable awareness and concern are driven by the Millennials, who need additional meanings and implications to their otherwise stereotypical and benign purchase (Kendall, 2010). These shifts in values and perceptions associated with luxury consumption are reflected in the creation of new product trends for travel and hospitality industry, ranging from organic and slow food, healthy living, transcendental travel, isolated travel, nature mining to sustainable eco-resources. The Howard HTL Report stated in International

Luxury Travel Market (2011) that the "luxury traveler has become eager to discover experiences off the beaten track." Price tags are not restrictive for a luxury traveler in seeking experiential authenticity, connectivity to local culture, and commitment to the local environment.

The conventional constructs for perceived brand image of sustainable luxury–travel brands are being replaced by environmental concerns, namely, climatic shifts, resource depletion, genetic distortions, food and water scarcity, etc. The eco-mindedness of travelers is paving ways for neo-luxury branding constructs and the gradual diffusion of ostentation and superfluous elitism into an altruistic and constrictive living, marks the common ground for luxury and sustainability. The logo-buying propensity associated with luxury consumption is giving way to the "story-behind-the-creation" of luxury brands and the consumption is increasingly determined by the brands' ability to connect consumers with socioenvironmental and ethnocultural issues (Kendall, 2010). Steinhart et al. (2013) observed that an environmental claim could positively enhance consumers' perception of both utilitarian and hedonic (luxury) products and services. Researchers have also claimed that the luxury industry will gradually embrace a "sustainable luxury" positioning with consumers' growing consciousness about socioenvironmental and ethnocultural aspects (Ducrot-Lochard and Murat; 2011). There has been negatively polarized observations too as empirical evidence was drawn to understand that environmental commitment is the least significant factor to endorse luxury products/services (Joy et al., 2012).

However, the luxury industry suffers from the lack of visibility when it comes to promoting luxury brands on the ground of sustainable issues (Carrigan et al., 2013). Only a handful of luxury marketers promote their brands modeled on sustainable parameters (Adams et al., 2018; Hendriksz, 2018) and even responsible brands, namely, BMW and Armani tend to highlight their sustainable features as peripheral issues compared with their core-luxury values grounded on hedonic features. Positive Luxury (2018), an organization, has been working on the promotional issues of sustainable luxury and is issuing the "Butterfly Mark" to luxury brands that set standards for sustainable sourcing and manufacturing practices. A number of luxury travel brands (hotel, resorts, travel accessories, beverages etc.) have received the "Butterfly Mark," thereby confirming its genetic formation on the ground of sustainable issues, namely, Conservatorium Hotel, Salt of Palmar, Soneva Fushi Resort, Rusticae Hotels, Louis Vuitton, Biotherm, Christian Dior, Helen Kaminski, Nomaddisem, Veraroad, Krug, Ruinart, Hennessy, etc.

To mark a positive brand imprint in the minds of the consumers, it is only but essential that the marketers communicate their environmental concerns and cultural roots associated with the brand DNA (Kunz et al., 2020; Platania et al., 2019; Rolling and Sadachar, 2018; Sthapit et al., 2017; Nash et al., 2016). Arrigo (2018) observed that flagship outlets could enact as the "branded-space" to augment stakeholders' perception about brand commitment toward sustainability in the context of luxury consumption. Earlier, Ivanova et al. (2013) proposed a multisensory signal theory on sustainable luxury to create a subliminal metaphorical representation of luxury brands that transmutes the attitudinal responsiveness of the marketers toward environment, culture, and ethical buying. Use of aesthetics and personalized creativity were also used to propagate brand messages on sustainability (Arrington, 2017; Poldner et al., 2017)

Despite visible behavioral changes attuned to responsible and sustainable luxury consumption, the research foray has been significantly insignificant (Atwal et al., 2019). In addition, the bulk of the research focus on sustainable luxury has been on tangible product industry. Experiential luxury consumption associated with travel and hospitality industry is gaining gradual prominence in academic research (Turunen, 2017) and are identifying the cultural divide (D'Arpizio et al., 2016, 2017). This study focuses on developing a scale framework to assess and capture the notions of sustainability issues associated with luxury travel in the context of travel, tourism, and hospitality industry. The researcher, further nomenclates (in the form of acronym) the proposed scale as SusTraveLux (STL). The objective of the study is to reinforce the existing theoretical posits pertaining to luxury branding and to expand the realm to understand the emerging environmental hedonism governing luxury brand consumption in travel, tourism, and hospitality sector.

5.3 METHODOLOGY

The study focused on identifying constructs to capture the sustainability issues associated with the luxury branding and its consumption pattern in the context of travel and hospitality industry, and hence proposes to develop a scale framework, namely, SusTraveLux. Explicit measurement instruments are scarce, however, a number of studies attempted covert identification of parameters to project sustainability of luxury brands in general. The luxury brand consumption in travel and hospitality sector reflects the inherent hedonism and cognitive alignment toward experiential maximization of delight.

5.3.1 CONCEPTUALIZING THE SUSTRAVELUX BRANDING CONSTRUCT

The scale items for SusTraveLux branding were generated from the studies of Schroeder (2005), Joy et al. (2012), Koronaki (2017), Baumgarth et al. (2014), and Kapferer (2014) which primarily focused on luxury branding. The study also incorporated the nine-item scale to measure the attitude of consumers toward luxury consumption developed by Kapferer (2014). The perception of luxury, operationalized by the seven-item construct (outstanding quality, hedonism, price, exclusivity, selective distribution and associated services, rarity, and creativity) developed by De Barnier et al. (2012), was also studied. Since a brand framework contains covert layers of embedded motives to process the offer and perceive value, the relevant instruments of "perceived sustainable luxury value" were also studied. The models were proposed by Lundblad and Davies (2016) (sustainable luxury value motivators were identified as self-expression, responsibility, social justice, protect the planet, self-esteem, and sense of accomplishment), Cervellon and Shammas (2013) (identified three sets of values, namely, sociocultural centered, ego-centered and eco-centered), Hennigs et al. (2013) (sustainable value motivators- financial, personal, functional, and interpersonal), and (Song et al., 2013) (self and social norms). The dynamic value model identified by Schwartz and Rubel-Lifschitz (2009) brought the contradictory nature of human psyche while making decisions of luxury purchase. The model visualized the inherent conflict in human psyche (self-enhancement versus self-transcendence and openness-to-change versus conservation) for its alignment as pro-sociocultural/pro-environmental entity versus pure hedonistic consumption, which opened up avenues to explore the latent dimensions of a sustainable luxury travel brand and contributed in identifying critical scale items for SusTraveLux brand framework. Identification of factors responsible for sustainable luxury purchase did consider the Theory of Planned Behavior too (Jain, 2018). The study postulated "art and culture" as a possible construct of sustainable luxury branding, in the context of luxury travel, tourism, and hospitality industry. However, this construct remained untested for luxury travel. Apart from identifying and contextualizing scale items to support the instrument, namely STL brand framework, the study focused on a number of luxury brands to thematically infuse the "essence of luxury" elements (personification, anthropomorphization, positioning, USP, etc.) with the instrument. The choice of such brands used two criteria: (1) the consumption of such luxury brands should be in the context of travel, tourism, and hospitality-driven transactions and (2) such brands should have a preceding image of environmental and cultural concern. Since the study was carried out in Indian context (20), luxury travel brands, that travelers had opportunity to interact with during their luxury trips to India, were chosen. The luxury travel brands were chosen from a wide range of luxury travel and hospitality services, namely, accommodation (Kipepeo, Himalayan Ark, Oberoi Villas), dining (Wasabi, Bukhara, Yuuka, Olive Qutub), travel accessories (Luis Vuitton, Globe Trotter), train safari (Palace-on-wheels. Maharaja Express), road travel (Olevia, Volvo B11R), airliner (Emirates, Etihad), tour operator (The Blue Yonder, Luxoindia, Abercrombie and Kent) and spa and wellness (Park Hyatt).

Considering the complexity of the study which desired to capture not only the hedonic alignment in the process of luxury consumption, but also the logical drive to endorse sustainability, it was essential to delve into the psyche of the travelers. The shared uninhibited sentiments and emotions, expressed as user-generated content (UGC) over social media, in the context of responsible consumption of luxury travel brands were explored to add-up to the findings from literature. Purposive sampling approach was used in accessing social media as it exposes to information-rich cases that demonstrate the phenomenon of interest intensely (Patton, 2002). Three online luxury travel forums were identified based on their blog-followings over social media, namely, thesqua.re (Twitter follower-62.98K; Facebook followers-165.20K; Alexa Rank-28.56K), FlyerTalk (Twitter follower-58.73K; Facebook followers-35.57K; Alexa Rank-29.8K), and Nomadic Matt (Twitter follower-120K; Facebook followers-255.70K; Alexa Rank-27.21K). Tweets posted on the Twitter account of the three selected luxury travel forums were collected for a period of 6 months starting from August 2019 to January 2020. The units of analysis were the microblogs depicting luxury travel experience, contextual brand interactions, and aspirational fulfillment shared over the Twitter platform. The software NVivo-12 was used to analyze the tweets.

Tweets were captured using hashtags "traveling with sustainable luxury," "sustainable travel brands" and "responsible travel branding for luxurious experience" from the three identified luxury travel platforms, namely, thesqua.re, FlyerTalk, and Nomadic Matt. Tweets and retweets were captured using the NCapture feature of NVivo-12 for a period of 6 months (August 2019–January 2020.). A total of 11,724 tweets were streamed in. The tweets collected from Twitter were imported to the qualitative data analysis software (QDAS), namely, NVivo-12. Table 5.1 computed the details of Twitter data extracted.

TABLE 5.1 Details of Extracted Twitter Data.

Luxury travel platform	No. of Tweets from June, 2019 to December, 2019	No. of users (with unique ID)	Average Tweets per user during June, 2019 to December, 2019
Thesqua.re	5876	2456	2.39
FlyerTalk	3619	1762	2.05
Nomadic Matt	2229	1299	1.71

Following the process outlined by Tian et al. (2019) and Fan and Niu (2016), analysis on the texts and graphics, used in the tweets, was conducted using open, axial, and selective coding (Corbin and Strauss, 1990). The tweets were read and contextual themes were open-coded, followed by assigning interpretive labels in graphical elements of the tweets to frame the nodes. Axial coding was used to classify and cluster nodes based on the homogeneity of thematic content. Finally, selective coding was done to summarize the identified nodes based on the thematic expressions. Sentiment analysis was deployed to assess the cognitive alignment of the users with respect to their experiential manifestations of luxury consumption while traveling. Dictionary-based (Lexicon) method was used to compute the sentiment scores which ranged from '+1' (extremely positive) to '−1' (extremely negative). Machine algorithm used the following equation to calculate the sentiment score centered around expressed positivity/negativity via text's (T) and cross-referring it with dictionary (D):

$$S_D^T = \frac{\sum_{W f D} h_D(W) \times f^T(W)}{\sum_{W f D} f^T(W)} = \sum_{W f D} h_D(W) \times p^T(W)$$

where W denotes each of the phrase cross-referred with dictionary D; word-sentiment scores as $h_D(w)$; frequency of word as $f_T(w)$, and normalized frequency of w in T as $p^T(w) = f^T(w)/\Sigma$.

The phrase classification and clustering was done using the Affective Norms of English (ANEW) dictionary. Two clusters, namely, (1) valence and (2) arousal were identified as measures of emotions (Russell, 1980). While "valence" represented the direction of the emotion, "arousal" captured the quantum of physical response. Machine algorithm (supervised classification process using Naïve Bayes theorem) assigned score for each tweet for valence and arousal. Higher "valence score" implied greater mark-ups of tweets in positive direction (in this case perception of sustainability embedded in luxury travel brands interfaced with while traveling) and higher "arousal score" denoted enhanced intensity in expressed emotions (while interacting with interfaced brands). Summation of valence and arousal scores for each phrase of tweet was done to obtain the emotional scores. Five clusters, based on text-clustering process (Meyer et al., 2008), could be identified by the algorithm and were nomenclated based on thematic proximity of the tweets: experience (experience of association with the luxury travel brand), self-reflection, environmental consciousness, cultural absorption, and thrill and excitement. Figure 5.1 represented the clustered heatmap.

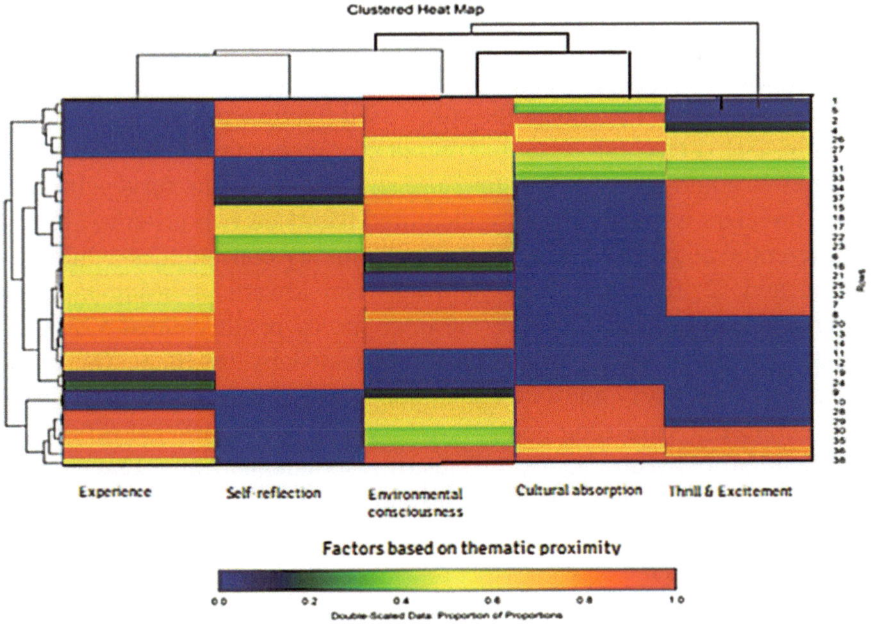

FIGURE 5.1 Heat map of thematic factors identified.

5.3.2 EXECUTION OF THE STUDY

Focus Group, comprised of eight members representing various level of stakeholders, was used to assess the content validity of the survey instrument based on subjective logic. Content validity index for Scale (S-CVI) was used to examine the survey instrument. The content of the survey instrument was finalized with 32 scale items to measure SusTraveLux branding construct. Piloting was done to assess the internal reliability of the scale. Exploratory factor analysis (EFA) was deployed to identify the underlying factor structure. IBM SPSS 26 was used to analyze the pilot data.

The study was carried out in three prominent tourist destinations of India, namely, Sikkim, Puducherry, and Uttarakhand. The choice of geographical locations to conduct the study was based on: (1) inbound tourists and their preferential alignment with luxury travel brands and (2) the effort to synch geomorphological vulnerability and sustainability in luxury service and product offers by the travel and hospitality marketers. The study was cross-sectional in nature spanning from November 2019 to March 2020.

Considering the quasi-quantitative nature of the study, which demanded quantification of hedonistic attitudes involved in perceiving and consuming luxury offers and decoding the offers for their underpinned sustainable agendas, the study used hermeneutic phenomenology (Kafle, 2013) research design. Appropriate intervening field agency was chosen to administer the questionnaire and to conduct the survey. Initially, 867 travelers were purposively identified based on two criteria: (1) their preferential alignment and engagement in luxury consumption while traveling and (2) their awareness and activism toward environmental and ethnocultural issues. However, the final survey result yielded 572 legitimate responses. Nonresponse bias was tested by assessing the differences between early and late respondents with regard to the mean of all the variables (Armstrong and Overton, 1977). No significant differences were found between the two groups of respondents. This suggested that response bias was not a problem in the study. The sample consisted of 51.92% male (297) and 48.08% female (275) with an average age of 47 years. The sample consisted of 311 (54.37%) overseas and 261 (45.63%) domestic tourists. The sample also predominantly had a high percentage of respondent having higher academic/professional qualification (85% graduation and above). A five-point interval scale was used for the study.

5.4 DATA ANALYSIS AND INTERPRETATIONS

Piloting was done through EFA and the computation of Cronbach's alpha was done to test the internal consistency of the measures and to examine scale items' performance within their respective dimension (Churchill Jr, 1995). EFA churned out a five-factor (22 item) instrument for the SusTravLux branding construct (factors loaded with > 0.650 were retained). The latent constructs were nomenclated and are based on the thematic agglomeration of factor loadings as "immersive experience" (six items); "ethnocultural rootedness" (six items); "socioenvironmental consciousness" (four items); "self-projection" (four items) and "passion and excitement" (two items). The Cronbach's alpha was found to be .907, indicating good reliability levels of the scale (Nunnally, 1978). The KMO value is .867, which indicated sample adequacy for the study. The Bartlett's test of Sphericity exhibited significant value (chi-square = 9951.813, df = 325, sig. = .000). This suggested that we may reject the null hypothesis that our correlation matrix is an identity matrix and will conclude that the variables are correlated highly enough to provide a reasonable basis for EFA. The total variance explained by the five factors was found to be 71.58 %, which was significant enough. Table 5.2 summarizes the results for EFA and Cronbach's alpha.

5.4.1 COMMON METHOD BIAS

As the study was rationally grounded on existing theories with contextual realism, EFA was also used to assess common method bias (CMB) (Arya et al., 2019). Considering the fact that validity issues might yield potentially misleading conclusion (Campbell and Fiske, 1959), the assessment of variance attributable to the measurement method (Podsakoff et al., 2003) was applied. Testing the biasness was critical for the study as the data were obtained through self-administered questionnaire (using online platform), and response to both the predictor and criterion variables were generated from the same respondent. The results revealed that the first factor explained the total variance of 18.32% (< 50%), which confirmed

TABLE 5.2 EFA Results.

Scale items	**Factors**				
	Immersive experience (IE)	**Ethnocultural rootedness (ECR)**	**Socioenvironmental consciousness**	**Self -projection**	**Passion and excitement**
Luxury travel brands explains my preferences in life (IE1)	0.810				
When I interact with my luxury travel brand, I forget everything else around me (IE2)	0.786				
Time flies when I am interacting with my luxury travel brand of choice (IE3)	0.758				
I feel satisfied while interacting intensely with my luxury travel brand (IE4)	0.757				
When interacting with my luxury travel brand, it is difficult to detach myself (IE5)	0.715				
My luxury travel brands galvanize mood change pleasantly (IE6)	0.685				
I prefer my luxury travel brands as they are reliable and consistent with my values. (ECR1)		0.782			

TABLE 5.2 *(Continued)*

Scale items	Factors				
	Immersive experience (IE)	**Ethnocultural rootedness (ECR)**	**Socioenvironmental consciousness**	**Self -projection**	**Passion and excitement**
I opt for particular luxury travel brands which embed and reflect local culture, traditions, festivals, and ethnicity (ECR2)		0.78			
I prefer my luxury travel brands as they promote social interaction (ECR3)		0.766			
I prefer my luxury travel brands as they aim to induce consciousness and change beliefs, attitudes, and behaviors (ECR4)		0.753			
I like the way my travel brands infuse luxury with ethnocultural existence of a destination (ECR5)		0.726			
My luxury travel brand connects me to the local culture and heritage (ECR6)		0.722			
My luxury travel brand/ s is/ are embarked on zero-waste philosophy (SEC1)			0.900		
My luxury travel brand/ s evoke/ s a sense of connectedness with the nature (SEC2)			0.885		

TABLE 5.2 *(Continued)*

Scale items	Factors				
	Immersive experience (IE)	**Ethnocultural rootedness (ECR)**	**Socioenvironmental consciousness**	**Self -projection**	**Passion and excitement**
My luxury travel brand is associated with environmental activism and makes me aware about its contribution to conserve natural environment (SEC3)			0.831		
My luxury travel brand evokes a sense of responsible traveling (SEC4)			0.796		
My luxury travel brand says a lot about the kind of person I would like to be (SA1)				0.905	
Association with my luxury travel brands lets me be a part of a shared community of like-minded consumers (SA2)				0.88	
My luxury travel brands fit well with my current stage/ status of life (SA3)				0.819	
My luxury travel brands connect with the part of me that really makes me going (SA4)				0.683	

TABLE 5.2 *(Continued)*

Scale items		**Factors**				
		Immersive experience (IE)	**Ethnocultural rootedness (ECR)**	**Socioenvironmental consciousness**	**Self -projection**	**Passion and excitement**
My relationship with my luxury travel brand has been exciting (PE2)						0.863
I am passionate about my luxury travel brands (PE3)						0.682
Cronbach's alpha		0.907				
KMO		0.867				
Bartlett's Sphericity	Chi-square	9951.813				
	df	325				
	Sig.	0.000				
Total Variance explained		71.583				

Extraction Method: Principal Component Analysis.

Rotation Method: Varimax with Kaiser Normalization.

[a]Rotation converged in 12 iterations.

the absence of common method bias. Further, it was established that the measures were free of common method variance (< 50%) and hence concluded to be an insensitive test which did not support the fundamental assumption of Herman's single-factor test (Podsakoff et al., 2003).

Once the scale reliability and factor structure were confirmed, confirmatory factor analysis (CFA) was deployed to assess the dimensionality and the convergent validity of the instrument, namely, SusTraveLux. Full information maximum likelihood (FIML) was used for CFA. The items were limited to load on its prespecified factors and the factors were allowed to correlate freely. All 22-scale items were retained in CFA. Figure 5.2 illustrates the measurement model with standardized regression weights of each item on the respective construct. Results reflect the unidimensionality and internal consistency of the measures.

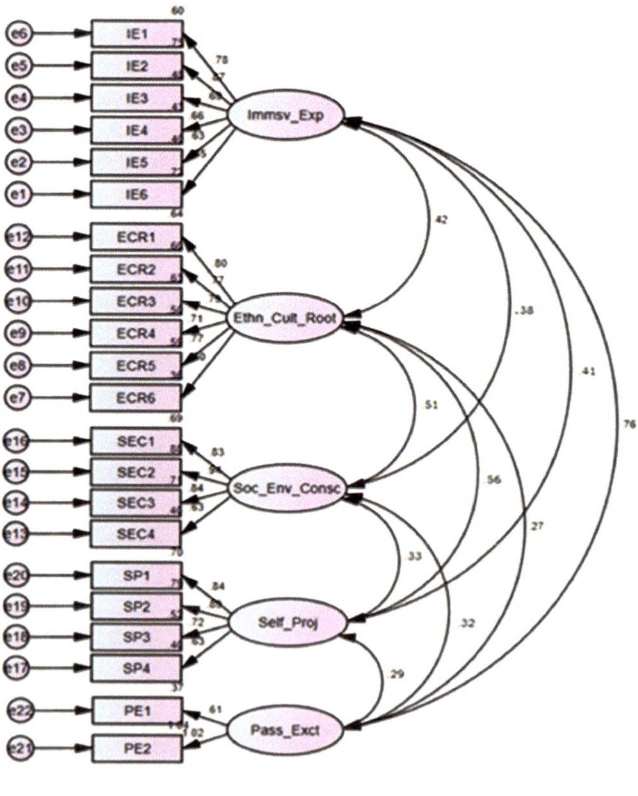

FIGURE 5.2 Zero-order measurement model.

The CMIN/DF was found to be 1.520 (ref. value: < 3, Arya et al., 2019), thereby establishing the adequate fit of the data to the proposed construct (Table 5.3). Chi-square estimation for the model was 494.126 (df = 325, $p = 0.000$). The chi-square statistic is sensitive to sample size. Therefore, the following fit indices (Table 5.4) were also considered: normed fit index (NFI) (0.947), comparative fit index (CFI) (0.939), Tucker-Lewis fit index (TLI) (0.933), goodness of fit index (GFI) (0.941) and RMSEA (< 0.04). All these indices presented acceptable values (Hair et al., 2010).

TABLE 5.3 Model Fit Analysis (CMIN/DF).

Model	NPAR	CMIN	DF	P	CMIN/DF
Default model	54	494.126	325	.000	1.520

NPAR, number of distinct parameters; DF, degrees of freedom; P, significance value; CMIN/DF, minimum discrepancy divided by degree of freedom.

TABLE 5.4 Model Fit Analysis (NFI, GFI, CFI, TLI, RMSEA).

NFI	GFI	TLI	CFI	RMSEA
0.947	0.941	0.933	0.939	0.038

NFI, normed fit index; GFI, goodness of fit index; TLI, Tucker-Lewis index; CFI, comparative fit index; RMSEA, root mean square error of approximation.

Composite reliability (CR) was found to be >0.7 for all constructs (Table 5.5) and hence considered to be significant. Average variance extracted (AVE) was > 0.50 (Fornell and Larcker, 1981), Convergent validity was established, as AVE was found to be greater than 0.5 for all constructs (Fornell and Larcker, 1981) and CR > AVE (Bagozzi, 1980). The maximum shared variance (MSV) was found to be consistently less than the AVE, which established discriminant validity (Hair et al., 2010). The analysis also reported Hancock's *H [MaxR (H)]* measure of reliability, which required fewer assumptions than Cronbach's α (Hancock and Mueller, 2006). Hancock's *H* reflected the minimum achievable or anticipated construct reliability associated with the measurement model of each factor (Hancock and Mueller, 2006). It ranged from .861 (immersive experience) to .977 (self-projection), and hence exceeded the recommended minimum value for of $H = .70$ (Hancock and Mueller, 2006) (Table 5.5).

TABLE 5.5 Construct Validity (CR/AVE/MSV/MaxR(H)).

	IE	ECR	SEC	SP	PE	CR	AVE	MSV	MaxR(H)
IE	**0.726**					0.806	0.518	0.044	0.861
ECR	0.004	**0.801**				0.908	0.628	0.172	0.939
SEC	0.064	0.415	**0.797**			0.858	0.608	0.172	0.954
SP	0.210	−0.001	0.025	**0.811**		0.900	0.698	0.044	0.977
PE	0.091	0.017	0.034	0.059	**0.752**	0.872	0.61	0.129	0.931

IE, immersive experience; ECR, ethnocultural rootedness; SEC, socioenvironmental consciousness; SP, self-projection; PE, passion and excitement; Diagonal elements show square roots of AVEs (average variance explained).

The SusTraveLux branding construct was further tested to assess the performance at a higher order level (Nobre and Simoes, 2019; Dabholkar et al., 1996). The higher order measurement model (Fig. 5.3) retained the five zero-order factors with all the scale items having significant loading value.

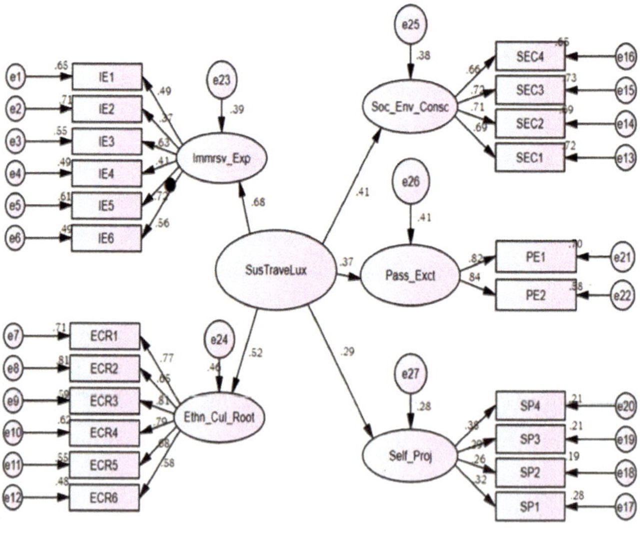

FIGURE 5.3 Higher order measurement model.

The higher order model achieved adequate fit as the CMIN/DF was found to be 1.969 (ref. value: < 3, Arya et al., 2019) (Table 5.6) and considered to be significant. Chi-square estimation for the higher order model was 636.269 (df = 323, p = 0.001). The baseline comparisons were also presented in Table 5.6. The GFI (0.911), NFI (0.908), CFI (0.918), and the TLI (0.904) were found significant (Hair et al., 2010). The RMSEA value (0.046) was also found < 0.05 and hence significant (Hair et al., 2010). The standardized regression weights for the retained higher order factors were significant (immersive experience: 0.68, $p < 0.01$; ethnocultural rootedness: 0.52, $p < 0.01$; socioenvironmental consciousness: 0.38, $p < 0.01$; self-projection: 0.28, $p < 0.05$; and passion and excitement: 0.37, $p < 0.05$).

TABLE 5.6 Goodness of Fit Indices for the Proposed Model and RMSEA Value.

		CMIN				**Baseline comparisons**				**RMSEA**
Model	**NPAR**	**CMIN**	**DF**	**P**	**CMIN/DF**	**NFI**	**GFI**	**TLI**	**CFI**	
Default model	78	636.269	323	0.001	1.969	0.908	0.911	0.904	0.918	0.046

NPAR, number of distinct parameters; DF,degrees of freedom; P, significance value; CMIN/DF, minimum discrepancy divided by degree of freedom; NFI, normed fit index; GFI, goodness of fit index; TLI, Tucker-Lewis index; CFI, comparative fit index; RMSEA, root mean square error of approximation.

CR was found to be >0.7 for all higher order factors and hence was significant (Hair et al., 2010). Convergent validity was established as the AVE was found to be greater than 0.5 for all higher order factors and CR > AVE. The MSV and the average shared variance (ASV) were found to be consistently less than the AVE, which established discriminant validity (Hair et al., 2010) (Table 5.7).

TABLE 5.7 Construct Validity (CR/AVE/MSV/MaxR(H)).

	IE	**ECR**	**SEC**	**SP**	**PE**	**CR**	**AVE**	**MSV**	**MaxR(H)**
IE	**0.789**					0.836	0.558	0.042	0.856
ECR	0.005	**0.776**				0.911	0.598	0.162	0.943
SEC	−0.044	0.395	**0.792**			0.823	0.618	0.152	0.899
SP	0.197	−0.041	0.038	**0.803**		0.897	0.638	0.038	0.909
PE	0.108	0.032	0.020	0.059	**0.829**	0.918	0.60	0.176	0.929

IE, immersive experience; ECR, ethnocultural rootedness; SEC, socioenvironmental consciousness; SP, self-projection; PE, passion and excitement; Diagonal elements show square roots of AVEs (average variance explained).

5.5 CONCLUSION AND DISCUSSION

The study focused on developing a framework construct to explain the dichotomy of responsibility and consumption embedded in sustainable luxury travel brands. Contemporary travelers perceive luxury travel to be an aspirational experience encapsulated within the realm of emotional exclusivity and embedded uniqueness. Travelers interact with luxury travel brands more frequently as the luxury travel brands continued to be affordable and accessible. The shifting dynamism of traveler relationship with luxury travel brands, therefore, needs to be explored. The existing body of literature postulated luxury consumption in the context of travel motives (Nobre and Simoes, 2019; Brandão et al., 2019; Berrozpe et al., 2017; Kaufman et al., 2016). This study expanded on the findings of Nobre and Simoes (2019) which extrapolated luxury consumption in the context of tangible consumption to a hypothesized luxury brand, in the field of luxury travel with responsible consumption pattern.

Theoretical realm of luxury branding contradicts its compatibility with sustainability (De Barnier et al., 2012; Vigneron and Johnson, 2004; Dubois et al., 2001) as the luxury brands were conceived to commit subjective exclusivity and perceived to deliver hedonic pleasure. However, the neo-luxury ideation expands the notion of hedonism with a sense of responsibility (Bendell and Kleanthous, 2007). Activism to support ecological sanctity and ethical standards of production had a strong impact on a number of luxury brands to alter their attitude.

The major contribution of this study could be attributed to the development and validation of SusTraveLux branding construct that captures the infused alignment of sustainable consumption wrapped in luxury offers in the context of travel, tourism, and hospitality industry. The study drew inspiration from the theories of subjective and hedonic consumption associated with experiential traveling (Bian and Forsythe, 2012; Kapferer, 2014; Vickers and Renand, 2003). At the same time, the study panned on the theories governing structural aspects of luxury branding (Kapferer, 2015; Adams et al., 2018; Atwal et al., 2019; Kunz et al., 2020, etc.).

The study established a five-dimensional construct for the SusTraveLux instrument, namely, "immersive experience," "ethnocultural rootedness," "socioenvironmental consciousness" and "self-projection" and "passion and excitement." *"Immersive experience"* entailed the spontaneous and uninhibited cognitive attachment of the traveler with the luxury travel brands. The intense relationship between the traveler and the luxury travel brands was reflected in the scale items ("*When I interact with my luxury travel brand, I forget everything else around me*"; "*When interacting with my luxury travel brand, it is difficult to detach myself*" *and* "*My luxury travel brands galvanize mood change pleasantly*") and emphasized the nature of aspirational values that drove this relationship. Earlier studies had empirically identified hedonism to impart brand meaning (Torelli et al., 2010).

The dimension of "*ethnocultural rootedness*" captured the concept of cultural symbolism and inherent intent of traveler, with motive of luxury consumption, to be an integral part of the brand's cultural and ethnic fabric. It revealed the responsible nature of luxury consumption with ethnocultural rootedness explaining brand meaning and the acculturation process emphasizing on brand affinity. The shift in perceived luxury brand value to absorb the concern for cultural integration reflected in the dimension ("*I opt for particular luxury travel brands which embed and reflect local culture, traditions, festivals and ethnicity*" *and* "*I prefer my luxury travel brands as they aim to induce consciousness and change beliefs, attitudes and behaviors*"). There has been a dearth in extant literature in support of ethnocultural integration with the fundamental luxury brand architecture except for some sporadic studies made on artification and artifact embellishment (Koronaki et al., 2017; Joy et al., 2012). However, artification of travel brands embarked on philanthropic and experimental collaboration (Baumgarth et al., 2014; Kapferer, 2014) for sustainable luxury embededness in culture. The dimension, therefore, provided the necessary clarity to understand the perceived branding facets and reinforced the theoretical underpinning of sustainable luxury consumption based on conscious and responsible endorsement of ethnic, cultural, and heritage assets.

The growing concern for social and environmental decay and the subsequent activism to oppose environmentally destabilizing acts by the organizations has led the brand marketers to rethink the luxury branding strategy. This notion consolidated with the emergence of responsible travel and tourism. Kapferer and Michaut (2017) observed diminishing contradiction between luxury and sustainability as travelers and tourists opted for more eco-friendly generic products (namely, bicycles, recyclable batteries, biofood, etc.). At the same time, a growing apathy can be observed in connection with luxury brands (often used as travel accessories), which, at the production level, used biotic inputs (animals, trees). This surrogated feel of animal genocide (e.g., crocodiles for skin, beavers, rabbits for fur, etc.) and ecological imbalance has been repulsive of "perceived branded luxury." In this context, the dimension of *"socioenvironmental consciousness"* aptly captures the shift in travelers' (consumers') perception of luxury travel brands. The specific scale items, namely, *"My luxury travel brand/ s is/ are embarked on zero-waste philosophy" and "My luxury travel brand is associated with environmental activism and makes me aware about its contribution to conserve natural environment"* resonated with this underlying shift in luxury perception.

The dimension of *"self-projection"* reflected the theory that posited that consumers responded more positively to brands that were aligned with their concept of "self" (Sirgy, 1982), thus predicting consumers' attitudes and purchase intentions (Aaker, 1999; Hong and Zinkhan, 1995). Retained scale items, namely, *"Association with my luxury travel brands lets me be a part of a shared community of like-minded consumers,"* conveyed the meaning of self-expression, while statements, namely, *"My luxury travel brand says a lot about the kind of person I would like to be" and "My luxury travel brands fit well with my current stage/status of life"* underlined the aspirational self of the travelers that evoked a sense of relationship with the luxury travel brands. This consolidated the proposition of value co-creation (Merz et al., 2018) on the basis of perceived brand value (Payne et al., 2009) predisposed from cherished relationships and abstractions (Heine and Phan, 2011). The findings of the study further re-established the transition of luxury concept based on perceived social images driven by exclusivity to aspirational immersive experience grounded on intimate and intensive subjective encounters. The travel sector might just provide the ideal kind of platform to experiment with luxury travel brands with scope for personalization to infuse greater sense of self-projection.

Passion instills excitement when it comes to luxury consumption in the context of travel behavior *("My relationship with my luxury travel brand has been exciting" and "I am passionate about my luxury -travel brands")*. The findings supported the aspect of mass-consumed luxury brands, possessing the ability to induce passion and excitement, to bond with consumers (Nobre, 2015; Nobre and Simoes, 2019). The retained passion and excitement scale items linked luxury travel to the more aspirational experience and subjective evocation that luxury travel brands might propagate.

The study had implications for the brand designers and the brand marketers too. The "submerged-eco-experience designers" are conceiving luxury travel brands, both offered through syndicated and isolated ventures, as a layered entropy that can trigger both hedonic and conscious energy. Contemporary luxury travel brands have widened their scope of consumption to get liberated from the perennial essence of cultural symbolism, self-reflection, aristocracy, and elitism. The surfacing of personalized travel brands encapsulates the balance between luxury and sustainability. Jaipur Virasat Foundation by Pradip Krishen, Nirmana Hotels by Aman Nath, Konyak Tea Retreat by Phezin Konyak, Roots Ladakh by Muzammil Hussain and Tafazzul Hussain, and Kalpavriksh by Ashish Kothari are some of the ventures that scaled-up their branded offers by fusing the feel of luxury with the responsibility and accountability toward society and environment. The SusTraveLux instrument will also assist in ideating brand differentiators and design brand communications. It can also identify the specific brand connectors given the nature of travel and tourism products to be marketed.

The study was limited to three specific tourist destinations of India and was cross-sectional in nature. Nonprobabilistic sampling technique was used to capture the experiential nature of the study. Therefore, the study cannot be generalized over a wider geographical and sociodemographical spread. The concept of sustainability in the context of luxury consumption is still debatable and not devoid of contradictions. Therefore, further exploration is required to examine the apprehended dichotomy and the gradual synch taking place. Further extrapolations of the study could be made by incorporating additional constructs that would explain specific behavioral patterns (cognitive, affective, and conative) representative of embracing sustainability in luxury offer.

KEYWORDS

- **luxury travel**
- **SusTraveLux**
- **sustainability**
- **brand experience**
- **ethno-cultural**

REFERENCES

Aaker, J. L. The Malleable Self: The Role of Self-Expression in Persuasion. *J. Market. Res.* **1999,** *36* (1), 45–57. https://doi.org/10.2307/3151914

Aaker, J.; Fournier, S.; Brasel, S. A. When Good Brands Do Bad. *J. Consumer Res.* **2004,** *31* (June), 1–16.

Adams, R.; Jeanrenaud, S.;Bessant, J.; Denyer, D.; Overy, P. Sustainability-Oriented Innovation: A Systematic Review. *Int. J. Manage. Rev.* **2018,** *18*, 180–205.

Adjei, J. K. The Role of Social Media in National Discourse and Mobilization of Citizens. *Proc. Int. Conf. Collab. Technol. Syst.em, Orando, Florida, USA* **2016,** *1*, 559–566.

Aguirre-Rodriguez, A. Cultural Factors That Affect Brand Personification Strategy Effectiveness. *Psychol. Market.* **2014,** *31* (1), 70–83.

Armstrong, J. S.; Overton, T. S. Estimating Nonresponse Bias in Mail Surveys Relationships. *J. Market. Res.* **1977,** *16* (Aug), 396–400.

Arrigo, E. The Flagship Stores as Sustainability Communication Channels for Luxury Fashion Retailers. *J. Retail. Consumer Serv.* **2018,** *44*, 170–177.

Arrington, D. W. Ethical and Sustainable Luxury: The Paradox of Consumerism and Caring. *Fashion, Style Popular Culture* **2017,** *4* (3), 277–285.

Arya, V.; Sethi, D.; Paul, J. Does Digital Footprint Act as a Digital Asset?—Enhancing Brand Experience through Remarketing. *Int. J. Info. Manage.* **2019,** *49*, 142–156.

Atwal, G.; Williams, A. Experiencing Luxury. *Admap* March **2007**, 30–32.

Atwal, G.; Williams , A. Luxury Brand Marketing—The Experience Is Everything! *J. Brand Manage.* **2008,** *16* (5/6), 338–346.

Atwal, G.; Williams, A. Luxury Brand Marketing: The Experience Is Everything! *J. Brand Manage.* **2009,** *16*, 338–346.

Atwal, N.; Wells, V. K.; Carrigan, M. and Henninger, C. E. Sustainable Luxury Marketing: A Synthesis and Research Agenda. *Int. J. Manage. Rev.* **2019,** *21, 405–426*

Bagozzi, R. P. *Causal Models in Marketing*; New York: John Wiley, 1980.

Baksi, A. K. Destination Bonding: Hybrid Cognition Using Instagram. *Manage. Sci. Lett.* **2016,** *6* (1), 31–46. doi: 10.5267/j.msl.2015.12.001.

Baksi, A. K. Craftourism: Development and Validation of a Scale to Measure Visitors' Behavioural Pattern Based on Identified Travel Motives. *Tourismos* **2017,** *12* (1), 101–125.

Baumgarth, C.; Lohrisch, N.; Kastner, O. L. Arts Meet Luxury Brands. In *The Management of Luxury*; Berghaus, B., Müller-Stewens, G., Reinecke, S., Eds.; Kogan Page Publishers, 2014.

Becker, K.; Lee, J. W.; Nobre, H. *J. Asian Finance, Econ. Busi.* **2018,** *5* (3), 51–63.

Bengtsson, A.; Bardhi, F.; Venkatraman, M. How Global Brands Travel with Consumers: An Examination of the Relationship between Brand Consistency and Meaning across National Boundaries. *Int. Market. Rev.* **2010,** *27* (5), 519–540.

Berrozpe, A.; Campo, S.; Yagüe, M. J. Understanding the Identity of Ibiza, Spain. *J. Travel Tour. Market.* **2017,** *34* (8), 1033–1046.

Beverland, M. Uncovering Theories-in-Use: Building Luxury Wine Brands. *Eur. J. Market.* **2004,** *38* (3/4), 446–466.

Bian, Q.; Forsythe, S. Purchase Intention for Luxury Brands: A Cross Cultural Comparison. *J. Busi. Res.* **2012,** *63*, 1443–1451.

Bijmolt, T. H. A.; Leeflang, P. S. H.; Block, F.; Eisenbeiss, M.; Hardie, B. G. S.; Lemmens, A.; Saffert, P. Analytics for Customer Engagement. *J. Serv. Res.* **2010,** *13*, 341–356.

Bizarris, F. S.; Stehlau, S.; Brandao, M. M. The Moderating Role of Brands for Low Income Luxury Consumers. *Independent J. Manage. Prod.* **2017,** *8* (3), 936–954.

Brakus, J.; Schmitt, B. H.; Zarantonello, L. Brand Experience: What Is It? How Is It Measured? Does It Affect Loyalty? *J. Market.* **2009,** *73* (May), 52–68.

Brandão, A.; Pinho, E.; Rodrigues, P. Antecedents and Consequences of Luxury Brand Engagement in Social Media. *Spanish J. Market.—ESIC* **2019,** *23* (2), 163–183. https://doi.org/10.1108/SJME-11-2018-0052

Campbell, D. T.; Fiske, D. W. Convergent and Ant Validation by the Multitraitmultimethod Matrix. *Psychol. Bull.* **1959,** *56* (2).

Canziani, B.; Watchravesringkan, K.; Yurchisin, J. A Model for Managing Service Encounters for Neo-Luxury Consumers. *Worldwide Hosp. Tour. Themes* **2016,** *8* (1), 41–52.

Carrigan, M.; Moraes, C.; McEachern, M. From Conspicuous to Considered Fashion: A Harm Chain Approach to the Responsibilities of Fashion Businesses. *J. Market. Manage.* **2013,** *29*, 1277–1307.

Chitturi, R.; Chitturi, P.; Raghavarao, D. Design for Synergy with Brand or Price Information. *Psychol. Market.* **2010,** *27* (7), 679–697.

Chun, H.; Diehl, K.; MacInnis, D. Savoring an Upcoming Experience Affects Ongoing and Remembered Consumption Enjoyment. *J. Market.* **2017,** *81* (3), 96–110.

Churchill, G. A.; Jr. *Marketing Research: Methodological Foundations*; Chicago: TheDryden Press, 1995.

Corbin, J.; Strauss, A. Grounded Theory Research: Procedures, Canons, and Evaluative Criteria. *Qual. Sociol.* **1990,** *13* (1), 3–21.

Cristini, H.; Kauppinen-Raisanen, H.; Barthod-Prothade, M.; Woodside, A. Toward a General Theory of Luxury: Advancing from Workbench Definitions and Theoretical Transformations. *J. Busi. Res.* **2017,** *10* (1), 101–107.

D'Arpizio, C.; Levato, F.; Kamel, M. A.; de Montgolfier, J. Luxury Goods Worldwide Market Study, Fall-Winter 2017. *Bain Report*, 2017. http://www.bain.de/Images/BAIN_REPORT_Global_Luxury_Report_2017.pdf (accessed 9 May 2020).

Dabholkar, P. A.; Thorpe, D. I.; Rentz, J. A Measure of Service Quality for Retail Stores: Scale Development and Validation. *J. Acad. Market. Sci.* **1996,** *24* (1), 3–16.

de Kerviler, G.; Rodriguez, C. M. Luxury Brand Experiences and Relationship Quality for Millennials: The Role of Self-Expansion. *J. Busi. Res.* **2019,** *102*, 250–262.

Dion, D.; Arnould, E. Retail Luxury Strategy: Assembling Charisma through Art and Magic. *J. Retail.* **2011,** *87*(4), 502–520.

Dubois, B.; Duquesne, P. The Market for Luxury Goods: Income versus Culture. *Eur. J. Market.* **1993,** *27* (1), 35–44. https://doi.org/10.1108/03090569310024530

Ducrot-Lochard, C.; and Murat, A. *Luxury and Sustainable Development: The New Alliance*; Eyrolles: Paris, 2011.

Dumoulin, D. What Is Today's Definition of Luxury? *Admap* March 2007, 27–30.

Ekinci, Y.; Yoon, T. H.; Oppewal, H. An Examination of the Brand Relationship Quality Scale in the Evaluation of Restaurant Brands. *Adv. Hosp. Leisure* **2005,** *1*, 189–198.

Escalas, J. Self-Identity and Consumer Behavior. *J. Consumer Res.* **2013,** *39* (1), 1–6.

Fan, Y.; Niu, R. H. To Tweet or Not to Tweet? Exploring the Effectiveness of Service Recovery Strategies Using Social Media. *Int. J. Oper. Prod. Manage.* **2016,** *36* (9), 1014–**1036**. https://doi.org/10.1108/IJOPM-10-2013-0461

Fletcher, G. J. O.; Simpson, J. A.; Thomas, G.; Giles, L. Ideals in Intimate Relationships. *J. Personal. Soc. Psychol.* **1999,** *76* (1), 72–89.

Fog, K.; Budtz, C.; Munch, P.; Blanchette, S. *Storytelling: Branding in Practice*, 2nd ed.; Springer: Copenhagen, Denmark, 2010.

Fornell, C.; & Larcker, D. F. Evaluating Structural Equation Models with Unobservable Variables and Measurement Error. *J. Market. Res.* **1981,** *18* (February), 39–50.

Fournier, S. Consumers and Their Brands: Developing Relationship Theory in Consumer Research. *J. Consumer Res.* **1998,** *2* (4), 343–373.

Fung So, K. K.; King, C.; Sparks, B. Customer Engagement with Tourism Brands: Scale Development and Validation. *J. Hosp. Tour. Res.* **2012,** *20* (10). doi: 10.1177/1096348012451456.

Gefen, D.; Straub, D.; & Boudreau, M.-C. Structural Equation Modeling and Regression: Guidelines for Research Practice. *Commun. Assoc. Info. Syst.* **2000,** *4* (1), 7.

Hair, J.; Black, W.; Babin, B.; and Anderson, R. *Multivariate Data Analysis*, 7th ed.; Prentice-Hall, Inc.: Upper Saddle River, NJ, USA, 2010.

Han, H.; Hyun, S. S. Role of Motivations for Luxury Cruise Traveling, Satisfaction, and Involvement in Building Traveler Loyalty. *Int. J. Hosp. Manage.* **2018,** *70*, 75–84.

Hancock, G. R.; Mueller, R. O. Structural Equation Modeling. In *The Reviewer's Guide to Quantitative Methods in the Social Sciences*; Hancock, G. R., Stapleton, S. L., Mueller, R. O., Eds.; Taylor and Francis, 2006; pp 371–383.

Heine, K.; Phan, M. Trading-up Mass-market Goods to Luxury Products. *Austr. Market. J.* **2011,** *19* (2), 108–114.

Hendriksz, V. Kering, H&M, Bestseller & More CEOs to Tackle 7 Crucial Sustainable Priorities, Fashion United, 2018. https://fashionunited. uk/news/business/kering-h-m-bestseller-more-ceos-to-tackle-7-crucial-sustainable-priorities/2018032728851 (accessed 30 July 2020).

Hollebeek, L. D. *Demystifying Customer Engagement: Toward the Development of a Conceptual Model.* Paper Presented at the ANZMAC 2009 Conference, Monash University, Melbourne, Australia, November–December, 2009.

Holt, D. B.; Quelch, J. A.; Taylor, E. L. How Global Brands Compete. *Harv. Busi. Rev.* **2004,** *82* (9), 68–75.

Hong, J. W.; Zinkhan, G. M. Self-concept and Advertising Effectiveness: The Influence of Congruency, Conspicuousness, and Response Mode. *Psychol. Market.* **1995,** *12* (1), 53–77. https://doi.org/10.1002/mar.4220120105

Hospitality Net. *Website of Hospitality Net.* Horwath HTL Report at the International Luxury Travel Market (ILTM) Identifies 'The Future of Luxury Travel' Published on 16 June 2011. http://www.hospitalitynet.org/news/4051855.html (accessed 23 May 2020).

Huang, Z. J.; Cai, L. A. Modeling Consumer-based Brand Equity for Multinational Hotel Brands–When Hosts Become Guests. *Tour. Manage.* **2015,** *46* (2), 431–443.

Ivanova, O.; K. and Poldner, & Branzei, O. Touch and feel. *J. Corp. Citizenship* **2013,** *52*, 102–130.

Jain, V.; Roy, S.; Ranchhod, A. Conceptualizing Luxury Buying Behavior: The Indian Perspective. *J. Prod. Brand Manage.* **2015,** *24* (3), 211–228.

Joy, A.; Sherry Jr.; J.; Venkatesh, A.; Wang, J.; Chan, R. Fast Fashion, Sustainability, and the Ethical Appeal of Luxury Brands. *Fashion Theor.* Sept **2012**.

Kafle, N. P. Hermeneutic Phenomenological Research Method Simplified. *Bodhi: Interdiscipl. J.* **2013,** *5*, 181–200.

Kapferer, J. N. The Two Business Cultures of Luxury Brands. In *Brand Culture*; Schroeder, J., Salzer-Mörling, M., Eds.; Taylor and Francis: London, 2006; pp 67–76.

Kapferer, J. N. The Artification of Luxury: From Artisans to Artists. *Busi. Horiz.* **2014,** *57* (3), 371–380.

Kapferer, J. N.; Laurent, G. Where Do Consumers Think Luxury Begins? A Study of Perceived Minimum Price for 21 Luxury Goods in 7 Countries. *J. Busi. Res.* **2016,** *69* (1), 332–340.

Kapferer, J. N.; Michaut, A. Is Luxury Compatible with Sustainability? Luxury Consumers' Viewpoint. *J. Brand Manage.* **2017,** *21* (1), 1–22.

Kaufman, H. R.; Loureiro, S. M. C.; Manarioti, A. Exploring Behavioural Branding, Brand Love and Brand Co-creation. *J. Prod. Brand Manage.* **2016,** *25* (6), 516–526.

Kendall, J. Responsible Luxury–A Report on the New Opportunities for Business to Make a Difference. *The World Jewellery Confederation*, 2010. Published in 2010. http://www.cibjo.org/download/responsible_luxury.pdf (accessed 13 July 2020).

Kim, A. J.; Ko, E. Do Social Media Marketing Activities Enhance Customer Equity? An Empirical Study of Luxury Fashion Brand. *J. Busi. Res.* **2012,** *65*, 1480–1486.

Kleanthous, A.; Peck, J. *Let Them Eat Cake. Satisfying the New Consumer Appetite for Responsible Brands.* WWF UK, 2006. Published in 2006. http://www.wwf.org.uk/filelibrary/pdf/let_them_eat_cake_abridged.pdf (accessed 18 July 2020).

Koronaki, E.; Kyrousi, A. G.; Panigyrakis, G. G. The Emotional Value of Arts-based Initiatives: Strengthening the Luxury Brand–Consumer Relationship. *J. Busi. Res.* **2017,** *85*, 406–413.

Kotler, P.; Kartajaya, H.; Setiawan, I.; Wiley InterScience (Online service). *Marketing 3.0: From Products to Customers to the Human Spirit*; Wiley: Hoboken, NJ, 2010.

Kunz, J.; Stephanie, M.; Schmidt, H. J. Sustainable Luxury: Current Status and Perspectives for Future Research. *Busi. Res.* **2020,** *56*, 541–601.

Lee, E. J. A Prototype of Multicomponent Brand Personality Structure: A Consumption Symbolism Approach. *Psychol. Market.* **2013,** *30* (2), 173–186.

Lee, J.; Hwang, J. Luxury Marketing: The Influences of Psychological and Demographic Characteristics on Attitudes toward Luxury Restaurants. *Int. J. Hosp. Manage.* **2011,** *30* (3), 658–669.

Liang, Y.; Ghosh, S.; Oe, H. Chinese Consumers' Luxury Value Perceptions—A Conceptual Model. *Qual. Market Res. Int. J.* **2017,** *20* (2), 247–262.

Liao, J.; Wang, L. Face as a Mediator of the Relationship between Material Value and Brand Consciousness. *Psychol. Market.* **2009,** *26* (11), 987–1001.

Ling, Y. J.; Ariff, M. S. B. M.; Zakuan, N.; Tajudin, M. N. M. Brand Personality, Brand Loyalty and Brand Quality Rating in the Contact Lens Perspective. *Rev. Integr. Busi. Econ.* **2014,** *3* (2), 433–448.

Liu, M. T, Wong, I. A.; Tseng, T.; Chang, A. W.; Phau, I. Applying Consumer-based Brand Equity in Luxury Hotel Branding. *J. Busi. Res.* **2017,** *81*, 192–202.

Macey, W. H.; Schneider, B. The Meaning of Employee Engagement. *Indust. Org. Psychol.* **2008,** *1*, 3–30.
Merz, M. A.; Zarantonello, L.; Grappi, S. How Valuable Are Your Customers in the Brand Value Co-creation Process? The Development of a Customer Co-creation Value (CCCV) Scale. *J. Busi. Res.* **2018,** *82* (January), 79–89.
Meyer, D.; Hornik, K.; Feinerer, I. Text Mining Infrastructure *in R. J. Statist. Softw.* **2008,** *25* (5), 1–54.
Miller, G. and Real, N. Postmodernity and Popular Culture. In *The Post-Modern Presence*; Berger, A. A., Ed.; London: Sage, 1998.
Munezero, M. D.; Montero, C. S.; Sutinen, E.; Pajunen, J. Are They Different? Affect, Feeling, Emotion, Sentiment, and Opinion Detection in Text. *IEEE Trans. Affect. Comput.* **2014,** *5* (2), 101–111.
Nash, J.; Ginger, C.; Cartier, L. The Sustainable Luxury Contradiction: Evidence from a Consumer Study of Marine-Cultured Pearl Jewellery. *J. Corp. Citizenship* **2016,** *63*, 73–95.
Nobre, H. *The New-Luxury Brand Relationship Scale: Measuring Relationships with Mass-Consumed Luxury Brands* (working paper); University of Aveiro: Portugal, 2015.
Nobre, H.; Simoes, C. New Lux Brand Relationship Scale: Capturing the Scope of Mass-Consumed Luxury Brand Relationships. *J. Busi. Res.* **2019,** *102*, 328–338.
Novotná, M.; Kunc, J. Experience Matters! Luxury Consumption Patterns and Motivation of the Czech Affluent Society. *Deturope* **2019,** *11* (2), 121–142.
Novotna, M.; Kunc, J. Luxury Tourists and Their Preferences: Perspectives in the Czech Republic. *Tourism* **2019,** *67* (1), 90–96.
Nunnally, J. C. *Psychometric Theory*; McGraw-Hill: New York, 1978.
Okonkwo, U. *Luxury fashion Branding: Trends, Tactics, Techniques*; Palgrave Macmillan: Basingstoke, Hampshire, UK, 2007.
Patterson, P.; Yu, T.; de Ruyter, K. *Understanding Customer Engagement in Services.* Paper presented at the ANZMAC 2006: Advancing Theory, Maintaining Relevance, Brisbane, Queensland, 4–6 Dec 2006.
Patton, M. Q. Two Decades of Developments in Qualitative Inquiry: A Personal Experiential Perspective. *Qual. Soc. Work* **2002,** *1* (3), 263–283.
Payne, A.; Storbacka, K.; Frow, P.; Knox, S. Co-creating Brands: Diagnosing and Designing the Brand Experience Relationship. *J. Busi. Res.* **2009,** *69* (3), 379–789.
Pine, B. J.; Gilmore, J. H. Welcome to the Experience Economy *Harv. Busi. Rev.* July/Aug 1998, 97–105.
Pine, B. J.; Gilmore, J. H. *The Experience Economy*; Harvard Business School Press: Boston, MA, 1999.
Platania, S.; Santisi, V.; Morando, M. Impact of Emotion in the Choice of Eco-Luxury Brands: The Multiple Mediation Role of the Brand Love and the Brand Trust. *Qual.—Access Success* **2019,** *20* (2), 501–506.
Podsakoff, P. M.; MacKenzie, S. B.; Lee, J.-Y.; Podsakoff, N. P. Common Method Biases in Behavioral Research: A Critical Review of the Literature and Recommended Remedies. *J. Appl. Psychol.* **2003,** *88*, 879–903.
Poldner, K.; Dentoni, D.; Ivanova, O. Aesthetic Mediation of Creativity, Sustainability and the Organization. *J. Clean. Prod.* **2017,** *140* (3), 1936–1947.

Positive Luxury. Our Story, Positive Luxury, 2018. https://positiveluxury.com/about/story (accessed 30 June 2020).

Prahalad, C. K.; Ramaswamy, V. Co-creation Experiences: The Next Practice in Value Creation. *J. Interact. Market.* **2004,** *18* (3), 4–9.

Rolling, V.; Sadachar, A. Are Sustainable Luxury Goods a Paradox for Millennials? *Soc. Respons. J.* **2018,** *14* (4), 802–815.

Roper, S.; Caruana, R.; Medway, D.; Murphy, P. Constructing Luxury Brands: Exploring the Role of Consumer Discourse. *Eur. J. Market.* **2013,** *47* (3/4), 375–400.

Russell, J. A. A Circumplex Model of Affect. *J. Personal. Soc. Psychol.* **1980,** *39* (6), 1161–1178.

Schroeder, J. E. The Artist and the Brand. *Eur. J. Market.* **2005,** *39*, 1291–1305.

Schroeder, J. E. The Cultural Codes of Branding. *Market. Theor.* **2009,** *9* (1), 123–126.

Seo, Y.; Buchanan-Oliver, M. Luxury Branding: The Industry, Trends, and Future Conceptualisations. *Asia Pacific J. Market. Logist.* **2015,** *27* (1), 82–98.

Shin, H.; Eastman, J. K.; Mothersbaugh, D. The Effect of a Limited-Edition Offer Following Brand Dilution on Consumer Attitudes toward a Luxury Brand. *J. Retail Consumer Serv.* **2017,** *38*, 59–70.

Sirgy, M. J. Self-concept in Consumer Behavior: A Critical Review. *J. Consumer Res.* **1982,** *9* (3), 287–300. https://doi.org/10.1086/208924

Sreejesh, S. Pillai.; Sarkar, A.; Roy, S. Validating a Scale to Measure Consumer's Luxury Brand Aspiration. *J. Prod. Brand Manage.* **2016,** *25* (5), 48–65. http://dx.doi.org/10.1108/JPBM-06-2014-0647

Steinhart, Y.; Ayalon, O.; Putermen, H. The Effect of an Environmental Claim on Consumers' Perceptions about Luxury and Utilitarian Products. *J. Clean. Prod.* **2013,** *53*, 277–286.

Sthapit, A.; Jo, G. Y.; Hwang, Y. Y. Communicating Responsible Luxury Brand: The Role of Luxury-CSR Fit and Dispositional Consensus on Brand Evaluation. *J. Distr. Sci.* **2017,** *15* (2), 7–14.

Sumbly, Y.; Siraj, S. A Conceptual Framework of Brand Experience for Luxury Brands. *J. Manage. Res. Analy.* **2019,** *6* (1), 60–64.

Sung, Y.; Choi, S. M.; Ahn, H.; Song, Y. A. Dimensions of Luxury Brand Personality: Scale Development and Validation. *Psychol. Market.* **2015,** *32* (1), 121–132.

Tian, X.; He, W.; Tang, C.; Li, L.; Xu, H.; Selover, D. A New Approach of Social Media Analytics to Predict Service Quality: Evidence from the Airline Industry. *J. Enterprise Info. Manage.* **2019,** *33* (1), 51–70. https://doi.org/10.1108/JEIM-03-2019-0086.

Torelli, C.; Keh, H. T.; Chiu, C. Y. Cultural Symbolism of Brands. *Brands Brand Manage.: Contemp. Res. Perspect.* **2010,** 113–132.

Townsend, C.; Sood, S. Self-affirmation through the Choice of Highly Aesthetic Products. *J. Consumer Res.* **2012,** *39* (2), 415–428.

Turunen, L. L. M. *Interpretations of Luxury: Exploring the Consumer Perspective*; Palgrave Macmillan: Cham, Switzerland, 2017.

Tutek, E.; Roth, C.; Simatic, A. Rethinking Luxury in Hospitality. Industry Report, 2018. https://www.howarthhtl.com (accessed 16 July 2020).

Tynan, C.; McKechnie, S.; Chhuon, C. Co-creating Value for Luxury Brands. *J. Busi. Res.* **2010,** *63* (11), 1156–1163.

Vargo, S. L.; Lusch, R. F. Evolving to a New Dominant Logic for Marketing. *J. Market.* **2004,** *68* (1), 1–17. https://doi.org/10.1509/jmkg.68.1.1.24036

Vickers, J. S.; Renand, F. The Marketing of Luxury Goods: An Exploratory Study—Three Conceptual Dimensions. *Market. Rev.* **2003,** *3*, 459–478.

Vigneron, F.; Johnson, L. W. Measuring Perceptions of Brand Luxury. *Brand Manage.* **2004,** *6* (2), 484–506.

Wilcox, K.; Kim, H. M.; Sen, S. Why Do Consumers Buy Counterfeit Luxury Brands? *J. Market. Res.* **2009,** *46* (2), 247–259.

Yang, F. X.; Lau, M. LuXurY Hotel Loyalty—A Comparison of Chinese Gen X and Y Tourists to Macau. *Int. J. Contemp. Hosp. Manage.* **2016,** *27* (2), 1685–1706.

Yang, W.; Mattila, A. S. Why Do We Buy Luxury Experiences? *Int. J. Contemp. Hosp. Manage.* **2016,** *28* (9), 1848–1867.

CHAPTER 6

INFLUENCE OF SOCIAL MEDIA ON TOURIST BEHAVIOR: INDIAN PERSPECTIVE

RASHMI[1,*], GUNJAN BHAYANA[2], and SAWITRI DEVI[2]

[1]*Management Department, Delhi Global Institute of Management, Faridabad, Haryana, India*

[2]*Institute of Management Studies & Research, MD University, Rohtak, Haryana, India*

**Corresponding author. E-mail: mbahr006@gmail.com*

ABSTRACT

The tourism industry of India is an emerging sector and playing a significant role in the growth of "services sector" in our country. Further, the tourism in India has high potential due to its rich, diverse culture and heritage, ecological variation from mountains to beaches, and natural scenic beauty creating huge "tourism product portfolio" catering to the requirements of diverse tourist population. Furthermore, the exponential growth of mobile phone and internet users has also boosted the growth of the tourism industry and the technology-enabled social media platform has brought significant shift in the awareness, outlook, and the expression of the tourist behavior by means of social media as people continue to share their holiday photographs, touring experience which in turn influence other tourists. These user-generated contents are considered most credible

Tourist Behavior: Past, Present, and Future. Narendra Kumar, PhD, Bruno Barbosa Sousa, PhD & Swati Sharma, PhD (Eds.)

by the online communities. Hence, this empirical study assesses the preferences of various social media and its influence on the tourist behavior. The study concludes that Facebook emerged as the most preferred platform and tourist decisions are significantly influenced by different social media contents with respect to choosing their travel destination.

6.1 INTRODUCTION

The tourism industry is highly information-intensive (Sheldon, 1997; Werthner and Klein, 1999). Hence, it requires the understanding of recent technological trends and latest channels used for distribution of the information that influences the consumer behavior and their decision-making specifically in the tourism domain. Nowadays, many internet-based "social media" platforms are trending around, that carry a lot of user-generated information which includes their travel-related experiences. And, this information is accessed and spread by various other users as the social media provides an application to the users to share the "post," "hash-tag," "tag," "digg," etc. Hence, these contents are becoming the sources of "online information" intending to educate other consumers with regard to the "products, brands, services and issues" (Blackshaw and Nazzaro, 2006). Further, "user-generated content" (UGC) such as "facebook, twitter, instagram, Pinterest, blogs, virtual communities, wikis, social networks, collaborative tagging, and media files shared on sites, such as YouTube and Flickr, have gained substantial popularity in online travelers' use of the Internet" (Gretzel, 2006; Pan et al., 2007). Hence, we can say that social media has drastically transformed the landscape of the tourism industry with regard to the distribution of the information which impacts the tourism system significantly (Buhalis and Law, 2008). Further, it has become essential for tourism players to understand the drivers of tourism and choice pattern of the tourist in order to develop and launch "successful marketing campaign and create better information ecosystems to enhance user experience." Although the use of social media is growing exponentially, however, it lacks the empirical evidences with regard to the impact of the social media on tourist behavior in the Indian context. Hence, the present study offers empirical evidences of the preferences of users with regard to various social media and their impact on their tourist behavior.

Objectives of the Study

- To explore the extent of usage of social media for planning a tour by tourist.
- To know the most preferred social media sources used for planning a tour by tourist.
- To understand how much social media influence the tourist decision for a tour/trip

6.2 LITERATURE REVIEW

Xiang and Gretzel (2010) conducted a study titled "Role of Social Media in Online Travel Information Search," and highlighted the importance of social media with respective to search engine. Throughout this study, it was concluded that there were nine U.S. tourist destination names which were widely searched by tourist players.

Potis et al. (2012), in the research paper titled "Social Media Use and Impact During the Holiday Travel Planning Process," highlighted the relevancy of social media in making holiday travel plans. From the findings, the result found that there is a strong correlation exists between level of influence and alteration level done in holiday plans.

Leung et al. (2013), in the paper titled "Social Media in Tourism and Hospitality: A Literature Review," highlighted the uses of social media in decision-making process. To conduct this study, content analysis was used and it was observed that from the consumer point of view, usage of social media becomes significant and from the supplier point of view, promotional and management strategy becomes significant.

Rashad and Lim (2014), in the paper titled " Impact of Social Networking Sites on Hospitality and Tourism Industries," stressed on the relevancy of social networking sites. This paper throws lights upon both the aspects, that is, positive and negative, such as effectiveness level of the consumers in terms of time and cost, building of company good image, etc. be like.

Živković et al. (2014), in the paper titled "The Impact of Social Media on Tourism," highlighted the dependency on ICT Sector. In this paper, by using the concept of market segmentation, it was easily observed that tourist packages are offered and are recognized through modern digital media, It also concluded that most of the travelers use online travel applications for searching the best destination and avail vacation offers.

Gururaja (2015) conducted a study on the topic "Impact of Social Media on Tourism and Hospitality," highlighted the significance of social media with respect to tourism industry. This paper further determines the effect of social media also. A sample size of 500 respondents including individuals and hotels were considered and it was found that most of the travel players takes the advantage of social media efficiently.

Sahoo and Mukunda (2017), in the paper titled "Role of Social Media in Promoting Tourism Business—A Study on Tourism Promotion in Odisha," highlighted that how social media extends the business of tourism sector successfully. In addition to this, social media helps to connect people socially and boosts the interaction process. It was also observed that people of Odisha city widely used social media.

Hur et al. (2017), in the research paper titled "An Exploration of the Factors Influencing Social Media Continuance Usage and Information Sharing Intentions Among Korean Travelers," examined the factors of social media on the users. By using SEM, it was observed that travel triggers provide information sharing services to users efficiently.

Han et al. (2018), conducted a study on "Evaluating User-Generated Content in Social Media: An Effective Approach to Encourage Greater Pro-Environmental Behavior in Tourism," signified that social media brings changes in pre-environmental behavior of tourism industry. It further stressed upon the UGC norms laid down with respect to social media, how effectively it improves the communication skills.

Pejman et al. (2019), in the paper "Place Branding and Moderating Role of Social Media," investigated the effect of social media in producing brand image for the tourists. It was found that sharing pictures and videos did not attract visitors. Hence, it is recommended that pre-environmental behavior should be tested.

Tolani and Pushkar (2019), in the paper "Study on the Impact of Social Media on Tourism Industry in India with Reference to Population of Uttar Pradesh," highlighted the popularity of social media among the population of city Uttar Pradesh with respect to tourism industry, In this, it was considered that online travel agencies gained a lot of popularity with the help of social media platforms.

Gretzel (2019), in the article titled "The Role of Social Media in Creating and Addressing Overtourism," highlighted the ground of social media which helps to meet psychological, social, and functional needs of the consumer. It further highlighted the terminology, that is, overtourism, it becomes combating when social media-based solutions are used effectively.

Chatzigeorgiou and Christou (2020), in the article named "Adoption of Social Media as Distribution Channels in Tourism Marketing: A Qualitative Analysis of Consumers' Experiences," highlighted the experiences of consumers with regard to social distribution services. And, by conducting this kind of study, in-depth interviews were done, and from that it was found that by adopting online technology, consumers developed both positive and negative attitudes simultaneously.

6.3 METHODOLOGY OF THE STUDY

The research design used for present study is descriptive cum exploratory. The questionnaire has been prepared using Google form for collecting data. The questionnaire has been distributed using emails and Whatsapp groups. The data of total 256 respondents have been found usable and analyzed for the purpose of this study. The data have been analyzed using percentage, bar, and pie charts in MS Excel 2010.

6.4 DATA ANALYSIS

The first objective of the study is "to explore the extent of usage of social media for planing a tour by tourist." The survey results revealed that around 225 (88%) respondents make use of various social media platforms to gather the information about their travel plan and destination.

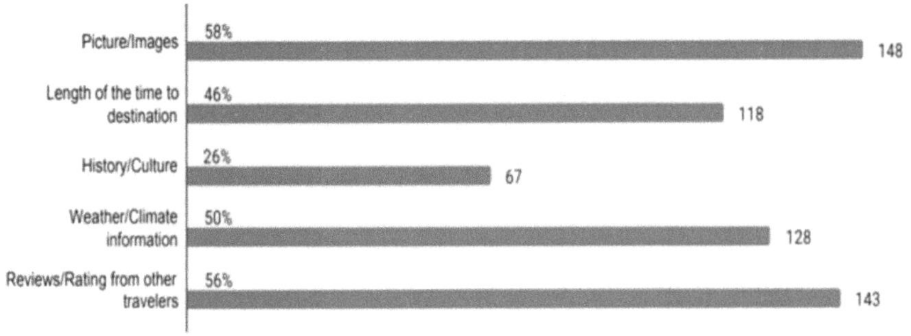

FIGURE 6.1 Most preferred social media contents.
Source: Survey results.

Figure 6.1 shows the percent and count of the responses of respondents on the most preferred contents with respect to their travel destination. As per the responses, 148 (58%) respondents preferred picture/images followed by Reviews/Rating from other travelers 143 (56%), Weather/ Climate information 128 (50%) and Length of the time to destination 118 (46%) and least 67 (26%) respondents preferred History/Culture of the travel destination. The respondents provided their opinion by giving multiple options with regard to the most preferred contents about travel destination, hence, the overall percent distribution exceeded 100%.

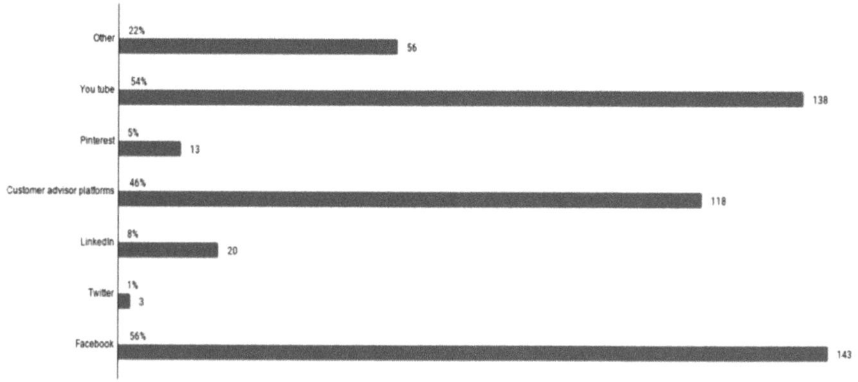

FIGURE 6.2 Most preferred social media sources.
Sources: Survey results.

Figure 6.2 shows the percent preference and count of respondents with regard to different social media sources while deciding their travel destination. The social media platform Facebook is most preferred social media sources among all as per the survey results. As per the responses, 143 (56%) respondents mentioned it as their first choice closely followed by 138 respondents were influenced by Youtube (54%), customer adviser platform was preferred by 118 (46%) which is the third highest preference followed by other sources 56 (22%), that is, instagram, Whatsapp, etc. Furthermore, Linkedln is also substantially preferred by 20 (8%) users, followed by Pinterest 13 (5%) and Twitter 3 (1%).

Figure 6.3 shows the responses of the respondents, weather social media influences their travel destination choices. Out of the total 256 respondents 246 (96%) respondents agreed that their travel destination choice gets influenced with the information available on social media.

There are 108 (42%) respondents agree that they get reasonably influenced, 82 (32%) slightly influenced, 56 (22%) significantly influenced by social media while deciding their travel destination.

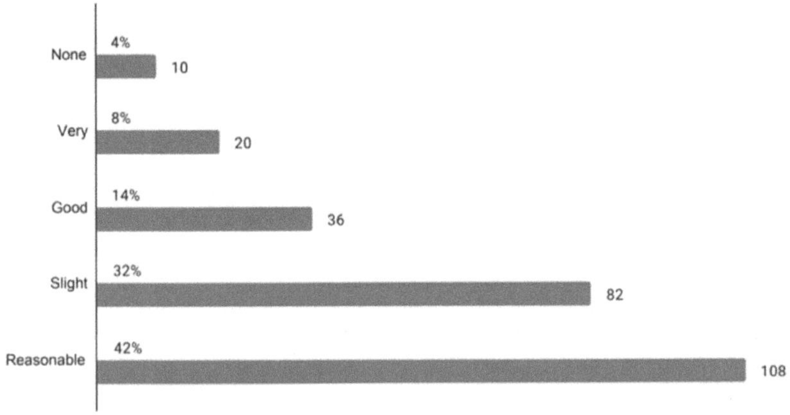

FIGURE 6.3 Influence of Social media on travel destination of tourist.
Sources: Survey results.

6.5 CONCLUSION

The present study gives insights into increasing use of social media platform in the tourism domain. Due to the advent of technology, various social media platforms have cropped up. These platforms have empowered the consumers to distribute the knowledge, information, and experiences as a user by producing and posting related contents specifically on their personal social media pages in order to share with others. In other words, social media is key channel of fueling "the creation and consumption of content" specifically travel-related. Undoubtedly, the social media users acts as motivator for other to travel. Hence, it creates sentiments that bring the desire to travel among others, therefore, give rise to travel trends. Further, the latest "experiential information available on social media influences the decision-making processes. Furthermore, the ratings and reviews are used to create the hit list of the travel destinations that assist the information seekers, thereby exert social influence" as referring "these lists have become an important pursuit for some social media users, as they allow them to build reputation, feel closely connected to the users and their audience, and to obtain approval from their own social network." Besides, social media has foster travel to particular places, specifically "city tourism" compared with rural as city offers connectivity required by the tourists to fulfil their "social media posting needs" thus will lead to concentrated tourism at certain places. However, "social media platform should be used strategically to promote certain behavior over others. These platforms should be subjected to strong social norms and being policed by social media users and subsequently integrated into the self-disciplinary gazes of social media users." However, the marketers in the tourism domain can harness these findings and accordingly plan their marketing strategies to popularize their tourism space in order to increase the footfall. They can also customize their tourism offerings as per the choices of the users in order to enhance user experiences.

KEYWORDS

- **thermal tourism**
- **cross-border destination**
- **relationship marketing**
- **wellness**

REFERENCES

Blackshaw, P.; Nazzaro, M. *Consumer-Generated Media (CGM) 101: Wordof-Mouth in the Age of the Web-fortified Consumer*; NielsenBuzzMetrics: New York, 2006.

Buhalis, D.; Law, R. Progress in Information Technology and Tourism Management: 20 Years on and 10 Years after the Internet: The State of eTourism Research. *Tour. Manage.* **2008,** *29* (4), 609–623.

Chatzigeorgiou, C.; Christou, E. Adoption of Social Media as Distribution Channels in Tourism Marketing: A Qualitative Analysis of Consumers' Experiences. *J. Tour. Herit. Serv. Market.* **2020,** *6* (1), 25–32.

Ebrahimi, P.; Hajmohammadi, A.; Khajeheian, D. Place Branding and Moderating Role of Social Media. *Curr. Iss. Tour.* **2020,** *23* (14), 1723–1731.

Fotis, J. N.; Buhalis, D.; Rossides, N. Social Media Use and Impact during the Holiday Travel Planning Process; Springer-Verlag, 2012; pp. 13–24.

Gretzel, U. Consumer Generated Content—Trends and Implications for Branding. *e-Rev. Tour. Res.* **2006,** *4* (3), 9–11.

Gretzel, U. The Role of Social Media in Creating and Addressing Overtourism. In *Overtourism: Issues, Realities and Solutions*, Vol. 1; De Gruyter: Berlin/Boston, MA, 2019; pp. 62–75.

Gururaja, R. *Impact of Social Media on Tourism and Hospitality*; MSRUAS: Bangalore, 2015.

Han, W.; McCabe, S.; Wang, Y.; Chong, A. Y. L. Evaluating User-generated Content in Social Media: An Effective Approach to Encourage Greater Pro-environmental Behavior in Tourism? *Journal of Sustainable Tourism* **2018,** *26* (4), 600–614.

Headquarters, U. N. W. T. O. Travel and Tourism under Pandemic Conditions, 2009.

Hur, K.; Kim, T. T.; Karatepe, O. M.; Lee, G. An Exploration of the Factors Influencing Social Media Continuance Usage and Information Sharing Intentions among Korean Travellers. *Tour. Manage.* **2017,** *63*, 170–178.

Leung, D.; Law, R.; Van Hoof, H.; Buhalis, D. Social Media in Tourism and Hospitality: A Literature Review. *J. Travel Tour. Market.* **2013,** *30* (1–2), 3–22.

Pan, B.; MacLaurin, T.; Crotts, J. C. Travel Blogs and Their Implications for Destination Marketing. *J. Travel Res.* **2007,** *46* (1), 35–45.

Sheldon, P. J. *Tourism Information Technology*; CAB International: Oxon, UK, 1997.

Tolani, M.; Pushkar, M. B. K. Study on Impact of Social Media on Tourism Industry in India with Reference to Population of Uttar Pradesh. *Int. J. Sci. Res. Rev.* **2019,** *7* (04), 812–815.

Werthner, H.; Klein, S. *Information Technology and Tourism: A Challenging Relationship*; Springer: Vienna, 1999.

Xiang, Z.; Gretzel, U. Role of Social Media in Online Travel Information Search. *Tour. Manage.* **2010,** *31* (2), 179–188.

Yazdanifard, R.; Yee, L. T. Impact of Social Networking Sites on Hospitality and Tourism Industries. *Global J. Hum.-Soc. Sci.: Econ.* **2014,** *14* (8), 1–5.

Živković, R.; Gajić, J.; Brdar, I. The Impact of Social Media on Tourism. *Singidunum J. Appl. Sci.* **2014,** 758–761.

CHAPTER 7

MEDIATING ROLE OF TOURIST MOTIVATION AND PARTICIPATION WITH SATISFACTION AND REVISIT INTENTION IN RISHIKESH, INDIA

LAKHVINDER SINGH[1,*] and DINESH DHAKHAR[2]

[1]*Department of Tourism Management, Government College, Kaithal, Haryana, India*

[2]*Department of Tourism & Hotel Management, Kurukshetra University, Kurukshetra, Haryana, India*

**Corresponding author. E-mail: lakhvindersingh.kuk@gmail.com*

ABSTRACT

The present chapter is concerned with understanding how tourist motivation and tourist participation interact, and at the same time, contribute toward a satisfying tourism experience. The chapter focuses on Rishikesh in Uttarakhand, India as a case of destination and randomly selected the participant through a field survey. By applying factor analysis, three dimensions of attributes, such as destination uniqueness, attractiveness, and facilities and services were identified and were found to contribute significantly to tourist experience satisfaction, with destination attractiveness contributing the most. The chapter discovered the extent of participation as influencer of tourist experience with satisfaction. Additionally, the findings indicated no direct effect of motivation to satisfaction, rather,

Tourist Behavior: Past, Present, and Future. Narendra Kumar, PhD, Bruno Barbosa Sousa, PhD & Swati Sharma, PhD (Eds.)

with the mediating effect of destination attribute assessment and tourist participation with the destination, the contribution of motivation toward satisfaction became significant. In other words, motivation can positively influence destination satisfaction only in the case of tourist participations of an experience were met and tourists were actually participated in various activities at destination. The findings of chapter contribute both conceptually and practically to tourism destination planning and management to create a satisfying tourist experience.

7.1 INTRODUCTION

The phenomena of tourism in modern world rising rapidly with successive growth during past few years and accepted as an important catalyst for economic development through revenue generation, foreign exchange, and employment opportunities (UNWTO, 2018). The World Travel & Tourism Council (WTTC, 2017) also estimated its direct contribution to GDP to be US$ 2306.0 billion and total contribution as US$ 7613.3 billion during the year 2016 and expected to increase rapidly in next decade. So, tourism industry has grown as an important development mean worldwide.

In a country like India, tourism has become an important agent of development in different perspectives, including economic, social, and regional development. Sofield and Li (1998) showed that the improvement of the travel industry is bolstered by rich wealth of verifiable and social assets, and the administration underpins the travel industry. Couple of research (Dasgupta, 2011), hence declared tourism development in the country is enriched by its rich cultural and natural resources, facilities and services, and along with adequate institutional mechanism. As a result, more than 10 million international and 1652 million tourists visited India during 2017 (Ministry of Tourism, Government of India, 2018).

The increasing size of tourist footfalls resulted in outcome of mounting demands and expectations among tourists for enhanced facilities and services. Resultantly, the destination planners strained to recompense such multifunctional needs, motives, and expectation of the tourists (Rinschede, 1992). This paradigm shift in tourism influences the behaviors of tourists and has influenced the choice of destination and product preferences (Wright, 2008). Therefore, it is imperative to systematically analyze the behavior and attitude of tourists at the destination (Sharma et al., 2011) to ensure the success and survival of tourist destinations in future. Hence, the concerned stakeholders should focus on pleasing the tourists by effectively designing and offering different facilities and services in tourism.

The handling of tourist demands and would assist with advancing an appealing goal picture to pull in likely sightseers and keep up rehash guests (Huh et al., 2006). In light of this foundation, past examination has additionally perceived different predecessors that influence travel experience including travel inspiration (Devesa et al., 2010), tourist perception (Donlon et al., 2010), performance of different attributes or elements (Komppula and Gartner, 2013), and tourist' participation and involvement in different activities at destination also influence travelers' intention for involving more frequently at destination (Gross and Brown, 2008). But, despite the importance of the phenomena, diminutive concentration of researcher and academicians was reported by literature toward investigation of travelers' satisfaction all the way through overall experience toward destination. Hence, there exists a research gap in this research line. Hence, against this background, the present chapter investigates the interaction among travel motives, attributes' quality, and tourist participating in different activities, as well as to examine the extent to which these constructs contribute to the contentment of tourists in Rishikesh as a destination.

7.1.1 *AIMS AND OBJECTIVE*

The study is designed to investigate tourist experience in Rishikesh in different folds, that is, (1) motivating factors behind traveling to Rishikesh, (2) satisfaction arise from different attributes' quality by that matching expectation, (3) extent of tourists' interaction and participation in different activities, and (4) what contributes to tourists' overall experience with satisfaction and revisit intentions.

7.2 LITERATURE REVIEW

7.2.1 *SATISFACTION*

Satisfaction is a collective evaluation of individual experiences from consumption of different products and services. In terms of tourism sector, tourist satisfaction is also imperative as it influences the use of various services during the visit and the satisfied tourist tends to be more willing to revisit the destination and recommend the destination to other tourists (Lee and Beeler, 2009; Kim and Brown, 2012).

Although different methods have been developed to study tourist satisfaction, but theory developed by Oliver (1980) know as expectancy disconfirmation paradigm (EDP) is the approach that gained sufficient consideration from researcher in the field of satisfaction (Yen and Lu, 2008). The theory proposed that consumer satisfaction is a function of expectation and expectancy disconfirmation. In the purchasing process, consumers compared the actual performance with their expectation of a product, and the gap between the two determines satisfaction. The theory was also commonly applied in the study of tourist satisfaction, which was explained as the result of the discrepancy between pre-travel expectation and post-travel perception (Chen and Chen, 2010; Lee and Beeler, 2009). Nevertheless, Tse and Wilton (1988) anticipated customer satisfaction as definite experience and emphasize expectations as not essential to measure satisfaction. Divergently, Yoon and Uysal (2005) indicated that the actual performance of destination elements in assessing tourist satisfaction can be only measured by comparing pre-travel expectations of different elements or facilities and services. Hence, it is an important parameter to identify satisfaction in several dimensions.

7.2.2 DESTINATION ATTRIBUTE AND SATISFACTION

The performance of destination attributes refers to the perceived performance of the functional aspects of a destination, such as service quality, location, and attractiveness (Ahn et al., 2011). The significant role of attribute performance in shaping satisfaction has been extensively discussed in the literature of Voon and Lee (2009), and it has been concluded that there are associations between experience quality, perceived value, satisfaction, and behavioral intentions. This calls for providing better products and services at destination to increase the satisfaction index of the destination.

Voon and Lee (2009) also accomplished tourist satisfaction as outcome of various destination elements, such as accessibility, amenities, access, cleanliness, nature, safety, food and beverages, culture, people, and local experience. Huh et al. (2006) also incorporated that positive tourist' perceptions toward destination attributes, such as attraction, shopping and souvenir, and information accessibility as predictor of enhanced satisfaction and better overall experience. Whereas, existence of underprivileged facilities and services resulted into discontent among tourists (Kozak and Rimmington, 2000). Therefore, attention should be given to identification of attributes contributing to tourist satisfaction at destination (Caber et al., 2012). The authors of the present chapter investigated the connection of tourists' perception of destination attributes with satisfaction and revisit intention.

7.2.3 TOURIST PARTICIPATION, SATISFACTION AND REVISIT INTENTION

The term participation indicates the association of tourists in different activities at the destination (McQuarrie and Munson, 1987). The notion has been illustrated in the literature (Laurent and Kapferer, 1985) and referred to a psychosomatic affirmation toward curiosity, motive, and excitement to particular course of action. Hence, it has been commonly described as major component of the decision-making (Chen and Wu, 2010). In the field of tourism, the interrelationship among tourist sites and traveler is taken into consideration in different perspectives, such as place affection, place uniqueness, and place acquaintance. The overemphasized use of such concepts resulted into diverse description of the phenomena among scholars (Prayag and Ryan, 2012).

Although the study of tourist participation is among the emerging topics of research, and there is scarcity of standardized instrument to measure the concept (Prayag and Ryan, 2012). However, a couple of scholars, Zaichikowsky (1985) applied a one-dimensional measurement using 20 statement parameters to assess the term. Laurent and Kapferer (1985) anticipated a tool using four facets to measure the attributes, including perceived importance, perceived pleasure, perceived sign, and perceived risk (Chen and Wu, 2010; Prayag and Ryan, 2012). In a practical approach, tourist participation comprises different aspects like self-identity, satisfaction, attention, menace, and conformity (Lee et al., 2009).

As a predictor of consumer behavior, Hwang et al. (2005) demonstrated the tourist participation as instantaneous predecessor to predict satisfaction and positive recommendation and revisit intentions. Lee and Beeler (2009) in their study also sustained the higher level of participation of tourists is associated with better satisfaction and behavioral intentions. Thus, higher tourist participation could also be regarded as a positive feature toward the development of a destination (Gross and Brown, 2008). Nevertheless, Prayag and Ryan (2012) in their study unable to notice any significant effect of tourist participation to satisfaction and behavioral intentions, thus described other dimensions, including motivations and tourist as interferer between participation and overall satisfaction.

7.2.4 MOTIVATION, SATISFACTION, AND REVISIT INTENTION

Tourist motivation is an important constraint of tourist behavior and considered as predictor of satisfaction and repeat visit behavior toward the destination. It therefore influences both cognitive perception and participation in travel (Kim and Brown, 2012).

Travel motivation, which has been studied on a multidisciplinary base, plays a focal role in tourism research since mass tourism began to thrive (Chen et al., 2011). Yoon and Uysal (2005) described motivation as psychosomatic and factual desires that are responsible for behavioral overview of tourists. Though diverse descriptions were noticed in the literature, and motivation is elaborated as the "driving force" that stimulates visitors for participation in several activities at the destination (Crompton, 1979).

Tourist motivation can be internal forces which mean people's desire to travel as well as the external forces related to destination attributes (Ritchie et al., 2010). Applying this framework, Crompton (1979) identified factors, such as, "escape from a perceived environment," "exploration and evaluation of self," "relaxation," "prestige," "regression," "enhancement of kinship relationships," and "facilitation of social interaction" as internal factors which motivate people to travel, and factors, such as "novelty," "recreation," "business," "education," "entertainment," "well-being," "escape from routine" as factors that attract people to experience a vacation. Hence, travelers' motivation toward a particular destination resulted in variation and changes due to diverse demands and desires (Kay, 2009; Ozel and Kozak, 2012).

Tourist motivation has also been theoretically related toward tourists' experience at the destination (Devesa et al., 2010). Motivation has also been elaborated by tourists as wish or imagination formed during pre-travel phase and ultimately recognized as a crucial element of tourist satisfaction and behavioral intentions (Meng et al., 2008). A study found that visitors with different motivations would perceive different outcomes of satisfaction and behavioral intentions. Thus, travelers with strong desire to visit a particular destination, likely to be found with satisfactory exposure and positive behavioral intention (Yoon and Uysal, 2005).

7.2.5 THEORETICAL MODEL AND ASSUMPTION

In view of the previous literature, cognition and involvement at destination were identified as interpreters of tourist satisfaction. Since motive behind travel chiefly occurred before traveling, and its influence on cognition and involvement ought to likewise be noted. Current study takes the interpretation of cognition as perceived by tourists and their participative at visited destination. The underlying influence of motivation toward the estimation of attribute performance, participative behavior of tourists at destination, and satisfaction and repeat visit behavior has also been affirmed by existing literature (Kim and Brown, 2012; Weaver et al., 2007).

The present chapter investigates about tourists' motivation, assessment of attributes' perception, and participation of tourists at destination were noticed as mediator of experiential satisfaction along with repeat visit intention. The association of three constructs to satisfaction was also considered and reported by previous researches (Kim and Brown, 2012; Lee and Beeler, 2009). Nevertheless, scarcity of attention from academician and researcher has been noticed regarding the investigation of interaction concurrently as an overall travel experience. Hence, a research gap is identified in this area. Moreover, researches in the field of tourism, probably intend to investigate the association between tourists' motivational factors leading toward perception of destination attributes, participation in different activities, satisfaction and revisit intention. Therefore, in an attempt to fill this gap, authors proposed to assess the mediating theoretical model proposed to develop and tested in the present chapter (Fig. 7.1).

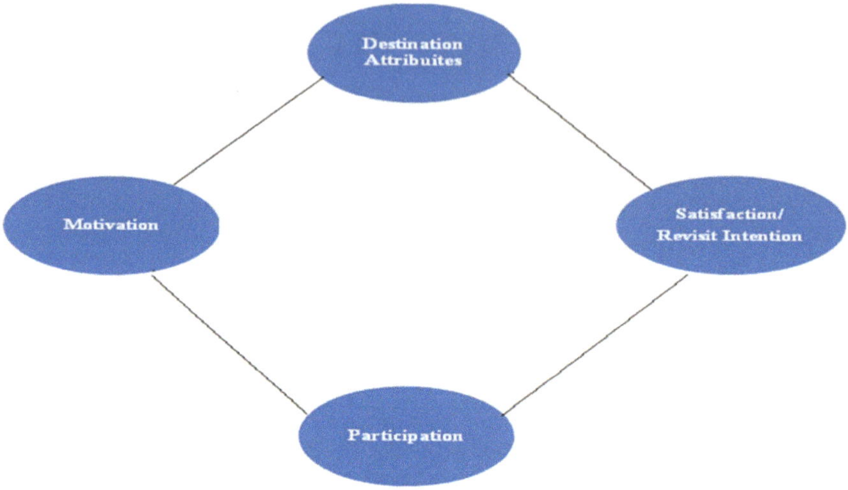

FIGURE 7.1 Theoretical model explaining the interrelationship of four selected variables.

Current researcher therefore developed a theoretical model on the basis of extensive review of literature and proposes an assumption to investigate the mediating role of tourist motivation and participation toward tourist satisfaction and repeat visit intention in Rishikesh.

7.3 METHODOLOGICAL APPROACH

The present chapter used survey method to collect the responses and opinions of tourists toward their visit in Rishikesh, India. Different tourist sites were identified in Rishikesh for the collection of data including attraction area, hotels, resorts, restaurants, guest houses, tourist information centers, and the office of tourism authority in the area. The data were collected for more than a year ranging from June 2018 to October 2019. The tourists for the survey were randomly selected to participate in the research and they were informed in advance about the purpose of the survey. The tourists were contacted only at the day after observing the suitability of the tourists. The questionnaires were distributed to 600 tourists, but only 282 tourists have completely filled the questionnaire with response rate of 47.0%.

The study, in order to test the proposed hypothesis developed from an extensive review of literature, used a self-constructed survey instrument in the form of questionnaire. The questionnaire was designed in different parts focusing to measure background information, motive toward visit, perception of destination's attributes, participation at destination, overall satisfaction, and repeat visit intention. In the first part, the questions, such as gender, age, nationality, education, occupation, income, etc. were asked from the tourists. The second part focuses on the measurement of motivation of tourists to visit Rishikesh based on 15 question statements. The third part of the survey form was designed to assess the experience of tourists with the help of 14 selected destination's attributes. The fourth section of the questionnaire was based on asking the question related to tourists perception about their participation in different activities. Tourists' extent and perception of participation in various activities at destination was evaluated in fifth section on the basis of four question statements. The sixth part of the survey instrument evaluated the behavioral intentions of tourists and included questions about willingness to recommend the destination as well as intentions to revisit the destination in the near future. The question statements designed in second to sixth parts were based on a five-point Likert scale ranged from strongly agree (5) to strongly disagree (1) after considering the utility of the measurement scale.

The statistical data analysis was applied by using statistical package for social science (SPSS) in different phases to carry out the study. First of all, the descriptive analysis was applied through frequency distribution, percent assessment, average and standard deviation to measure the background information and average score of items of different constructs mentioned in the above paragraphs.

Thereafter, the dimension reduction technique was carried out to reconstruct the measurement items. The principal component analysis with a varimax rotation method was used to make the most of the justification of variation. The Kaiser-Meyer-Olkin measure of sampling adequacy and Bartlett's test of sphericity were tested which reported a considerable value (0.6) and acceptable score of a Bartlett test (p value less than 0.05) indicating the sampling adequacy and suitability of factor analysis conducted. The scale's reliable value was also identified through the application Cronbach's alpha test which explored more than the threshold level (0.7) describing a good measure of inherent consistancy of the data.

Finally, theoretical mediation model identified from a literature review was also tested by employing mediation analysis. The analysis assesses the impact of a third external variable (M) on the association between independent variable (X) and dependent variable (Y) in consultation with studies applied with similar approach (Fiedler et al., 2011; MacKinnon et al., 2007). Likewise, the apprehension coefficient analysis (Preacher et al., 2007), by existing researches, and the authors of present chapter also considered and applied unstandardized coefficients, standardized coefficients, and bootstrap analysis to report the result and provide an understanding of the relationship and interference among selected constructs, including motivation, participation, destination attributes, and satisfaction and revisit intentions.

7.4 RESULT ANALYSIS AND DISCUSSION

The result of the chapter is presented in different sections starting from discussion of demographic and background information of tourists followed by descriptive analysis of the constructs, that is, motivation, destination attributes, satisfaction, and revisit intentions. The study also presented the result of factor analysis for the identification of factors of destination attributes and finally presented the result of mediation analysis.

7.4.1 DEMOGRAPHIC AND BACKGROUND OF TOURISTS

The information is summarized on demographic information of the sample. As shown in the table, 57.1% male and 42.9% female tourists comprise the sample. Most of the respondents were between the ages of 18 and 35 (83%), with 85.8% of the respondents reporting that they had a higher than college-level education background, indicating that the overwhelming majority of the respondents are highly educated. The median household income fell in the ranges of INR 400,001–600,000 per year (27.3%). Over half of the respondents (55.1%) were representing India and thus were domestic tourists. The statistical results of the visitation level specified that over 53% of the participants were repeat visitors who were familiar with Rishikesh. Besides, only 6.1% of participants were traveling with organized groups. Others either traveled alone or with friends and family. The results implied that the respondents were able to make their own decision on where to visit and what activities to participate in during their visit to Rishikesh.

Local cuisine and exotic food along with culturally related souvenirs are regarded as important components of tourism. The findings indicated that a majority of the respondents had tasted local food and purchased souvenirs during their visit in Rishikesh. According to the statistical results, the most popular local food selected by tourists is the north Indian dishes (71.63%), which is famous all over the country. The souvenir that nearly half of the tourists purchased is the clothes, idols, and other stone items (45.74%) that can be found in Rishikesh. Regarding accommodations used by tourists, 46.10% of the respondents said that they were staying in budget class hotels, which dominated the main part of the accommodation, whereas 20.92% of respondents stayed with friends and relatives living in Rishikesh. The majority of the participants (57.7%) spent 3 days at destination, indicating that the respondents were qualified because they had received enough time to experience the destination, and they had an overall image of Rishikesh.

7.4.2 TOURIST MOTIVATION TO VISIT RISHIKESH

Table 7.2 presents the rating scores of tourist motivation to visit Rishikesh. Question number 5 of the survey asked about tourists' overall motivation to visit Rishikesh. Using a five-point Likert-type rating scale, the statement, "I was not at all motivated to visit Rishikesh" was assigned the score of 1, and, respectively, "I was highly motivated to visit Rishikesh" was assigned the score of 5. The mean score of the 282 valid answers is 3.79, with a standard deviation of 0.866. The frequency distribution indicated that 145 of the 282 respondents (51.4%) were moderately motivated to visit Rishikesh. Fifty-three of the respondents (18.8%) reported that they were highly motivated to visit Rishikesh. With a similar percentage (55%), 55 respondents thought it was hard to define their motivation, whereas, 29 respondents (10.28%) reported that they did not have a strong motivation before their visit. It was noted that no respondents stated that they were not motivated at all.

The mean score and standard deviation of the remaining 15 motivation items were computed by SPSS based on the five-point Likert-type scale test, where 1 represents "strongly disagree," and 5 represents "strongly agree." A higher score means that respondents were more motivated by the item. The results of the mean score indicated that the highest rated motivation items were "enjoy natural beauty of Rishikesh" (M = 4.27), "interest in Rishikesh's fame for religious and spirituality" (M = 4.22), and "knowing more about Rishikesh (M = 4.18). The lowest rated motivations were "business travel" (M = 2.42), "escaping from the routines of life" (M = 3.03), and "escaping from the stress of daily life" (M = 3.76). It was surprisingly noticed that the mean score of "to escape from the routines of life" was significantly smaller than other motivations. After conducting the frequency analysis, the results suggested that 22 of the 282 respondents strongly disagreed with the statement. This chapter also described about the aggregate average of different motivation scale items in the descriptive Table 7.2. This table highlights moderate (M=3.89, S.D.=0.84) and positive motivation rating among tourists visiting Rishikesh. Thus, it is ascertained that most of the tourists to the destination were highly motivated to visit and thus indicates higher rating about motivation items (Table 7.1).

TABLE 7.1 Mean Score and Standard Deviation of Tourist Motivation to Visit Rishikesh.

Items	Mean	Std. Dev.
I wanted to experience the rich culture in Rishikesh	4.17	0.82
I am interested in the famous temples and Ashrams in Rishikesh	4.11	0.82
I was interested in Rishikesh fame as a religious and spiritual destination	4.22	0.75
I wanted to participate in learning of yoga and meditation activities at the destinations	3.96	0.82
I felt that Rishikesh is a must visit destination for me	3.93	0.83
I wanted to learn the history of Rishikesh	4.15	0.67
I wanted to learn and know more about Rishikesh	4.18	0.72
I wanted to learn the history of Sages and penance in Rishikesh	4.09	0.74
I wanted to enrich my knowledge about current Rishikesh	4.09	0.69
I wanted to escape from stress in my daily life	3.76	0.89
I wanted to escape from the routines of life	3.03	1.15
I wanted to relax and rest while traveling	4.12	0.74
I wanted to be enjoyed natural beauty of Rishikesh to make me happy	4.27	0.66
I wanted to reunite with my family and friends	3.82	1
I am here because of business	2.42	1.24
Average Score	3.89	0.84

7.4.3 *EVALUATION OF ATTRIBUTE PERFORMANCE*

The mean scores of the 14 items of respondents' evaluation of destination attributes performance is presented in Table 7.2. A higher score represented that respondents agreed that the attribute stated above satisfied their requirements of travel. The results of the mean scores fell into the range from 3.59 to 4.22, suggesting that respondents tend to have positive attitudes toward destination attribute performance. To compare the mean scores of the 14 attribute items, participants reported that the best performing attribute of Rishikesh is "the natural attraction of Rishikesh is highly valued" (M = 4.22). This was followed by "the wellness and recreation opportunities are highly valued" (M = 4.01) and "the religious and spiritual attraction in Rishikesh" (M=4.00). As expected, the three items with the highest mean scores were the statements regarding the cultural attributes of Rishikesh, which is consistent with the literature. The value of heritage attributes of a destination has been discussed in the literature (Huh et al., 2006). Excluding these three attributes, all the remaining items scored below the mean value of 4. The "behavior of locals" earned the lowest score (M = 3.59), which meant that participants thought the behavior of hosts needed to be interpreted in a more representative way. Conclusively, the aggregate score of respondents' evaluation about destination attributes was measured more than moderate, indicating 3.84 as mean score with good measure of standard deviation. Thus, it indicates that tourist participants were reported higher side rating of attributes' perception scale.

TABLE 7.2 Mean Score and Standard Deviation of the Respondents' Evaluation of Destination Attributes.

Items	Mean	Std. Dev.
The natural attractions in Rishikesh are highly valued	4.22	0.73
The natural resources are well conserved in Rishikesh	3.82	0.79
The tourist sites are unique compared with other destinations	3.74	0.87
The wellness and recreational opportunities are highly valued	4.01	0.78
The religious and spiritual sites of Rishikesh are attractive to me	4	0.71
The culture of Rishikesh is very interesting to me	3.74	0.78
The behavior of locals well interpreted in Rishikesh	3.59	0.83
The dressing and language of locals reflect the identity of Rishikesh	3.88	0.68

TABLE 7.2 *(Continued)*

Items	Mean	Std. Dev.
The ethics and moral values are well conserved in Rishikesh	3.78	0.75
I felt safe when I visited Rishikesh	3.73	0.74
Rishikesh is easily accessible to me	3.84	0.78
The transportation is convenient within Rishikesh	3.73	0.82
I can get local exotic food at Rishikesh	3.85	0.78
I have the opportunity to buy ethnic souvenirs from Rishikesh	3.78	0.8
Average Score	3.84	0.77

7.4.4 TOURIST PARTICIPATION IN DIFFERENT ACTIVITIES

Table 7.4 presents the results of descriptive statistics of tourists' participation in different activities to enrich experience of visit. A higher score reported by respondents represented that they felt emotionally bonded with Rishikesh during or after their visit. On the contrary, a lower rating suggested that the respondents did not feel strongly attached to Rishikesh. The mean scores ranged from 3.59 to 3.96, indicating that a marked majority of the respondents were not enthusiastically participated in different activities (all the mean scores of the items were below point 4). The lowest mean score (3.59) was given to visit as part of holiday package. The highest score was loaded at the pleasure obtained from their participation (3.96) in recreational and thrilling activities, including river rafting, camping, bathing, trekking, yoga, and meditation. In the end, the descriptive Table 7.4, also portrayed the average mean score at moderate point (M=3.79) falling more toward high perception index along with a good measure of standard deviation (Table 7.3).

TABLE 7.3 Mean Score and Standard Deviation of Tourists' Participation at Destination.

Items	Mean	Std. Dev.
Rishikesh is a very special destination to me	3.7	0.75
I got pleasure from participation in recreational activities (river rafting, boating, bungee jumping, kayaking, photography, trekking, camping, cliff diving, and bathing in waterfalls and river banks)	3.96	0.67

TABLE 7.3 *(Continued)*

Items	Mean	Std. Dev.
I felted religiously and spiritually attached to Rishikesh	3.89	0.68
I have a lot of interest to visit Rishikesh to learn yoga and meditation	3.9	0.67
Visiting Rishikesh to participation and experience a special event	3.79	0.71
I gave myself pleasure by getting involved in purchasing souvenirs and crafts in Rishikesh	3.72	0.78
I would not substitute any other destination for the types of things that I did at Rishikesh	3.59	0.8
Average Score	3.79	0.72

7.4.5 SATISFACTION AND REVISIT INTENTION

The highest mean score of behavioral intention was loaded at the statement of "intention to revisit" (M = 4.06), followed by the statement of "recommendation to relatives and friends" (M = 4.04). Consistent with the literature, the willingness to revisit and recommend to others are intensely associated with the satisfaction level (Kim and Brown, 2012; Lee and Beeler, 2009). The statement that "my experience at Rishikesh was exactly what I expected" earned the lowest mean score of 3.70. Tourists' satisfaction of a destination or a tourism site is linked to their pre-visit expectation and after-visit experience (Huh et al., 2006). The low score might explain that a partial number of respondents thought that the performance of Rishikesh did not meet their expectation. Although the mean value of 3.70 is the lowest of the eight items, it is still greater than the median value of 3. The mean scores of satisfaction items basically suggested that respondents were satisfied with their overall experience at Rishikesh. In response to average score, the tourists indicated moderate satisfaction with high rating toward satisfactory experience. As far as revisit intention was concerned, it highlighted high perception among tourists regarding their revisit intention toward Rishikesh. Thus, it indicates a moderate satisfaction and high loyalty among tourists for Rishikesh.

TABLE 7.4 Mean Score and Standard Deviation of Tourists' Satisfaction and Revisit Intention.

Satisfaction items	Mean	Std. Dev.
Rishikesh is one of the best destinations	3.84	0.72
My choice to visit Rishikesh was a wise one	3.96	0.63
I think I made the right decision to visit the destination	3.93	0.68
I am satisfied with my overall experience during my visit	3.94	0.66
My experience at Rishikesh was exactly what I expected	3.7	0.71
Satisfaction Average Score	3.87	0.68
Revisit Intention Items	**Mean**	**Std. Dev.**
My experience at Rishikesh is satisfactory and force me remain loyal	3.98	0.61
If I got another chance, I would be pleased to revisit the destination	4.06	0.66
I would like to recommend Rishikesh to my relatives and friends as a travel destination	4.04	0.76
Revisit Intention Average Score	4.03	0.68

7.4.6 FACTOR ANALYSIS

By using the principle component method with a varimax rotation, the variables were reduced to a smaller group of factors.

7.4.7 FACTOR ANALYSIS FOR DESTINATION ATTRIBUTE EVALUATION

After performing the principal component factor analysis, all 14 items were remained, and a three-factor structure was obtained. Of the total variance, 60.91% was explained by the three factors (see Table 7.5). The KMO (0.878) and Bartlett's ($p < .000$) test and eigenvalues (all greater than 1) verified the acceptance of the factor structure. The reliability was verified by the Cronbach's alpha test (0.785, 0.832, and 0.804). With regards to the content, the three factors were denoted as "representativeness" (factor 1), "attractiveness" (factor 2), and "facilities and services" (factor 3). The three components were consistent with the conclusion from the literature review. Factor 1 represented the four items with regard to the interpretation of items that represent the Rishikesh also determine that tourist could enjoy an authentic experience at the destination. Factor 2 included the five items related to the evaluation of the attractiveness of attributes. As expected, identified attributes were treated as core elements of the destination. The two factors respectively accounted for 21.419% and 20.313% of the total variance. The factor of "facilities and services" represented supporting facilities at Rishikesh that could satisfy visitors' basic needs. Factor 3 explained 19.176% of the total variance. Although it is lower than the the other two factors, it still plays a role in the model.

TABLE 7.5 Factor Analysis of Evaluation of Destination Attributes in Rishikesh.

Factors	Factor loading		
	1	**2**	**3**
Factor 1: Representativeness			
The behavior of locals well interpreted in Rishikesh	0.778		
The heritage of Rishikesh is authentic	0.742		
The dressing and language of locals reflect the identity of Rishikesh	0.619		
The ethics and moral values are well conserved in Rishikesh	0.583		
Factor 2: Attractiveness			
The natural sites at Rishikesh are highly valued		0.763	
The religious and spiritual sites of Rishikesh are attractive to me		0.752	
The wellness and recreational opportunities are highly valued		0.713	
The tourist sites are unique compared to other destinations		0.605	
The culture of Rishikesh is very interesting to me		0.531	
Factor 3: Facilities and services			
Rishikesh is easily accessible to me			0.758
The transportation is convenient within Rishikesh			0.729
I can get local cuisine exotic food at Rishikesh			0.724
I have the opportunity to buy ethnic souvenirs in Rishikesh			0.664
I felt safe when I visited Rishikesh			0.561
Eigenvalues	5.944	1.519	1.064
Variance (%)	21.419	20.313	19.176

TABLE 7.5 *(Continued)*

Factors	Factor loading		
	1	2	3
Cumulative variance (%)	21.419	41.732	60.908
Reliability alpha	0.785	0.832	0.804
Number of items (total 14)	4	5	5

7.4.8 MEDIATION ANALYSIS

The main purpose of the study was to develop a theoretical model to reveal the relationships among motivation, experience, and satisfaction. Based on the literature review, the relationship between motivation and overall satisfaction was mediated by the effect of attributes performance evaluation and tourists' participation at the destination. Given the results of factor analysis, motivation was factored as religious and spiritual motivation and relaxation and recreation and entertainment motivation. The evaluation of destination attributes was generated into three factors, such as representativeness, attractiveness, and facilities and services. To perform the mediation analysis, the SPSS was utilized as the statistical software.

7.4.9 MEDIATION EFFECTS OF MOTIVATION AND PARTICIPATION WITH SATISFACTION

As shown in Figure 7.2, motivation significantly affected participation ($b=30.84$, $SE = 2.56$, $p < 0.01$, 95% CI [25.79, 35.89]). A higher level of participation led to better satisfaction with significant coefficient ($b = 0.58$, $SE = 0.05$, $p <0.01$, 95% CI [0.47, 0.68]). The effect of motivation is significant on representativeness ($b = 0.18$, $SE = 0.06$, $p < 0.01$, 95% CI [0.07, 0.29]), ct attractiveness ($b = 0.54$, $SE = 0.05$, $p < 0.01$, 95% CI [0.55, 064]), and other facilities and services ($b = 0.29$, $SE = 0.06$, $p < 0.01$, 95% CI [0.18, 040]). Their connection with satisfaction was also illustrated by the statistic results. The effect size on intention of representativeness was 6.47 ($b = 6.47$, $SE = 2.33$, $p < 0.01$, 95% CI [1.88, 11.06]). The most significant effect on intention of attractiveness was also observed ($b=9.7$, $SE = 2.77$, $p < 0.01$, 95% CI [4.24, 15.16]).

Clearly, better performance on attractiveness could lead to better satisfaction. Besides, the effect of performance of facilities and services of destination on satisfaction is also significant (b = 7.85, SE = 2.37, $p < 0.01$, 95% CI [3.17, 12.52]).

The total effect of motivation on satisfaction was significant (b = 31.16, SE = 2.70, $p < 0.01$, 95% CI [25.83, 36.48]). The indirect effect of participation was significant (b = 17.84, Bootstrapping SE = 2.38, Bootstrapping CI [13.56, 23.05]). The mediating effect of representativeness is presented with the effect size of 1.16 (b = 1.16, Bootstrapping SE = 0.67, Bootstrapping CI [0.18, 2.89]). Attractiveness also had a positive indirect effect (b = 5.23, Bootstrapping SE = 1.64, Bootstrapping CI [2.22, 8.79]). In the same way, the attribute of facilities and services had a significant indirect effect on the relationship (b = 2.27, Bootstrapping SE = 0.81, Bootstrapping CI [0.91, 4.08]). The total indirect effect of the three attributes and participation of the relationship between motivation and satisfaction was significant (b = 26.5, Bootstrapping SE = 2.97, Bootstrapping 95% [20.75, 32.56]). However, the direct effect was not significant (b = 4.66, SE = 2.77, $p = 0.09$, 95% CI [-0.79, 10.10]), which was reflected in the literature review. The ratio of the indirect to total effect was 0.85, which means that the mediators (attribute performance and participation) account for 85% of the total variance between motivation and satisfaction.

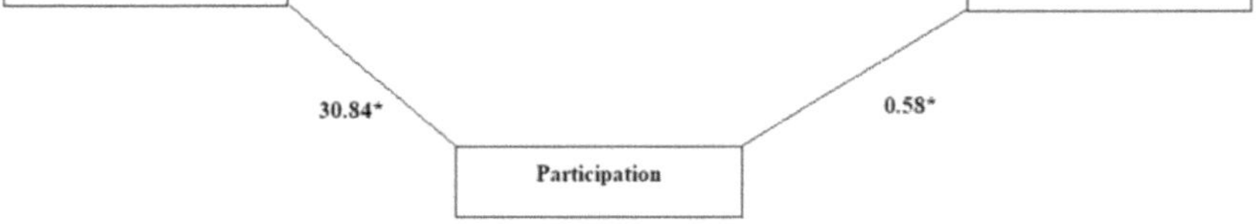

FIGURE 7.2 Attribute performance evaluations and participation mediation of the relationship between motivation and satisfaction.

Note: C' denotes the direct effect. C denotes the total effect. * indicates $p < 0.01$.

7.4.10 MEDIATION EFFECTS OF MOTIVATION AND PARTICIPATION TO REVISIT INTENTION

As shown in Figure 7.3, motivation significantly predicted participation (b = 13.65, SE = 2.56, p < 0.01, 95% CI [8.60, 18.69]). The effect on representativeness was significant (b = 0.25, SE = 0.06, p < 0.01, 95% CI [0.14, 0.36]). Increased motivation could cause better evaluation on attractiveness (b = 0.23, SE = 0.05, p < 0.01, 95% CI [0.13, 0.32]) and facilities and services (b = 0.13, SE = 0.06, p < 0.05, 95% CI [0.02, 0.25]). The effect of the mediators on revisit intention was the same as illustrated in the first model.

The total effect of motivation was significant (b = 12.7670, p < 0.01, 95% CI [7.44, 18.09]). Similar to the first model, the direct effect of motivation was not significant (b = 0.007, SE = 2.22, p > 0.05, 95% CI [-4.36, 4.38]). The indirect effect size of participation on revisit intention is 7.89 (b = 7.89, Bootstrapping SE = 1.91, Bootstrapping CI [4.34, 12.08]). The indirect effect of representativeness was significant (b = 1.62, Bootstrapping SE = 0.77, Bootstrapping CI [0.39, 3.45]). The indirect effect of attractiveness had a greater effect size than the other two attributes (b = 2.20, Bootstrapping SE = 0.89, Bootstrapping CI [0.84, 4.44]). And the mediating effect of facilities and services was tested to be significant (b = 1.05, Bootstrapping SE = 0.64, Bootstrapping CI [0.09, 2.68]). The total indirect effect of the four mediators proved to be significant (b = 12.76, Bootstrapping SE = 2.37, Bootstrapping CI [8.38, 17.77]). The results indicate revisit intention would change by 12.76 in a positive path through the mediation of participation and the evaluation of the attribute performance when tourists' motivation changed.

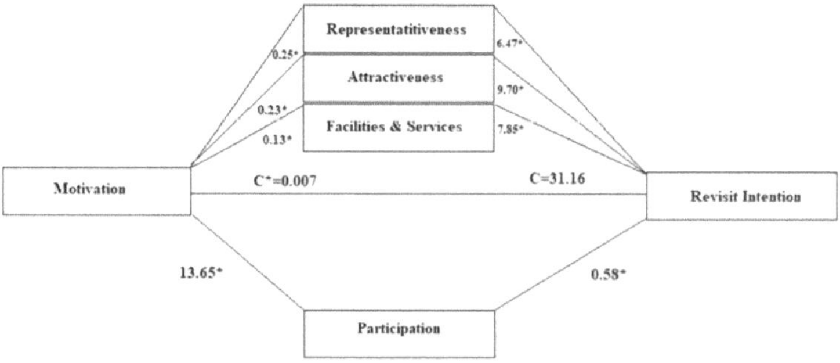

FIGURE 7.3 Attribute performance evaluation and participation mediation of the relationship between motivation and revisit intention.

Note: C' denotes the direct effect. C denotes the total effect. * indicates $p < 0.01$.

Hence, from the mediation analysis, the study assessed that tourist motivation and participation significantly mediating the tourist satisfaction and revisit intention in Rishikesh. Hence, it can be ascertained that tourist satisfaction and repeat visit intention to Rishikesh among participant tourists were significantly and positively influenced by their motivation behind visit and participation behavior at the destination.

7.5 DISCUSSION OF MAJOR FINDINGS

From the empirical analysis, tourists responded positive agreement for most of the selected destination attributes in Rishikesh. The study supported three major factors of destination attributes, namely, representativeness, attractiveness, and facilities and services. The descriptive table highlights the attractiveness of Rishikesh as most positively perceived factor by tourists in Rishikesh. The factor facilities and services focused on supporting facilities and services to satisfy tourists' basic needs,namely, the needs of accessibility, transportation, eating, souvenir purchasing, and safety. As expected, all three dimensions significantly contribute to satisfaction and revisit intentions. Among the three dimensions, attractiveness contributed the most in determining the satisfaction and revisit intention, implying that attractive and interesting tourist sites and activities affected tourists' experience most directly and significantly. Another important discussion point is the role of supporting facilities and services. The mediating effect of "facilities and services" attribute on satisfaction and revisit intention is even stronger than the "representativeness" attribute. The results brought much attention to the importance of providing supporting facilities and services to meet tourists' needs during travel. The attributes might not be considered the main motivational factors that attract tourists, but a negative performance could decrease their overall travel experience.

The findings indicated that most of the respondents expressed that the participation of tourists in different activities in Rishikesh during the visit motivated them. The results of the descriptive statistics presented that people undergo more participation at the destination when achieve contentment from the travel experience. The study also supported and indicated positive result regarding motivation variables related to recreation and pleasure activities, such as participation in river rafting, camping, trekking, bathing, and boating, and also reported with the highest mean score.

The findings certainly illustrated the necessity of studying arousing participation as an important component of a tourist experience at a destination. As stated from empirical result, the mediating model indicates that tourists who are more motivated by the destination elements tend to feel highly participated at the destination and travel activities, respectively, and tend to be more satisfied with high revisit intention. In addition, the inclusion of the level of participation in the theoretical model assists in the understanding of tourists' emotional change during the travel process.

They discovered that the motivating factors behind choosing Rishikesh as a travel destination were related to culture, religion, and spirituality to visit Rishikesh. However, tourist also reported that recreation and relaxation motivations, such as wellness, yoga, meditation, and adventure tourism activities act as major force behind their visit to study area. Furthermore, the present study investigated that the mediation analysis failed to report that travel motivation has a direct effect on satisfaction. However, with the mediating effect of attribute performance and participation with a destination added, motivation becomes a significant factor in predicting revisit intention. The result of the study also indicated that between the two mediators, motivation is more bonded with the mediating effect of participation. It is probably because that participation is the emotional and affective interaction that occurs between tourists and a destination. On the other hand, destination performance obviously has a stronger direct effect than participation on determining satisfaction and revisit intention. This could be explained by tourists who were not strongly motivated to visit a destination could be impressed and satisfied with a good performance of the destination, which will lead to a positive revisit intention.

7.6 CONCLUSION

The main purpose of the study was to reveal the relationships among motivation, participation, destination attribute performance, satisfaction, and revisit intentions through a proposed theoretical model. The study also estimates the effect of assumption among these constructs. The study was designed to understand how a destination could provide a satisfying tourism experience through its attractions and different activities and facilities. Both the cognition and the affect derived from a travel experience were taken into consideration in the study to investigate tourists' evaluation of a destination and participation in different activities at the destination.

The research questions proposed in the study were sufficiently answered by the results and revealed the dimensions of destination attributes and tourist participation at the destination, that provided a baseline measure for understanding the connection between motivation, satisfaction, and revisit intention. The results supported the influential effects of motivation, attribute performance, and participation on overall experience at destination. The summary of proposed assumptions illustrated that the main constructs of the study variables assist in the understanding of the holistic travel experience that tourists have at the destination.

The findings of the study provided strategic implication for destination planning and management to create an unforgettable experience for tourists. It also has drawn implication regarding the identification of different dimensions of a destination, creatively interpret them into tourism products, and present and deliver them to a target market.

Although, the study has contributed to developing the conceptual model of predictors of satisfaction and providing suggestions to destination marketing. However, there are still limitations of the study. The study used primary data collected through survey undertaken over a long time. Though the survey time was controlled with weekends and during daytime, the procedure was still considered as a convenience sampling. To avoid the potential bias, a comprehensive approach for collecting of data during all weekdays and during night also needs to be considered in future studies. In future, qualitative study based on in-depth face-to-face interview survey instrument is suggested for future studies to investigate more about the real perceptions of respondents.

KEYWORDS

- **motivation**
- **participation**
- **attributes**
- **performance**
- **satisfaction**
- **revisit intention**

REFERENCES

Ahn, T.; Ekinci, Y.; Li, G. Self-Congruence, Functional Congruence, and Destination Choice. *J. Busi. Res.* **2011,** *66*, 719–723.

Caber, M.; Albayrak, T.; Matzler, K. Classification of the Destination Attributes in the Content of Competitiveness (by Revised Importance-Performance Analysis). *J. Vacation Market.* **2012,** *18* (1), 43–56.

Chen, A. H.; Wu, R. Understanding Visitors' Involvement Profile an Information Search: The Case of Neimen Song Jiang Battle Array Battle Festival. *Event Manage.* **2010,** *13*, 205–222.

Chen, C.-F.; Chen, F.-S. Experience Quality, Perceived Value, Satisfaction and Behavioral Intentions for Heritage Tourists. *Tour. Manage.* **2010,** *31*, 29–35.

Chen, Y.; Mak, B.; McKercher, B. What Drives People to Travel: Integrating the Tourist Motivation Paradigms. *J. China Tour. Res.* **2011,** *7*, 120–136.

Dasgupta, D. *Tourism Marketing*; Pearson Education: India, 2011.

Devesa, M.; Laguna, M. and Palacios, A. The Role of Motivation in Visitor Satisfaction: Empirical Evidence in Rural Tourism. *Tour. Manage.* **2010,** *31* (4), 547–552.

Donlon, J. G.; Donlon, J. H.; Agrusa, J. Cultural Tourism, Camel Wrestling, and the Tourism 'Bubble' in Turkey. *Anatolia: Int. J. Tour. Hosp. Res.* **2010,** *21* (1), 29–39.

Fiedler, K.; Schott. M.; Meiser, T. What Mediation Analysis Can (Not) Do. *J. Exp. Soc. Psychol.* **2010,** *47*, 1231–1236.

Gross, M. J.; Brown, G. An Empirical Structural Model of Tourists and Places: Progressing Involvement and Place Attachment Into Tourism. *Tour. Manage.* **2008,** *29*, 1141–1151.

Huh, J.; Uysal, M.; McCleary, K. Cultural Heritage Destinations Tourist Satisfaction and Market Segmentation. *J. Hosp. Leisure Market.* **2006,** *14* (3), 81–99.

Hwang, S. N.; Lee, C.; Chen, H. J. The Relationship among Tourists' Involvement, Place Attachment and Interpretation Satisfaction in Taiwan's National Parks. *Tour. Manage.* **2005,** *26*, 143–156.

Kay, P. L. Cultural Experience Tourist Motives Dimensionality: A Cross-Cultural Study. *J. Hosp. Market. Manage.* **2009,** *18*, 329–371.

Kim, A. K.; Brown, G. Understanding the Relationships between Perceived Travel Experiences, Overall Satisfaction, and Destination Loyalty. *Int. J. Tour. Hosp. Res.* **2012,** *23* (3), 328–347.

Komppula, R.; Gartner, W. C. Hunting as a Travel Experience: An Auto-Ethnographic Study of Hunting Tourism in Finland and the USA. *Tour. Manage.* **2013,** *35*, 168–180.

Kozak, M.; Rimmington, M. Tourist Satisfaction with Mallorca, Spain, as an Off-season Holiday Destination. *J. Travel Res.* **2000,** *38* (3), 260–269.

Laurent, G.; Kapferer, J. Measuring Consumer Involvement Profiles. *J. Market. Res.* **1985,** *22* (1), 41–53.

Lee, C. H.; Lee, B.; Bernhard, B. J.; Lee, T. K. A Comparative Study of Involvement and Motivation among Casino Gamblers. *Psych. Invest* **2009,** *6*, 141–149.

Lee, J.; Beeler, C. An Investigation of Predictors of Satisfaction and Future Intention: Links to Motivation, Involvement, and Service Quality in a Local Festival. *Event Manage.* **2009,** *13*, 17–29.

MacKinnon, D. P.; Fairchild, A. J.; Fritz, M. S. Mediation Analysis. *Annu. Rev. Psychol.* **2007,** *58*, 593–614.

McQuarrie, E. F.; Munson, J. M. The Zaichkowsky Personal Involvement Inventory: Modification and Extension. *Adv. Consumer Res.* **1987,** *14,* 36–40.

Meng, F.; Tepanon, Y.; Uysal, M. Measuring Tourist Satisfaction by Attribute and Motivation: The Case of a Nature-Based Resort. *J. Vacation Market.* **2008,** *14* (1), 41–56.

Ministry of Tourism, Government of India. *India Tourism Statistics-2014 at a Glance*. Published by Ministry of Tourism, Government of India, 2018.

Oliver, R. L. A Cognitive Model of the Antecedents and Consequences of Satisfaction Decisions. *J. Market. Res.* **1980,** *17* (4), 460–446.

Oliver, R. L. Cognitive, Affective, and Attribute Bases of the Satisfaction Response. *J. Consumer Res.* **1993,** *20*, 418–430.

Ozel, C. H.; Kozak, N. Motive Based Segmentation of the Cultural Tourism Market: A Study of Turkish Domestic Tourists. *J. Qual. Assur. Hosp. Tour.* **2012,** *13* (3), 165–186.

Prayag, G.; Ryan, C. Antecedents of Tourists' Loyalty to Mauritius: The Role and Influence of Destination Image, Place Attachment, Personal Involvement, and Satisfaction. *J. Travel Res.* **2012,** *51* (3), 342–356.

Preacher, K. J.; Rucker, D. D.; Hayes, A. F. Addressing Moderate Mediation Hypotheses: Theory, Method, and Prescriptions. *Multivariate Behav. Res.* **2007,** *1*, 185–227.

Rinschede, G. Forms of Religious Tourism. *Ann. Tour. Res.* **1992,** *19* (1), 51–67.

Ritchie, B. W.; Tkaczynski, A.; Faulks, P. Understanding the Motivation and Travel Behavior of Cycle Tourists Using Involvement Profiles. *J. Travel Tour. Market.* **2010,** *27*, 409–425.

Sharma, V.; Bhadula, S.; Joshi, B. D. Assessment of Religio-Touristic Activities in Haridwar City of Uttarakhand State (India): A SWOT Analysis. *Report Opin.* **2011,** *3* (7), 25–29.

Sofield, T. H. B.; Li, F. M. S. *Rural Tourism in China*; Routledge; The Business of Rural Tourism: London, 1996.

Sofield, T. H. B.; Li, F. M. S. Tourism Development and Cultural Policies in China. *Ann. Tour. Res.* **1998,** *25* (2), 362–392.

Tse, D. K.; Wilton, P. C. Models of Consumer Satisfaction: An Extension, 1998.

United Nation World Tourism Organization. *World Tourism Barometer, 2016 (January)*. Published by World Tourism Organization, Madrid Spain, 2018.

Voon, B-H.; Lee, N. Identifying Dimensions of Tourist Satisfaction for a Cultural Destination: The Case of Longhouses in Sarawak. *Int. J. Busi. Soc.* **2009,** *10* (2), 65–82.

Weaver, P. A.; Weber, K.; McCleary, K. W. Destination Evaluation: The Role of Previous Travel Experience and Trip Characteristics. *J. Travel Res.* **2007,** *45*, 333–344.

World Travel and Tourism Council. *World Travel and Travel Economic Impact Report-201t.* Published by World Travel and Tourism Council, London, 2017.

Wright, K. *The Christian Travel Planner: Christian Traveler's Guide*. Executive Director, World Religious Travel Association, 2008. http://www.christiantravelplanner.com/pdf/Christian%20Travel%20Planner%20Excerpt.pdf (accessed 29 April 2012).

Yale, P. *From Tourist Attractions to Heritage Tourism*; Elm Publications: Huntingdon, 1991.

Yen, C-H.; Lu, H-P. Effects of e-Service Quality on Loyalty Intention: An Empirical Study in Online Auction. *Manag. Serv. Qual.* **2008,** *18* (2), 127–146.

Yoon, Y.; Uysal, M. An Examination of the Effects of Motivation and Satisfaction on Destination Loyalty: A Structural Model. *Tour. Manage.* **2005,** *26*, 45–56.
Zaichikowsky, J. L. Measuring the Involvement Construct. *J. Consumer Res.* **1985,** *12* (3), 341–352.

CHAPTER 8

CHANGING TOURIST BEHAVIOR WITH GROWING TECHNOLOGY USAGE: THE ENTRY OF GENERATION Z TOURISTS

RICHA MISHRA*

Amity University Uttar Pradesh, India

*E-mail: *richa1188@gmail.com*

ABSTRACT

Technology has become an integral part of our lives. We are extensively using technological mediums for our day-to-day activities. Tourism industry is also highly affected by the technological advancements. Earlier large part of the business was based on face-to-face interactions but now the scenario has completely changed. Tourists these days are super tech-friendly unlike the earlier generations, which were less dependent on technology. Today's generation Z tourists believe in doing everything over the internet. They tend to trust the information available online more and believe it to be more reliable. They go back to internet even for cross checking the information given to them by the tourism service providers. Tourists are using internet for everything from information gathering to bookings to even feedback. They also tend to like service providers with good customer friendly websites and prompt replies via a chat or an e-mail. Tourism providers are also trying to catch-up to this trend are using

Tourist Behavior: Past, Present, and Future. Narendra Kumar, PhD, Bruno Barbosa Sousa, PhD & Swati Sharma, PhD (Eds.)

technology for their business from marketing to final bookings. The tourists and the tourism service providers are using technology for their specific purposes. The service providers use technology to reach their prospective clients, to promote their products and services, to facilitate travel and to carry out various pre- and post-travel formalities, are using technology. Hiving a smart phone and laptop is just normal in today's times and so is the use of internet. It is important for tourism service providers to keep themselves updated with technology looking at the tourist's preference toward tech-friendly options.

8.1 INTRODUCTION

We say tourism is happening when people move out from their homes, cities, and comfort zone for reasons, such as business, leisure, or pilgrimage etc. Tourists are the center of all tourism activities. It is the tourists who define the fate of a tourism destination. Tourism is all about experiences and how well a tourism destination can provide these experiences to its tourists defines that destination. Tourism industry is ever evolving and so are the tourists. With every generation, we see a different set of tourists. There is a change in their likes and dislikes, tastes, preferences, spending pattern, destination choices, technology usage, and duration of travel, etc. The growing usage of technology among the tourists has increased the exposure of the industry, everything and anything is available in one click. All good or bad information reach the tourists easily. Tourists are using internet and social media platforms to gather information, for reviews about a destination or a service provider and also to plan their entire itinerary. The use of technology has seen a steady rise in travel planning and it starts right from choosing the destination to hotel selection to mode of transportation and also while selecting the attractions to visit. The decision to travel is influenced by the information they get on the internet. Technology has certainly changed the face of tourism worldwide. From virtual tours to online payments, zoom meetings to form filling, everything is done on the internet. The tourists are well prepared in advance before contacting a travel agent. A travel agent or a service provider will have to accept the fact that today's tourists already have enough information with them and are only looking for final confirmation or something extra or unique which they cannot get on the internet. Tourists also prefer everything to be done online and minimizing their physical hassle. They return to internet for information, reviewing, and cross-checking. They also expect the service providers to use technology to the maximum. The industry has to be on toes all the time to keep pace with the changing tourist's behavior. Tourism industry can only survive when they are able to meet the expectation of the tourists, and for this understanding, the prospective tourists and their wants are primary. The tourists have been differentiated based upon their traveling pattern into Baby Boomers, Generation X, these are the past generation tourists, then came the Millennial also known as Generation Y, these are the present generation tourists and then came the Generation Z also called Geni, these are the next generation tourists. It becomes the responsibility of tourism service providers to transform and evolve as per the changing tourists. The tourists of each generation differ a lot in their characteristics. Baby Boomers out of all are least reliant on technology and prefer physical interaction, whereas Generation X and the Millennial are learning to use technology and getting comfortable with it. The Generation Z rely totally on the technology, and each decision of theirs is influenced by the social media. The tourism industry is now getting comfortable with Baby Boomers and trying hard to grasp the Millennial fully and nowthe next generation is catching up called the generation Z. The task of the tourism service providers has become all the more difficult with all these generations co-existing. The group of tourists will include Baby Boomers and Millennial along with Generation Z. A tourism professional will have to learn to cater to the needs of all groups of tourists. However, it is the Generation Z tourists who are becoming the decision makers, even if the payments are made by a Millennial, but all other decisions related to travel are taken by these Generation Z tourists. Therefore, these tourists are quickly becoming an important segment for the tourism policy makers and the service providers. Understanding these Generation Z tourists, the changes in their behavior from the past generations, and as to how the tourism industry worldwide should get ready to cater to these new set of tourists is important. Certainly, the Generation Z tourists are more advanced, more informed, and tech-friendly and are continuously looking for unique traveling experiences. The tourism service providers in order to satisfy this set of tourists has to be one step ahead. Tourists these days are more aware, they are well prepared, cautious, not willing to compromise and are looking for a perfect holiday without any hassles.

The following objectives are covered in this chapter as these are important to understand the impact of technology on the changing behavior of tourists:

- To assess how the Generation Z is different from Millennial and Baby Boomers.
- To understand the likes and dislikes of Generation Z tourists.
- To analyze the areas to be improved in order to satisfy the current generation tourists.
- To study the impact of Covid-19 situation on the present generation tourists.

8.2 LITERATURE REVIEW

Today's tourists, also referred as Generation Z, are full tech savvy. They like to speak and write in technological language, such as emojis and stickers. These tourists are constantly looking for real-time information through videos and pictures which they can further use and share. The tourism service providers and the tour operators need to learn and fully understand their language, it to give messages and to interact with them. Smartphone is something without which no Generation Z tourist can survive, and also these tourists expect that the service providers should take their demands seriously (Kamboj and Sharma, 2016). The motivation to travel for today's tourists is to do something new and have fun. The excitement of doing something different is the main push. They tend to use technology from the time they decide to travel and till the time they return home. They are inclined toward those tourism products which are attractive along with being convenient (Setiawan et al. 2018). The destination management organizations should pay attention to the expectations to the current tourists. The tourists consider the websites operated by destination management organization as an important source of information and a lot of traveling decisions are taken based on these websites, however, it is seen that that the websites run by destination management organization are still not up to the mark and need to be improved a lot. Even the most popular destinations have poorly operated websites. This should be a concern for the suppliers and the service providers (Hamed, 2017). Today, tourism is not restricted to a particular segment of the society. The introduction of low-cost travel has made traveling possible for everyone. The era of digitalization and technological advancement has further added to the ease of travel for everybody and the young generation is most benefited by this phenomenon. The easily available and exposure to large information has lead to the changing needs and demands of the tourists (Monaco, 2018). Online media has taken a center stage in advertisement and promotions for the millennial and post-millennial tourists because of the trust they have on the online media. Although they are influenced by the word of mouth, but still largely rely on the internet for everything. The information available online gets updated continuously and new features are added regularly. Experience is the most important in travel for these tourists. They prefer adventure over comfort. These technology-driven tourists always keep in mind the prices, structure while planning their trip (Maitra and Singh, 2019). Generation Z tourists are soon becoming the decision makers in the family. Their views are considered and followed. They are in the center of very decision-making. They are tourists who soon will be paying the travel industry for their services. The tourism industry along with infrastructure development should also focus on understanding the next generation tourists. It is also seen that the budget requirements will also increase in catering to the needs of these tourists (Robinson and Schänzel, 2019). Generation Z tourists do not like complexities in general. They prefer easy access to readily available information. Their destination choice is also influenced by how easilythey get information about that tourist site. These tourists like traveling within the country. These tourists do not prefer traveling with group tours escorted by travel agents. These tourists can also be termed as drifter and explorers. The motivation for travel lies in exploring good food, engaging in activities, gain new experiences, and going away from the daily routine and environment for some time. They believe in value for money and rely more on social media and internet for information rather than the word of mouth (Wiweka et al., 2017). Home stays and guesthouses are preferred accommodation types of Generation Z tourists. These tourists not only travel alone but also travel with families. These tourists like communicating with the locals and know more about their culture, lifestyle, and food habits. They are looking for fun, rest, and relaxation while traveling. Their use of technology and internet is unlimited. They are always online for collecting information, looking for recommendations, comparing options, sharing experiences, etc. They generally look for options which are convenient and attractive (Setiawan et al., 2018). The present day tourists tend to like authentic and unique destinations and activities either modern or culturally inclined. They also look for entertainment activities. These tourists are willing to pay more price if it is worth. They do not go for usual traditional and low-priced stuff. Internet remains their main source of information and for making bookings. They use online reservation options frequently. They also rely on travel blogs. The difference between the older set of tourists and these new generation tourists is quiet evident. Tourism service providers are working hard to gain their trust and confidence (Baltescu, 2019). Fulfilling the needs of the tourists and focusing on client satisfaction are important for tourism industry as its success revolves entirely around the tourists and how happy they are. Generation Z tourists are socially influenced, conscious tourists looking for superior quality, affordable, and accessible products. Generation Z has changed the traveling pattern of tourists, as they prefer traveling alone, having their own specific preferences and therefore their behavior will surely impact the coming future generations of tourists (Kishore, 2019). The Generation Z is more careful in their choices over the internet than the previous Generation Y. Generation Y used internet as it is without any restrictions and filters. While Generation Z are specific in their choices and they will look only those options on the internet which they like. They also keep their social accounts filtered, and posting selfies is really important to them. Generation Z also look for emotional quotient of the destination along with enchantment and pleasure. These are the important factors which they take into consideration while making a choice of a particular destination. The tourism marketers and service providers should keep in mind these qualities of the Generation Z in order to attract and earn their loyalty. They are more modest than their previous generations (Hamed and Christine , 2018). The Generation Z tourists differ from their past generation in the way they make decisions, loyalty, values and beliefs, satisfaction, motivations, personality, perceptions, expectations, attitude, self-concept, and trust. The three factors which have influenced Generation Z widely are Technology, Generation Y, and their concern over ethics that they should follow while making traveling choices (Cohen et al., 2014). The facilities like Wifi are more important for Generation Z than a swimming pool and a gym. Also prices being charged for accommodation and other services are also a matter of concern for this Generation Z because they give value for money. This Generation prefers going to a destination of sea and sand rather than a historical, religious, or a cultural destination. In order to attract Generation Z, the service providers need to focus on innovating their products and creating new products, positioning their brand by highlighting the products they have, creating new and advance infrastructure, create online tour planning and destination-guiding applications, and try to provide services at minimal prices and charge lesser taxes (Tavares et al., 2018).

8.2.1 THEORETICAL APPROACH

Understanding the needs, wants, and desires of a tourist is important. In order to understand the behavior pattern of tourists, it is important to know about their likes and dislikes, background, education, income, age, gender, country of origin, traveling history, etc. A service provider or a travel agent can only provide total satisfaction to their tourist if they understand them well. A service provider will have to understand the motivation level of a tourist and what exactly a tourist desires. The gap between expected, perceived, and actual delivery of services should be as narrow as possible. Understanding the past, present, and coming generation of tourists is also a job of a tourism service provider. Tourism industry must evolve and develop according to the needs of the tourists whether it is in terms of technology usage, tourism infrastructure, or overall tourism products and services.

8.3 MATERIALS AND METHODS

8.3.1 DEFINITIONS

- **Baby Boomer:** These are the people born in the baby boom period from 1945 to 1964. They had very little exposure to technology and still face difficulties with it. They do not trust technology and prefer face-to-face communication.
- **Generation X:** This generation has seen the rise of technology and media platforms. They are the people born during 1965–1981. Although they have witnessed the evolution and rise of technology, internet, and this whole digital space came into existence in front of them, but still they are less-dependent on technology. The use of smartphones is limited to very basic and when necessary.

- **Generation Y or The Millennials:** These are the people born during 1982–1994. Although these people were not born with technology but they have surely made their way through it. Using technology is necessary for their personal and professional life. They have made themselves comfortable with various technological media and are using them on their daily basis.
- **Generation Z or The Geni:** These are the people born during 1995–2010. This generation is born with a smartphone and tablet around them. For them, using Internet, Facebook, or any other social media platform is just normal and so common. These people are using technology for everything. From socializing to education, everything is done digitally. All their decisions are influenced by the information they get over the internet and the reviews they read on social media.

8.3.2 CONCEPTS

- **Tourist Motivation:** It consist of the factors that determine the behavior of a tourist. These factors inspires a tourist to plan a trip. There are push and pull factors that make a tourist like or dislike a specific tourism product. These are the guiding forces or the motivators behind a tourist's desire to travel.
- **Tourist Satisfaction**: It is the state in which a tourist finds himself fully content with the tourism products and services, he is happy and feels satisfied at the end of the travel. All his expectations from the travel are fulfilled and he has received what he desired.
- **Tourist Behaviour**: It is the overall decision-making process followed by a tourist. How tourists behave during and after their travel. The means they use to plan their trip. The likes and dislikes, destination choice, accommodation preferences, mode of travel, food preferences, technology usage, motivational factors, willingness to spend, pax preferences, expectations, overall attitude, and duration of travel together constitutes the behavior of the tourists.
- **Technology in Tourism**: There is large use of Information and Communication technologies in tourism. Customer relationship management is entirely based on technological solutions. Travel Bookings these days are totally dependent on technology and

cannot be done otherwise. Use of internet, social media, online reviews, and digital platforms is enormous.

Incorporating all the above definitions and concepts gives us an idea of the changing tourist behavior. Without understanding the above definitions, it will be hard to understand the process of change. Behavior change is a complex process and it does not happen due to a single factor. However, the impact that each factors creates on behavior change varies from one individual to another. We see a difference in the way tourists behave due to reasons, such as their age, family backgrounds, educational qualification, spending capabilities, gender etc. Apart from these, the factors that also play an important part in shaping these tourists are the technology usage, their life experiences, economics of their country, society they live in, etc. Therefore, it is important to take all these concepts into consideration in order to fully understand the impact that the technology has created on tourist behavior and in what ways it has changed their decision-making process.

8.4 RESULTS AND DISCUSSION

The use of technology is everywhere and it has certainly helped in simplifying the process and minimizing human efforts. In today's times, we prefer doing things sitting at home to the extent possible and we only move out only when it is really necessary. This has become possible only due to technological inclusion in our day-to-day lives. Tourism industry is also dependent on technology for all its pre-travel and post-travel procedures. Not only the tourism industry professionals but also the tourists are using technology to its maximum. The tourist earlier use to rely on travel agents and service providers for all their travel-related queries, but now they have internet. They have all the information well in advance and the service providers are contacted only to confirm and do the final bookings. Tourists prefer using technology even for the final bookings, payments are done online, documents are sent and the final booking are received via an e-mail. The human contact has been reduced to bear minimum. Technology is impacting a tourist's decision to the extent that the tourists only wants to travel to places whose information is available over the internet and the bookings for which can be done online. They feel safe using only those services about which they have read somewhere on the internet or on some social media platform. Digital presence has become the key in today's times. Tourists in the earlier days used to trust travel agents on the information they give for itinerary, and all the planning was done by the travel agents and the tourists used to follow as instructed, but now there is a role reversal. The tourists these days have their own itinerary prepared, they have a list of attractions, accommodation, and activities, and they already have a plan as per their choice. Present day tourists are specific in their choices and the instructions are given by them in a comprehensive manner and the travel agent is expected to follow as it is. All this has happened due to extensive use of technology in our daily lives. The tourists have so much of information at their disposal to use for planning and preparing their travel plan. The earlier generations however were restrictive in their technological usage but the generation Z is all out. They are depending on technology for everything. Their day begins with social media and ends with social media. Technology is in their blood and they cannot help but google for everything. The people of this generation are overly specific in their choices, they do not like paying more, and most importantly, they are not open to surprises and like things to go as per the plan. They need things to be exactly in the way it has been promised. The reason for their travel is not just rest and relaxation but also due to the societal pressure. They desire to have a unique and picturesque traveling experience which is worth sharing on Facebook or Instagram. They want a story which can be shared with their friends and social groups. These tourists just cannot go out of network, wifi is their necessary requirement apart from the situation when they specifically opt for such places. The technology usage does not stop even during their travel, they are constantly on their phone checking and cross-checking. This generation tourists like to make decisions themselves whether they are traveling with family, with friends, or alone. The earlier generations usedto trust the travel agent and the service providers for large part of their travel, but the Generation Z keeps all decision-making to themselves and trust the internet over anything else. The tourism service professional certainly must keep them technologically updated. They must be aware of the latest technology in the market being used by their prospective clients and a well-functioning and user-friendly websites is a must for them. Having positive reviews on social media from their past tourists is very much required because the perception about a destination or a service provider these days is built on reviews and ratings they get on various social media platforms. As the physical contact with the clients is minimum, the service providers must strive for a positive and trustworthy online image. These days the tourists are relying on travel agents for their first-hand experience, for confirming the information available on the internet, and for the tasks which cannot be done online. Today's tourists also like promptness and cannot bear laid back attitude. They do not like to be bothered again and again for same things. Professionalism is also in their list. It is not like they do not want to pay at all but they always look for value for money. They might even pay more but only if it is really necessary. The COVID 19 pandemic will further add highest hygiene level and minimum physical contact to their list. They will look for less crowded places where social distancing can be maintained. Tourists will be more suspicious of their health and safety. Travel agents will have to bear that burden of ensuring the tourists safety and confirming that whatever services they will provide to the tourists from the start to the end of their travel will be safe. The role of travel agents will be more now as the tourists will look for someone reliable, whom they know personally and who can ensure their safety while traveling.

8.5 CONCLUSION AND SUMMARY

Technology has become an important part of our lives. We are dependent on technology even for the smallest of the things and travel planning is one of them. The tourists and the tourism service providers are using technology for their specific purposes. Technology has been used by the service providers for the reasons, such as to reach their prospective clients, to promote their products and services, to facilitate travel, and to carry out various pre- and post-travel formalities. The tourists however use technology for information gathering, reaching out to suitable service providers, to make decisions regarding the destination to travel, transportation to be used, accommodation, attractions, recreational activities, to support them during their travel and to make post-travel recommendations, and give reviews on the destination and the service providers. The earlier generations were less reliant on technology and were dependent on travel agents for all the information, planning, and decision-making, but the Generation Z is tech savvy and trust technology above anything else. They prefer making all their decisions based on the information they get on the internet. Using technology is just normal for them and they cannot imagine their life otherwise.

KEYWORDS

- **generation Z**
- **tourists**
- **technology**
- **tourism**
- **internet**
- **social media**
- **thermal tourism**
- **cross-border destination**
- **relationship marketing**
- **wellness**

REFERENCES

Baltescu, C.A. Elements of Tourism Consumer Behaviour of Generation Z. *Bull. Transilvania Unive. Braşov Ser. V Econ. Sci.* **2019,** *12* (61) No. 1.

Cohen, S.A.; Prayag, G.; Moital, M. Consumer Behaviour in Tourism: Concepts, Influences and Opportunities. *Curr. Iss. Tour.* **2014,** *17* (10), 872–909.

Hamed, H.; Christine, S. Generation Z and the Tourist Experience: Tourist Stories and Use of Social Networks. *J. Tour. Futures* **2018**.

Hamed, H. M. Marketing Destinations to Millennials: Examining the Compatibility between the Destination Marketing Organization Website and the Millennial Tourist Prospects. *J. Tour. Recreat.* **2017**.

Kamboj, R.; Sharma, P. Tourism for Tomorrow: Travel Trends across Generations: From Baby Boomers to Millennials. *Amity Res. J. Tour. Aviat. Hosp.* Jul–Dec 2016, *01* (02).

Kishore, S. Characterizing Generation Z and Its Implications to Booking Practices in the Hotel Industry, 2019.

Maitra, R.; Singh, M. Youth Tourism-A Qualitative Study on Travel Experience and Expectation of Young Earning Generation, 2019.

Monaco, S. Tourism and the New Generations: Emerging Trends and Social Implications in Italy. *J. Tour. Futures* **2018**.

Robinson, V. M.; Schänzel, H. A. A Tourism Inflex: Generation Z Travel Experiences. *J. Tour. Futures* **2019,** *5* (2), 127–141.

Setiawan, B.; Luh Putu Trisdyani, N.; Pramania Adnyana, P.; Nyoman Adnyana, I.; Wiweka, K.; Retno Wulandani, H. The Profile and Behaviour of 'Digital Tourists' When Making Decisions Concerning Travelling Case Study: Generation Z in South Jakarta. *Adv. Res.* **2018,** *17* (2), 1–13.

Tavares, J. M.; Sawant, M.; Ban, O. A Study of the Travel Preferences of Generation Z Located in Belo Horizonte (Minas Gerais – Brazil). *e-Rev. Tour. Res.* **2018,** *15* (2/3).

Wiweka, K.; Kurniawati, R. et al. The Profile and Behaviour of Smart Tourism (Generation Z) in Making Decision and Travelling, Case Study: Young Tourist in Jabodetabek, 2017.

CHAPTER 9

TOURISTS' ATTITUDES TOWARD CULTURAL HERITAGE: PRE- AND POST-VISIT EVALUATION

MUHAMMAD HASSAN MAHBOOB[1,*], MUHAMMAD ASHFAQ[1], ASAD AFZAL HUMAYON[2], and MUSTAFA KAN[3]

[1]*Institute of Agricultural and Resource Economics, University of Agriculture, Faisalabad, Pakistan*

[2]*COMSATS Institute of Information Technology, Vehari, Pakistan*

[3]*Kirsehir Ahi Evran University, Agricultural Faculty, Department of Agricultural Economics, Kirsehir, Turkey*

[*]*Corresponding author. E-mail: m.hassan1695@gmail.com*

ABSTRACT

Cultural heritage tourism is the branch of tourism where people travel to experience the place which authentically represents the stories of the past people. Cultural tourists were those who had concerned with buildings, monuments, landscapes, books, documentations, artifacts, and historical structuresof a country. There are several types of tourism which are almost equally important for the tourists to travel frequently. More than 1000 cultural heritages are officially listed by UNESCO in 167 countries. Majority of the cultural heritages are in the Europe and North America. Tourist's nature is a concept that denotes the fundamental dispositions and distinctiveness including ways of feeling, acting, accepting,

Tourist Behavior: Past, Present, and Future. Narendra Kumar, PhD, Bruno Barbosa Sousa, PhD & Swati Sharma, PhD (Eds.)

motives, and thinking that tourists are said to have naturally. Theere are seven attributes of cultural heritages that identified the purpose of the tourists. Tourists' traveling attitude divided into two different parts, the first one is the pre-tour and the other one is the post-tour attitude. There are multiple factors which play an important role in the pre- and post-attitude. Since 1950, the tourism industry has grown rapidly and surpassed many other industries in the world, but now the tourism industry is facing the worst crisis due to the COVID-19 pandemic. The above number was expected to increase in 2020. Tourists give importance to cultural heritage and to their visit; in this sense, they are more willing to pay for cultural heritage if money goes for cultural heritage conservation and for better facilities.

9.1 INTRODUCTION

The word '*tour*' came from Greek and Latin words, circle and the turn, which represents the activities of tourists circling away from homes and returning to homes. Ancient Greece and Roman were claimed as the origins of cultural heritage tourism and beach resorts because in the 18th century, they organized travel with the supporting infrastructure, essential destinations, and experiences of sightseeing as the parts of their tradition. On the other hand, it is quite difficult to define the conversion from privations of small groups of monks to recognizably tourist's practices. Although pilgrimage to Mecca is the site began more than 2000 years ago, pilgrimage to Buddhists is of similar antiquity.

9.1.1 TOURISM AND ITS TYPES

Tourism has been defined in many ways; the United Nations World Tourism Organization (UNWTO) created a common glossary of term tourism through a project from 2005 to 2007. It defined tourism is a social, cultural, and economic phenomenon that entails the movement of people to countries or places outside their usual environment for individual or business-professional purposes. These people are called visitors/tourists (which may be local or international) and tourism has to do with their activities, some of which imply tourism expenditure (United Nations World Tourism Organization, 2008).

There are several types of tourism which are given below in Figure 9.1. Through this chart, we can see that tourism is the movement of people for several purposes.

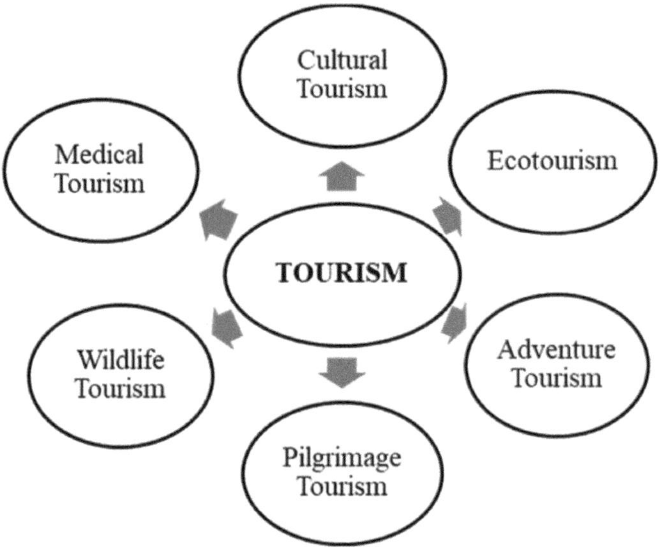

FIGURE 9.1 Types of tourism (UNWTO).

These forms of tourism allow tourists to do different activities in different places, but their attitude plays a vital role in choosing a destination for touring activities, so it can be cultural or non-cultural. The purposes of this chapter have a greater concern with those tourists who keen to visit cultural heritages or places of a particular country. The types of tourism and their characteristics are given below which are the most powerful motives of tourists.

- Cultural tourism (archeological sites, castles, museums, historical buildings, palaces, arts, etc.)
- Pilgrimage tourism (to achieve religious attitude and practices)
- Adventure tourism (to do some adventurers activities like hiking, skydiving, and scuba diving)
- Ecotourism (traveling to undisturbed areas of natural beauty)
- Wildlife tourism (eco and animal friendly, safari, hunting, etc.)
- Medical tourism (to visit for health care)

9.1.2 CULTURAL TOURISM

The basic purpose of this chapter is to evaluate tourist's attitude toward cultural heritage. Therefore, author is focusing only on cultural tourism and cultural tourists. The term cultural tourism and heritage tourism often written separately but both are related and overlapping phenomenons. Cultural tourism can be defined as a subset of tourism, which is concerned with a tourist's engagement within a country or outside a country's lifestyle of the people, history of the people, geographical areas, their art, religion (s), architecture, heritages, and other factors that help to shape their way of life. The tourism literature identifies the range of cultural tourism activities as including or watching, the use of such cultural heritages as archeological sites, castles, museums, historical buildings, palaces, ruins, famous buildings, arts, sculptures, festivals, galleries, events, theaters, music, primitive cultures, and other things that are based on culture (Jamieson, 1994).

9.1.3 CULTURAL AND HERITAGE TOURISTS (CHTS)

The word "tourists" first time used in 1772 and tourism in 1881, the word tour derived from old English turian, from Old French torner, from Latin tornare, which is itself from Ancient Greek torner (etymonline.com). There are several types of tourists according to *UNWTO*, domestic tourists who travel within a country, inbound tourists who travel in a given country, and outbound tourists (travel one country to another country). Cultural tourists were those who had concerned with buildings, monuments, landscapes, books, documentations, artifacts, and historical structures of a country. Cultural heritages have a significant impact on tourist's attitudes. On the other hand, heritage tourists keep their interest to look more at the art gallery, museums, buildings, historical structures, archaeological sites, and other cultural attractions.

9.2 WORLD CULTURAL HERITAGE SITES

Several heritage sites in the world are recognized by the United Nations Educational Scientific and Cultural Organization (UNESCO). These heritages have some special characteristics of being included in UNESCO because of their background which might be ancient ruins or historical structures, mountains, or wildness areas. According to UNESCO as of June 2020, a total number of heritage sites in the world are 1,121 in 167 countries, majority of the heritages are those who have cultural characteristics (77.5%), natural is 19% and only 0.34% are those heritages who have mixed features of cultural and natural. Cultural heritage tourists are very keen to visit different cultural heritages of the world which include the following attributes.

(1) Tangible culture (buildings, landscapes, work of art, and artifacts)
(2) Intangible culture (such as traditions, folklore, knowledge, and languages)
(3) Natural heritage (including culturally significant landscapes and biodiversity) (Fig 9.2)

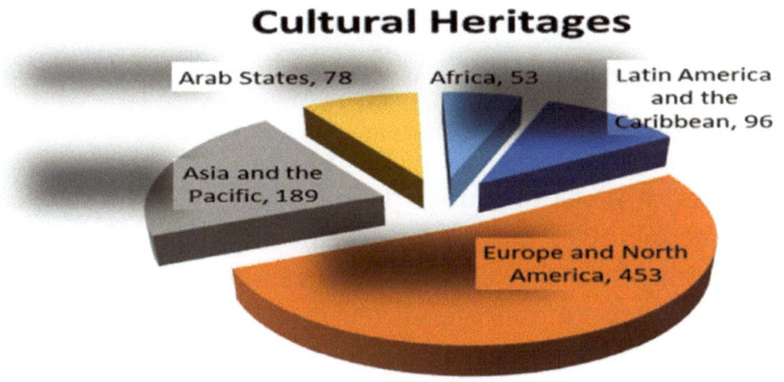

FIGURE 9.2 Officially recognized cultural heritages in the world.
Source: World Heritage List (2020); Total = 869/1121. https://creativecommons.org/licenses/by-sa/3.0/igo/

These region's heritages are officially listed at UNESCO. This figure shows that majority of the cultural heritages are in the Europe and North America 453 and few are in the Africa 53. These regions have international recognition and legal protection, and UNESCO gives the funds for the conservation. Almost 189 cultural heritages are in the region of Asia and the Pacific for the tourists in whom 55 are located only in China along with Italy. The Figure 9.3 shows the top 10 countries with most heritages in the world. Tourism-tourists directly related to each other, and domestic and international tourists visit a large number within a country or outside of the country every year.

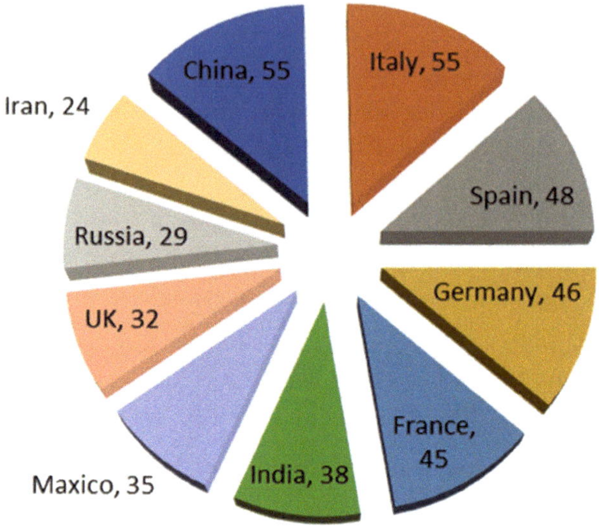

FIGURE 9.3 Countries with large number of heritages.
Source: UNESCO World Heritage Sites (2020).

Every year millions of tourists travel around the world for different purposes. The following chart 3.4 shows the top destinations where tourists visited during 2019 (WTO, 2019). France is the most attractive country in the world where the tourists visited in a large number almost 90.20 million in 2019, and other countries and tourists data have been given in Figure 9.4.

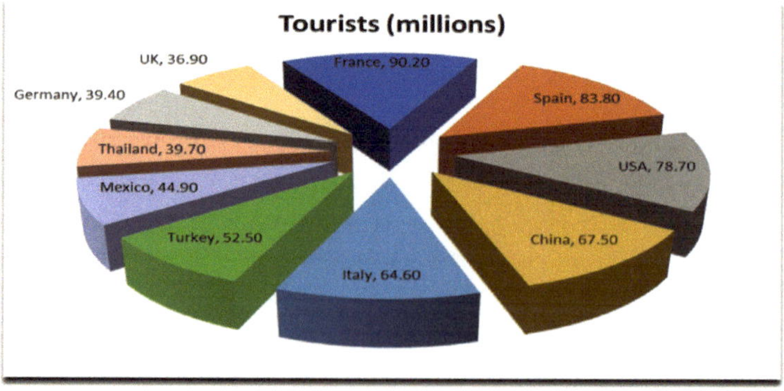

FIGURE 9.4 Countries in term of international visitor's arrival in 2019.
Source: WTO (2020).

9.3 FACTORS AFFECTING TOURIST'S ATTITUDE

Tourist's nature is a concept that denotes the fundamental dispositions and distinctiveness including ways of feeling, acting, accepting, motives, and thinking that tourists are said to have naturally. Recent studies about cultural and heritage tourism have focused on tourism development, management, as well as investigating tourist's attitude characteristics who visit cultural heritage places. Nowadays, the tourism industry has become the fastest-growing industry in the world; according to WTO, in 2018 almost 1.442 billion tourists visited different destinations and these numbers are increasing constantly. Tourist's satisfaction is also important in the success of destination marketing because it influences tourist's demand and consumption of products and services and the decision to return (Kozak and Rimmington, 2000). It is necessary to investigate those factors that encourage tourists to visit a cultural heritage and their satisfaction after visiting a destination. Tourists express their feelings in the form of satisfaction and dissatisfaction before and after visiting a cultural heritage. If tourists got satisfaction they will recommend them to their friends, family, etc. Now the main question raised here about determinates that influence or motivate tourists to visit a cultural heritage.

9.4 PRE-VISIT ATTITUDE

Attitude can be seen as statements toward objects, people, or events that include expressions that can be concluded as negative, positive, good-bad, either nice or not, and also beneficial or harmful. Attitude consists of emotional, cognitive, and behavioral elements. Several factors influence the tourists to visit cultural heritages. The following factors identify the purpose of this chapter that is based on a case study of Lahore, Pakistan.

- The socio-economic characteristics of the tourists
- Cultural heritage attributes
- Travel behavior
- Tourist expectations

9.4.1 THE SOCIO-ECONOMIC CHARACTERISTICS OF THE TOURISTS

The socio-economic characteristics of the tourists have a key role when researchers assess the tourist's satisfaction with cultural heritages. Tourist's socio-economic, demographic, and behavioral are the common factors in tourism research to profile by gender, age, marital status, income, occupation, education, etc. The study by Silberberg (1995) provided a common pattern of the tourists. In this study, he identified the tourists as one who has more money or income spends more on vocations and spends more time at cultural heritage. Moreover, educated tourists, female tourists, and older age categories were highly interested in a visit and came in large numbers. Educated tourists are well aware of the cultural history and cultural heritage, so they travel to gain knowledge about their own culture or other country's culture. These various socio-economic characteristics influence tourist's travel motivation toward cultural heritage connection between tourist's socio-demographic and destination image have been largely researched by Prayag (2012) and it has been claimed that socio factors greatly influence people's perception. Based on a case study, it has been seen that tourists who visited cultural heritage majority were male, educated, belong to the high-income level and their spending was also high for cultural heritage tourism.

9.4.2 CULTURAL HERITAGE ATTRIBUTES

This chapter attempts to evaluate the pre-visit attitude of the tourists toward cultural heritage; therefore, the attributes of the destination are important to satisfy tourists when they visit these destinations. Anderson et al., 1997 chose several attributes of cultural heritage, such as historical buildings, galleries, museums, festivals, places, sports, and old towns. They identified the important attributes of being a castle, museums, gardens, and historical buildings when tourists decided to visit cultural heritages. These attributes are also helpful for the marketing and development of cultural heritage tourism in a country. Cultural heritage's attributes (CHAs) significantly attract tourists. CHAs can be culture, history, festivals, traditional, folk arts, craftworks, architecture, historical sites, and museums. There are seven attributes of cultural heritages that identified the purpose of this chapter. These seven attributes are based on previous literatures that are similar to this study (Anderson et al., 1997; Sofield and Li 1998).

- History-Tradition
- Historical buildings
- Religious places
- Museums
- Galleries/Libraries
- Arts/Handicrafts
- The Outer Wall/Architecture

9.4.3 TRAVEL BEHAVIOR

Cultural heritage tourism is not only the type that generates economic benefits but also a tool that precedes interaction between different cultures (Mahboob et al., 2020). Cultural heritage tourism is a mind-opening experience, which defines that the world is not made up of a single model of living; even other models of living also exist. Many studies have demonstrated that experiences change the attitude of tourists (Uriely, 2005; ÇELİK, 2019). Tourists traveling behavior doesn't depend on a few things; there are so many factors that influence the tourist's behavior like, how they get information about a destination. Information can be attained through multiple sources, Newspapers, Television, internet or co-workers/ colleagues, etc. Tourists after getting information from several sources about a destination; after that they go on holidays for several reasons.

- To make this visit memorable
- To have an experience of cultural heritage
- To bring a change in the moods
- To reduce the stress level
- To explore a new place which they never visited
- To spend quality time with family or friends
- To meet new people, new cultures, experience a new thing, and do self-development
- Can buy well-known products, etc.

According to UNWTO (2018), almost 56% of the tourists visited for the only purpose of leisure, recreation, and holidays.

9.4.4 *TOURIST'S EXPECTATIONS*

All over the world, tourists travel in a large number. Tourist's expectations are now a major component in tourism. Although tourists are the bigger source of generating multiple benefits, but these benefits depend upon tourists. Tourist's expectations and then their spending are vital phenomenons in tourism. First of all, tourists travel toward those destinations that are safer and secure for the good experience of tourism rather than bad incidents; such incidents including a wide range of issues that could endanger tourist's lives and well-being. For instance, lack of proper sanitation, road issues, and hygienic standards may cause health concerns that have a dramatic negative impact on hosting destinations. Therefore, tourists avoid visiting that particular destination. Second is less expensive. Tourists visit in large numbers even many times which destination is less expensive, like traveling cost, accommodations, food and beverages, etc. And third has less environmental issues. The environmental issues due to expansion of tourism have been globally increased, such as pollution, waste, noise, sewage, and congested human vehicle traffic. Tourists look more for sustainable tourism rather than environmentally affected destination.

9.5 POST-VISIT ATTITUDE

The second part of this chapter evaluates pre-visit attitude of the tourists; it also evaluates post-visit attitude factors. National and international tourists observe many things during their visit toward cultural heritage. The positive feedback of the tourists much needed for the destination's popularity. Post-visit analysis based on the following factors:

- Perception toward improvements at the destination
- Tourist's willingness to pay
- Travel expenditures

9.5.1 *PERCEPTION TOWARD IMPROVEMENTS AT THE DESTINATION*

As we discussed that tourists face many issues before visit, but they also face many issues during their visit at any particular destination like, they observed that public facilities were not good enough which should be improved, toilet issues were also major concern for the tourists. Moreover, pedestrian paths, rubbish bin, maintenance of the heritage, and other facilities which should be improved for the tourists. If these issues are not treated on priorities, they will come again at this destination. These concerns of the tourists show that improved facilities at any destination attract the tourists to come again and again. Tourist's satisfaction is very important for the stakeholders that means tourists would like to visit again. Case study of Lahore shown that majority of the tourists were satisfied by visit and management, and they had a thought that both the destinations Badshahi Mosque and Royal Fort have the attraction for the tourists to visit these destinations, and were satisfied the way management cared for the cultural heritages. But they suggested plantation, hospitality, and cleanliness departments have to improve their status.

9.5.2 TOURIST'S WILLINGNESS TO PAY

The maximum amount that an individual is willing to pay for a product or service called willingness to pay (WTP). The certain point or amount at which the owner of a product or service provider is willing to accept a transaction with the buyer is willing to pay. Contingent valuation method brings out market valuation of a non-market good to compute total preservation value. Contingent valuation method (CVM) has been used that contains both constituents of use and non-use (Echeverria, 1995). The principle of contingent valuation can be used for gathering the data of tourist's willingness to pay for visitation of cultural heritage. Therefore, in the form of dichotomous-choice questions (Yes-No) asked to the tourists or a flat sum of money which would he/she pay for a product or services. There are many other researchers who had used this method (Randall et al., 1974; Bishop and Heberlein, 1979). There are a couple of models which used tourist's WTP, ordinary least square (OLS) model and binary logistic model. When tourists pay a flat sum of money like, 5$, 10$, etc., we use OLS and when the dependent variable is in the form of Yes-No then we employ binary logistic model (Alberini et al., 1997: Alberini, 1995; Kanninen, 1995: Mahboob et al., 2020). Meleddu and Pulina (2016) investigated the individual's willingness to pay for tourism and used it on quantitative data and applied logistic model. We also used OLS method where the values of an explained variable (Y) and other values of the explain variables (X). Past literature shows that there are many factors/ variables that affect tourists WTP. The most encouraging factor for tourists is the income level; we can say if the income level of the tourists tends to increasing or high income level, tourists would like to spend more for the visit. Moreover, if the tourists are more educated surly they are well aware to the cultural heritage tourism and will visit to different cultural heritages in the world. In this context, younger people try to visit more than older, distance from the home, total cost of the tour, and their satisfaction play an important role in the WTP. All these variables determine tourist's WTP for cultural heritage. There positive and significant relationships show that particular variable influence tourist's willingness to pay, and negative and insignificant relationship shows that variable doesn't influence WTP.

9.5.3 TRAVEL EXPENDITURES

Tourist's expenditures for visiting a cultural heritage have seen significant determinant of tourist's travel. Tourist's spending during their visit not only boosts local economy but also helps to boost the economic growth. A tourist directly and indirectly involves with many tourism firms, like tourism operators, agencies, product markets, etc. Travel expenditures can be transportation, food and beverages, accommodation, shopping, and other expenditures. Many tourists plan to spend their days that are economical in contrast of their budget. Across the different regions, Europe's combined tourist's spending dominant at $570b. Here are the top tourists hotspots countries and tourists spending for the year of 2019 by UNWTO (Table 9.1).

TABLE 9.1 Top International Destinations and Tourists Spending (2019).

Ranking	Country	International visitors (Millions)	Dollar spending (Billions-$)
1	Spain	83.5	79.7
2	USA	79.3	214.1
3	China	65.7	35.8
4	Italy	64.5	49.6
5	Turkey	51.2	29.8

TABLE 9.1 *(Continued)*

Ranking	Country	International visitors (Millions)	Dollar spending (Billions-$)
6	Mexico	45	24.6
7	Thailand	39.9	60.5
8	Germany	39.6	41.6
9	UK	39.4	52.7
10	Austria	61.9	22.9

Source: UNWTO (2019).

9.6 TOURISM IN 2020

International tourists have broken their visiting record in 2019. Almost 1.5 billion tourists visited different destinations which shows an increase of 3.8% per year (UNWTO). Since 1950, the tourism industry was growing rapidly and beaten up many other industries in the world but now, the tourism industry in the world is facing the worst crisis due to the COVID-19 pandemic. The above-given number was expected to increase in 2020. Unfortunately, the COVID-19 pandemic has caused a 22% fall in international tourists during the months of January–March. The world tourism organization was expecting a better situation for the remaining quarters of 2020, but they are expecting a huge fall in the tourist's number of almost 60–80% in 2020 if the outbreak and uncertainty continue to dominate till December. UNWTO secretary-general said, "The entire world is facing an unprecedented shock in the health and economic sector and tourism has been hit hard with millions of jobs risk." Tourism is one of the labor-intensive sector of the economy. Due to this pandemic, almost US$ 80 billion has been lost in exports. The UNWTO panel of expert's survey on the impact of COVID-19 on tourism is expecting that the domestic demands will recover faster than international demand. and this demand could be recovered in the last quarter of 2020, but mostly are expecting in 2021.

9.7 CONCLUSION

Culture is a convoluted structure that reflects the human psyche, manners, and communal environment. Globally, many developing countries are blessed with different heritages that are the asset for them to promote them and to earn sufficient benefits. The cultural heritage tourism based on many historical facts; therefore, it has become one of the finest attractions in the world for the tourists. In the world, different cultures prevail. Almost, 869 cultural heritages are officially listed by UNESCO out of 1121 total heritages. In the 1121 heritages there are combined number of cultural heritages, natural heritages, and mixed heritages, but a large number of 869 cultural heritages show that the cultural heritages are more prominent than other heritages, and tourists are more attracted toward cultural heritages. As we can see, cultural heritages are more in the world; therefore, tourists are more attracted toward cultural heritages, In 2019, 1.144 billion tourists visited different cultural heritages. Tourists attitude play an important role in the increasing numbers; their pre-visit attitude and post-visit attitude play a defining role in tourism industry. Cultural heritages demand is increasing day by day in many countries because of the improvements and facilities that are provided to the tourists. Destination's attraction is important but conservation and preservation is also an important factor. Tourist's socio-economic characteristics, heritage uniqueness, their positive attitude, and their expectations are valuable before visiting a particular destination. Cultural heritage tourism characteristics, development, and management have been identified in the recent studies. Tourist's socio-economic characteristics, their behavior, their attitude toward cultural heritage, and other related features of the tourists have been demonstrated in the past and recent studies (Mahboob et al., 2020). Tourists pre- and post-visit behavioral intentions are very important for the stakeholders. Different models can be used for estimating tourists WTP. Tourists come in large numbers and their spending is another key element of attitude and visit. Tourist's WTP is not just determined by their socio-economic or demographic characteristics but rather by their attitude, actions, and their motivation. Tourists give importance to cultural heritage and to their visit; in this sense, they are more willing to pay for cultural heritage if money goes for cultural heritage conservation and for better facilities. Most tourists were younger and educated; Tourism planners need to start tourism curriculum in schools and colleges to get involvement of national residents. Understanding to tourist's attitude is a useful tool in destination marketing decisions. Tourist's attitude toward cultural heritage especially with regards to Badshahi mosque and Royal fort depends on authorities. Tourist's spending during their visit; may encourage local industry (e.g., hotels, handicraft products, and transportation) to earn profit.

KEYWORDS

- **cultural heritage**
- **tourism**
- **pre and post visit**
- **tourists**

REFERENCES

Alberini, A. Efficiency vs Bias of Willingness-to-Pay Estimates: Bivariate and Interval-Data Models. *J. Environ. Econ. Manage.* **1995,** *29* (2), 169–180.

Alberini, A.; Kanninen, B.; Carson, R. T. Modeling Response Incentive Effects in Dichotomous Choice Contingent Valuation Data. *Land Econ.* **1997,** 309–324.

Anderson, V.; Prentice, R.; Guerin, S. Imagery of Denmark among\Visitors to Denish Time Arts Exhibitions in Scotland. *Tour. Manage.* **1997,** *18* (7), 453–464.

Bishop, R.; Heberlein, T. Measuring Values of Extramarket Goods: Are Indirect Measures Biased? *Am. J. Agric. Econ.* **1979,** *61* (5), 926–930. doi:10.2307/3180348

ÇELİK, S. Does Tourism Change Tourist Attitudes (Prejudice and Stereotype) towards Local People? *J. Tour. Serv.* **2019,** *10* (18), 35–46.

China's Liangzhu Archaeological Site Now a UNESCO World Heritage Site. *news.cgtn.com*. (accessed 6 July 2019).

Echeverria, T. W. Sea-Ice Conditions and the Distribution of Walleye Pollock (*Theragra chalcogramma*) on the Bering and Chukchi Sea Shelf. *Can. Spl. Pub. Fisheries Aqua. Sci.* **1995,** 131–136.

Jamieson, W. The Challenge of Cultural Tourism. *ICOMOS Can. Bull.* **1994,** *3* (3), 2–6.

Kanninen, B. J. Bias in Discrete Response Contingent Valuation. *J. Environ. Econ. Manage.* **1995,** *28* (1), 114–125.

Kozak, M.; Rimmington, M. Tourist Satisfaction with Mallorca, Spain, as an Off-Season Holiday Destination. *J. Travel Res.* **2000,** *38* (3), 260–269.

Mahboob, M. H.; Ashfaq, M.; Humayon, A. A.; Kan, M. Tourists Attitude towards Cultural Heritage and Halal Products: A Case Study of Lahore. *J. Org. Culture Commun. Conflict* **2020,** *24* (2), 1–8.

Meleddu, M.; Pulina, M. Evaluation of Individuals' Intention to Pay a Premium Price for Ecotourism: An Exploratory Study. *J. Behav. Exp. Econ.* **2016,** *65*, 67–78.

Online Etymology Dictionary. *etymonline.com*. (accessed 2 June 2016).

Prayag, G. Paradise for Who? Segmenting Visitors' Satisfaction with Cognitive Image and Predicting Behavioral Loyalty. *Int. J. Tour. Res.* **2012,** *14* (1), 1–15.

Randall, A.; Ives, B.; Eastman, C. Bidding Games for Valuation of Aesthetic Environmental Improvements. *J. Environ. Econ. Manage.* **1974,** *1* (2), 132–149.

Silberberg, T. Cultural Tourism and Business Opportunities for Museums and Heritage Sites. *Tour. Manage.* **1995,** *16* (5), 361–365.

Sofield, T. H. B.; Li, F. M. S. Tourism Development and Cultural Policies in China. *Ann. Tour. Res.* **1998,** *25* (2), 362–392.

The World Tourism Organization Reports the Following Ten Destinations as the Most Visited in Terms of the Number of International Travelers in 2019.

UNESCO World Heritage Centre. World Heritage List Statistics. *unesco.org*. Archived from the original on 14 May 2011.

UNWTO World Tourism Barometer and Statistical Annex, December 2020|World Tourism Organization.

Uriely, N. The Tourist Experience: Conceptual Developments. *Ann. Tour. Res.* **2005,** *32* (1), 199–216.

World Heritage List (Ordered by Region). UNESCO World Heritage Centre. Archived from the original on 5 July 2020.

CHAPTER 10

FOODIE TOURIST BEHAVIOR

NILGÜN DEMIREL[1,*] and EZGI KIRICI TEKELI[2]

[1]*Travel-Tourism and Leisure Services Department, Igdir Vocational School, Dr., Igdir University, Turkey*

[2]*Department of Tourism Guidance, School of Applied Sciences, Dr., Karamanoğlu Mehmetbey University, Turkey*

[*]*Corresponding author. E-mail: nilgundemirel90@hotmail.com*

ABSTRACT

With the dominance of individual tourism mobility, there have been changes in tourist motivation and interest in gastronomic tourism. There are advances in the variety of tours, such as culinary tours, gastronomy tours, festival tours, gourmet tours, cooking course tours that foodie tourists may prefer. Foodie tourists are divided into many subgroups according to their eating and drinking preferences, but in general they are interested in organic, natural products, and they like to taste ethnic or local foods. Foodie tourists want to learn about the traditions and customs of food culture, such as "what time the food is eaten, with whom, table manners, eating habits, which food and drink is compatible with food and drink". Foodies may prefer local, organic, and sustainable products also they adopt slow food habits. They engage in fishing and farm work in addition to sustainable agriculture. Authenticity is the decisive element in their restaurant preferences and local food are seen as major tourist attraction centers. Authenticity positively affects their destination selection. On the

Tourist Behavior: Past, Present, and Future. Narendra Kumar, PhD, Bruno Barbosa Sousa, PhD & Swati Sharma, PhD (Eds.)

other hand in their holidays, foodie tourists may both seek for familiar tastes for eating and drinking and may also be willing to try different tastes from their own familiar tastes. Entomophagous foodies are included in second group. They like eating edible insect species but few people prefer to eat insects. Some foodie tourists are simply fond of coffee, wine, or any famous drink and are motivated to travel just to try it. Foodies consume luxurious foods and they tend to pay higher prices during their travels. Especially they spend more money on their travels to abroad. Gastronomy festivals are other attraction element for foodies. They can socialize in these festivals as well as eating different foods. According to their digitalization choices, they can share the photo of their best foods. Most of them use social media actively and share what they eat and destination's local foods. Foodie tourists' destination choices and consumption habits are valuable source for development of gastronomy tourism. This chapter aims to detail different typologies related to foodie tourists. It is supposed to contribute to the body of literature both theoretically and practically on tourism sector, professional tourist guides and variety of tourism products in the sector.

10.1 INTRODUCTION

Food consumption constitutes an important area in modern consumption societies and reveals different cultural aspects of the consumption society. The food consumed varies according to the subcultures within a society. These cultural differences led to the emergence of the "foodie culture" and this concept, which expresses food consumption, has gained a more aesthetic value (Ambrozas, 2003). The foodie concept is a concept, highly related to travel and is considered and evaluated in the context of tourism. Because people who are fond of eating and drinking wonder and want to experience local flavors from different cultures. This situation leads to the emergence of the tourists who are passionate about food and whose main reason for traveling is food and drink. In the literature, these tourists are named as "foodie." In this context, the emergence of foodie culture leads to the formation of a tourist type.

In the future, it is thought that there will be many options for foodie tourists; therefore, consumerism will increase and a high entropy society related to eating and drinking will exist. As time goes by, the meaning of the concept of luxury changes and the importance of cultural and social capital increases. Depending on developments in information technologies in the world, it is easier for tourists to exchange ideas about destinations and their experiences in the relevant destination. This situation directly affects the destination preferences. These tourists, who make their own destination preferences and are fond of eating and drinking, emerge as a type of tourist who reflects their increasing desire and high-level expectations and wants to experience local various food (Yeoman and Beatte, 2016).

It is observed that the number of foodie tourists has increased recently. For foodie tourists who exhibit different attitudes and behaviors from other types of tourists, tasting local and traditional dishes or beverages such as tea, coffee and wine constitutes the main purpose of the travel. Therefore, they can talk about meals and restaurants throughout the day. They are particularly curious about recipes or the historical development of food. Sometimes, they like to eat in luxury restaurants and sometimes, prefer authentic, shabby places, and may pay high prices. In addition, it is very important that food and beverages are organic. For this reason, some foodie tourists deal with organic farming and animal husbandry. They prefer to buy local products in their destinations. For this reason, they are also interested in the agricultural products in the destination and occasionally they want to participate in the cultivation of agricultural products. One of their most common behaviors is to take pictures and share them on social media, as they see the food as an artistic product (Demirel and Kırıcı Tekeli, 2020).

All these attitudes and behaviors of foodie tourists are very effective in travel motivation and destination preferences. While food is just a source of nutrition for some people, on the other hand, for foodie tourists who love to taste different foods and seek different local flavors, "food" is the main reason to travel. For this reason, it is seen that food and beverages in general have a great influence on travel motivation and destination selection (Björk and Kauppinen-Räisänen, 2017).

In this chapter, primarily the concept of foodie is mentioned. The first emergence date of the concept has been mentioned and the foodie concept has been defined using different sources. Also, the characteristics of the foodie tourist are mentioned. Then, different typologies related to foodie tourists are included. Foodie tourist typologies were put forward by different researchers at different times. The emergence of different types of foodie tourists has led to some changes in the attitudes and behaviors of foodies, and the attitudes and behaviors of foodies are included in the relevant title. Finally, travel motivations and destination preferences of foodie tourists are mentioned.

10.2 THE CONCEPT AND SCOPE OF FOODIE

The foodie concept was used for the first time in an article published in the New York Times magazine. However, the popularity of the concept increased with the article titled "Cuisine Poseur" published in Harper's & Queen magazine in 1982. This concept was also mentioned in "The Official Foodie Handbook" published by Barr and Levy in 1984 (Dewitt, 2010; Poole, 2012, cited from Barr and Levy, 1984).

In recent years, it is seen that, several different concepts have been put forward to define people who are fond of eating and drinking. However, none of these concepts are seen as acceptable and useful alternatives to replace "foodie." These alternatives are "gourmet," "gastronom," "epicure," "connoisseur," and "gourmand." As can be seen, there is a concept confusion about the word "foodie." This concept, which is especially confused with the word "gourmet," was seen as equivalent in the USA in the 1960s and 1970s. A gourmet is described as "a food and wine expert skilled at recognizing good flavors and textures." This situation separates the concept of gourmet from the concept of foodie in terms of "having knowledge about food and beverage" (Collins, 2015). Another concept that is similar to gourmet is "şikemperver." Şikemperver is a Persian word, meaning "a person who is a hearty eater" (www.tdk.gov.tr), and "şikem" means stomach and "perver" means a person who loves nurturing and educating (Türkoğlu and Akoğlan Kozak, 2015).

The meanings of foodie in the dictionary are respectively "gourmet," "a person with a good sense of taste," "someone who is fond of food or is interested in world cuisine" (www.tureng.com). In another dictionary, its meaning is "someone who loves food and knows a lot about it" (www.dictionary.cambridge.org). A consensus has been reached on the definition of "foodie" concept which is discussed in this article as "people who are fond of eating and drinking."

Barr and Levy (1984) defined the people who are fond of eating and drinking as "people talking about food always, everywhere, dealing with food almost obsessively, talking about restaurants, recipes and even unknown fruits (such as radicchio) and not thinking that all of these are unnecessary and absurd." Additionaly, these people consider food, as an important field of art, such as painting and drama (Collins, 2015, cited from Barr and Levy, 1984). At the same time, these people see themselves as "people who are in love with all the processes of food, especially vegetables, from seed to plate" (Ambrozas, 2003).

Johnston and Baumann (2015) interviewed many people to identify people who are fond of eating and drinking, and during the interview, they received people's views on many issues such as food media, foodies' shopping, cooking and eating-out habits, authenticity in food and beverages, genuineness, etc. According to the results obtained, it was emphasized that the concepts of "education," "identity," "exploration," and "evaluation" are important while defining foodie. It has been stated that people who are fond of eating and drinking are highly educated about cooking, and generally have an interest and desire to get more knowledge about food. In terms of the self, foodies see food and beverages as a part of their "selves" and identify themselves with them, and even internalize them. Discovery often means that a foodie tries new recipes regularly. At the last stage, the evaluation of the food is perceived as a very pleasurable situation and interpreted as a gratification.

Even if they are not knowledgeable or adventurous about the food they are eating; nowadays, people who are enthusiastic about "food" think of themselves as people who are fond of eating and drinking. However, a person who is fond of eating and drinking should have a high degree of interest in getting knowledge about food as mentioned earlier. In addition, these people should be knowledgeable and well-equipped in subjects such as the historical origin of the food, its social history, flavor profile and food and wine pairing, etc. (Collins, 2015). Robinson and Getz (2016) searched and sized the characteristics of foodie tourists in their study. These dimensions are very important in defining yourself as a foodie. These four dimensions are as follows:

- For foodies, cooking is considered as an important act for their "selves" and social identities. First of all, foodie tourists love to cook so much. Cooking is the greatest pleasure in their life. In addition, it is observed that foodie tourists have a constant desire to learn.

- Foodie tourists care about the quality of the ingredients they buy for cooking. Quality is also an important issue for the food they buy. Naturally, this reflects in the expenditures they spend for their kitchens and the materials required for cooking as well. They especially prefer local and organic foods.
- Eating for a foodie is quite enjoyable. However, this is largely about socialization. Because foodie tourists love to please, entertain others, and accompany other people in a great food experience.
- Foodie tourists are fastidious. In other words, foodie tourists are considered to be demanding, meticulous people. They are very meticulous in purchasing and using food and beverages. They always keep the kitchen and the equipment they use, clean.

It seems possible for people carrying all these characteristics to define themselves as foodies. These characteristics show that not every person interested in food and drink should be called a foodie. Mohsen (2017) summarizes the characteristics of foodie tourists in five items: learning and exploration, pleasure and entertainment, inspiring effects, quality, and experiencing food and beverages individually or on the advice of others. As can be seen, the determined themes are very similar to the description made by Robinson and Getz (2016).

Although similar foodie characteristics have been put forward by researchers, there are different explanations about foodies. According to the first view, foodies reflect social, cultural, and environmental values. From the second point of view, being a foodie is understood as being a snob and elite (Yozukmazet al., 2017, cited from Getz et al., 2014). This issue is also discussed by Johnston and Baumann (2015). According to the first view, foodies are people who are democratic and have an interest in all kinds of food. In this regard, they prefer healthy food, local products, ecological, and organic foods. In other words, the foodie trend is a modern movement that anyone can join. According to the second view, foodies are snobbish, elite, and spoiled people who are influenced by consumption culture and belong to the upper class. As can be seen, there are some discussions about the concept of foodie. These characteristics proposed by researchers on foodie tourists have led to the emergence of different foodie typologies.

10.3 FOODIE TOURIST TYPOLOGIES

When the literature is examined, it is seen that different typologies related to foodie tourists have been suggested. The first typology for people who are fond of eating and drinking was discussed in 1984 by Barr and Levy and evaluated under seven titles. In the study of Sloan in 2013, six typologies were determined regarding people who are fond of eating and drinking. The typology of people who are fond of eating and drinking, discussed by Bourdain in 2013, was evaluated under five headings. In 2015, Getz et al. gathered this typology under three titles. Finally, Demirel and Kırıcı Tekeli determined a foodie tourist typology in 2020 from the perspective of professional tourist guides and discussed under five headings. These typologies are included in Table 10.1.

TABLE 10.1 Foodie Tourist Typologies.

Barr and Levy (1984)	Sloan (2013)	Bourdain (2013)	Getz et al. (2015)	Demirel and Kırıcı Tekeli (2020)
Whole-foodier Than Thou	Fans of Foreign/Spicy Food	Chefestants	Dynamic Foodies	Organic Consumption-Oriented Tourists
Squalor Scholar Cook	Enthusiastic Chefs	Coffeegeeks	Active Foodies	Tourists Who Don't Care about Paying More for the Food
Made in Paris	Gourmet Foodies	Dining Digerati	Passive Foodies	Tourists Who Wonder Eating and Drinking
Paris C'est un Dump	Organic Foodies	Locapours		The Tourists Who Consider The Food as An Artistic Product
Gorgeous East in Me	Healthy Foodies	Pizzaratti		The Tourist Who Post the Food Photos on Social Media
Foodies on Ice	Restaurant Foodies			
All-American				

Sources: Barr and Levy (1984), Bourdain (2013), cited from Green (2013), Sloan (2013), Getz et al. (2015), and Demirel and Kırıcı Tekeli (2020).

The first foodie typology was proposed by Barr and Levy (1984) and has been studied under seven titles. Whole-foodier Than Thou; give importance to natural nutrition by growing their own vegetables and fruits. They cut food and meat themselves and prefer simplicity and simpler ingredients when cooking meat. Squalor Scholar Cook; research and learn the history and recipes of food. They stick to traditional recipes. They also have sufficient academic and historical knowledge to distinguish one dish or one ingredient from another. Made in Paris; in order to progress toward becoming a foodie, they first learn basic cooking techniques in small restaurants in Paris. Paris C'est un Dump; have an upper class lifestyle. They spend most of their time in upscale restaurants, subscribe to food magazines, and are very selective about food. Gorgeous East in Me; are fond of ethnic and foreign food. They always want to try new flavors and experience and learn different cultures through food. Foodies on Ice; see food as an artistic material. For this reason, they try to impress people by making busts and sculptures out of ice, cake decorations or butter. All-American; is a small town foodie looking for local food and supplies. She tries to improve the products, and she grows in her own home in order to cook better.

Sloan (2013) in her article, titled "The Foodie Phenomenon," grouped foodies under six titles. Fans of Foreign/Spicy Food; constitutes 30% of the foodies. These foodies are fond of Italian, Spanish, Mediterranean, Indian, Mexican and Asian food. Enthusiastic Chefs; although they are more consumers than producers, they are eager to try new recipes. They are known to buy specialty food products and try new recipes. Gourmet Foodies; this gourmet group interested in high-class expensive food, spend most of their money trying new dishes in upscale restaurants. Organic Foodies; when choosing all foods, they pay attention to that they are natural, sustainable, organic and also environmental friendly. Healthy Foodies; primarily pay attention to the calories and nutritional values of the food. Restaurant Foodies; in the menus of the restaurants they visit, they prefer to try unknown dishes other than common ones. Such adventurous foodies also prefer products with a sour, bitter, and aromatic taste.

Bourdain (2013) examined foodies in five groups. Chefestants; are culinary contestants competing for fame and prize money. Usually, they stand out in television programs (reality shows) about the lives of ordinary people. Coffeegeeks; they are very fond of coffee and addicted to caffeine. They know coffee brewing techniques well. They are fond of coffee types and coffee making equipment. Dining Digerati; this foodie typology, which has become very popular recently, is mostly known as the types of gastronomes who communicate through blogs, tweets, social media accounts, and messaging. Locapours; consists of people who are only keen to drink wine, beer, and locally produced alcoholic beverages. Pizzaratti; these pizza snobs begin an endless quest for pizza perfection, and constantly debate their ability to produce a variety of pizza breads, cheeses, sauces, and pizza toppings.

Getz et al. (2015) examined the foodie typologies more broadly than other researchers and classified them into three groups. Dynamic Foodies; participating in food-related events, enjoying all kinds of food-related activities and traveling for food are the most important features of them. They are usually young or middle-aged, belong to the high income group and are well educated. The vast majority of them have traveled many times for food-related events. Food and drink is an important factor in the decision-making process for vacation for this kind of foodies. Participating in festivals, purchasing what the farmer produces and learning new things about food from the chefs, make them happy when they are back home from holidays. They generally prefer 4 and 5 star hotels, travel frequently and look for the "self catering" opportunity. This foodie typology is very knowledgeable and confident about the use of the internet, and they often prefer to book online and search online for information. Dynamic foodies often travel as couples or families, and this affects their decisions and opportunities. Active Foodies; although they do not travel much, eating and drinking is effective in the holiday purchasing decision process for them. Therefore, their travel satisfaction is not a criterion related to eating and drinking. Their preferred activity is to visit the farmers' markets or fish markets. Their average age is high. Holiday choices include choosing regional cuisine, buying fresh food from farmers' markets, attending food festivals, taking small trips to islands and villages, and choosing cheap air transportation. Their use of social media is low because they devote their time to family and friends. Before traveling, they check the websites of the destinations and make all their travel and accommodation processes online by booking. Passive Foodies; although they love food, few foodies travel to eat in this typology. Their preferences are more in favor of farmers markets which are especially closer to their homes. They attach importance to word of mouth communication and decide their travel by looking at the websites of the destinations based on the recommendation of friends and relatives. They demand their money's worth absolutely and want a good food experience. They travel; however, they prefer to express their passion for food mainly in their own home and with a group of close-knit friends or family.

Demirel and Kırıcı Tekeli (2020) evaluated the foodie tourist typology, which they determined using the observations of tourist guides, under five headings. Organic Consumption-Oriented Tourists; prefer organic food and beverages. Most of these tourists are engaged in organic agriculture and animal husbandry and prefer to buy local products directly from the producers during their holidays or in their daily lives. Tourists Who Don't Care about Paying More for the Food; prefer to pay high prices for a meal in an authentic place rather than eating in a luxury restaurants. Tourists Who Wonder Eating and Drinking; during their travels, they ask the tourist guides about places to eat, they are curious about the architectural design of these places, the recipes of the food and try to get information about the history of the meal. In addition, they want to get information about the traditions and customs associated with the food culture, such as what time and with whom the meal is eaten, table manners, eating and drinking habits, and food is in harmony with which food and drink. The Tourists Who Consider the Food as An Artistic Product; think that the food placed on the plate has a design and that this food has become as a piece of art, and they visualize the shape of the food on the plate by taking their photos. The Tourist Who Post the Food Photos on Social Media; this foodie tourist typology uses social media very effectively and they share their food photos on social media.

As can be seen, foodie typologies are referred to with different names in different cultures. However, when the common motivation of foodie tourists is evaluated, in terms of travel, it is observed that, they are young people using social media effectively and they can travel for tasting a kind of meal, and pay high fees for that meal. Additionally it is seen that they prefer to eat in authentic places and they are very keen to know the process of the food from the very beginning of cooking to the plate, as well as, they pay attention to organic consumption and buy products of the local farmers of local markets.

10.4 THE ATTITUDE AND BEHAVIOR OF FOODIE TOURISTS

The foodie community defines itself with its "highly cultured and chosen" characteristics, and this definition reveals their individuality in food and beverage. Accordingly, not everyone who likes to eat, cooks or tastes can become a foodie. Foodies have certain characteristics that distinguish them from others (Deleuze, 2014). Since there cannot be the same foodie typologies in every country or destination, it is possible to classify foodie tourists according to their attitudes and behaviors in terms of characteristics that can vary from person to person. The attitudes and behaviors of foodie tourists according to their general characteristics and typologies are examined under six headings.

10.4.1 FOODIES PREFERRING LOCAL, ORGANIC, AND SUSTAINABLE PRODUCTS

Traveling for food and drink has actually created a sustainable food movement for all people. (Deleuze, 2014). With the motto "you are what you eat," the local food consumption trend started, awareness about sustainability was created, healthy and organic eating patterns emerged and the foodie culture started to emerge after these trends (Vojnoviç, 2017). Foodies are considered to be the people who most adopt this trend. It is assumed that, foodies who take care to consume local and organic products grow products to meet their own needs, try to learn the details of cutting meat, engage in fishing, voluntarily engage in farm work, participate in cooking competitions, participate to food industry meetings, also participate to events and meetings related to sustainable agriculture. They are thought to support "the slow food" movement, try to make alcohol at home and join the dining clubs (Green et al., 2015). In particular, it is known that, while traveling, foodie tourists prefer slowly consumed, local and traditional foods and drinks instead of fast food (Demirel and Kırıcı Tekeli, 2020). Tourists' interest in healthy, safe, organic food and drink, local food and seeds has increased. Thus, tourists who are in need of consuming genuine and authentic local foods have started to form an important part of the travels. It is also seen that the image of the destinations will be positively affected by this interest in food and drink (Çorman, 2018).

10.4.2 FOODIES USING SOCIAL MEDIA

Nowadays, the connection between social media and food has increased as social media platforms, such as Instagram, have become a part of the culture. It is known that, social media platforms now include the foodie community and that especially photo sharing sites on eating and drinking have a control power in the society. Foodies are actively using social media with a sense of creating a new identity, expressing tastes and values, enriching and belonging to different online communities. With the help of blogs, revealing their talents and competencies and proving that they are different from others, is among the motivations of foodies (Vojnoviç, 2017). By using social media or blogs to demonstrate their professionalism and to lead opinions by choosing electronic word of mouth communication, especially regarding to the destinations they travel, contributes to the reputation of foodies (Watson, 2013).

10.4.3 FOODIES WHO CARE ABOUT AUTHENTICITY

While cooking, it is a part of authenticity to carefully prepare traditional and more different (exotic) ingredients with both modern and rustic cooking techniques. Although the concept of authenticity varies from person to person, authentic recipes become very attractive to tourists. Authenticity represents a cuisine that is both rooted in the past and revitalized with new techniques. This is not the daily, routine and ordinary reality in food and beverage at the heart of tourism, but a trend that allows to realize a different dream by escaping the ordinariness (Deleuze, 2014). Foodie tourists play an important role in revealing the authenticity of a culture, both in a political and cultural context. Foodies are especially in an effective envoy position to bring authenticity to the forefront and to promote ethnic structure in food culture (Oleschuk, 2017). Foodies learn all the values of the region due to their interest in eating and drinking, and by researching the history, origin and way of making food. For this reason, their interest in authentic food increases even more. Therefore, they do not avoid spending too much money for food and prefer to pay high fees for a meal in an authentic place rather than eating at luxury restaurants (Demirel & Kırıcı Tekeli, 2020).

10.4.4 FOODIES CONSUMING LUXURIOUS FOODS

The level of vacation expenditures of individuals in touristic destinations varies depending on preferences such as flights, accommodation in luxury hotels, and dining in luxury restaurants. Even if the cost of the holiday increases, the pleasure gained from food and drink develops in different ways for foodie tourists. Because foodie tourists do not hesitate to pay large amounts for eating and drinking (Green, 2013). While foodie tourists seeking luxury, do not cover so much place in domestic tourism demand, they spend more money on their experiences abroad. Foodie tourists consuming luxuries, aged 25–50, always want to get their money's worth (Deleuze, 2014).

10.4.5 ENTOMOPHAGOUS FOODIES

Entomophagy refers to the way of nutrition by eating small amounts of edible insect species (larvae or adult) rather than daily consumption. Insects can go through different culinary stages and cooked (e.g., fried, pickled, and roasted) or become components for processed food (e.g., insect flour). It is widely consumed across various cultures and regions in Central America, Africa and Asia. In recent years, interest in entomophagy has increased in western countries. Since 2014, some European countries (Belgium and the Netherlands) have started selling edible insects in supermarkets. As a food source, edible insects can be a supplement to the poor diets of developing countries, while also helping as a complementary food/additive in the diets of developed countries (Sogari et al., 2017). Edible insects are actually seen by some European consumers as future meat substitutes. It is known that young men interested in healthy and sustainable food were the first group to adopt entomophagy. Among women, prior knowledge and experience with edible insects can change the perspective on cooked insects. But young female foodies are known to be more neophobic and less adventurous (Megido et al., 2018). Although the majority of people in Western countries refuse to eat insects for cultural reasons; the attitude of accepting and adopting entomophagy has become common among young foodies in Italy, therefore, their motivation to experience has been explored. The results showed that almost all students tasted the product and were willing to try other edible insects in the future (Sogari et al., 2017).

10.4.6 SOCIAL FOODIES PARTICIPATING IN FESTIVALS

Food selection and preparation methods make people a member of a particular group and generally reflect their general relationship with society and themselves (Vojnoviç, 2017). While people are eating during the travel, they socialize, gain new experiences and get to know the culture of the region they travel to. Festivals, local events, and celebrations organized to promote the special food and food culture of a region, have an important effect on the selection of that region touristically and the experiences of the tourists traveling that region. These kinds of events create an environment where the participating individuals can experience the local culture and socialize with the local people (Çorman, 2018).

Foodies want to participate to festivals in the countries they are traveling to, they want to gather as a couple or in groups and attend home meetings, host their guests and share their meals with them. They love to discover and talk about new and different dishes by creating a meeting atmosphere in restaurants (Getz and Robinson, 2014). Yozukmaz et al. (2017) stated that the destinations that highlight local restaurants, local cuisine culture and local festivals are more preferred by foodie tourists. For this reason, destinations' organizations for food and beverage festivals intended for the foodie tourist typology gain importance in terms of marketing strategy (Kline et al., 2015).

10.5 TRAVEL MOTIVATIONS AND DESTINATION PREFERENCES OF FOODIE TOURISTS

It is stated that foodie tourists have a great passion for food and drinks. Reading and learning new things about food and drinks, talking about food and drinks, tasting or cooking new food and drinks, eating delicious food at home or restaurant, growing your own products, etc. show that foodie tourists socialize more within this framework. This is a lifestyle for foodie tourists and affects their travel motivation and their preferred destinations (Ambrozas, 2003).

Especially big cities are preferred in the choice of destination and their source of motivation is to taste different local flavors. It is also seen that they travel frequently to taste the flavors of different cultures (Ambrozas, 2003; Karim and Chi, 2010). During the holiday, foodie tourists may both seek for familiar tastes for eating and drinking and may also be willing to try different tastes from their own familiar tastes (Karim and Chi, 2010).

Anderson et al. (2016) investigated the travel experiences of foodie tourists in their study. According to the results of the research, some differences have been reached in the travel preferences of foodie tourists. It is observed that foodie tourists, who are more experienced in eating and drinking and travel frequently, more prefer food and beverage festivals and cooking courses. It is stated that those who are less experienced, both travel less and prefer more common and fun trips that include natural and cultural heritage. Their common point is that their source of motivation is food and drink. In this context, many destinations give particular importance to culinary culture as a marketing tool and promote their destinations to appeal to foodie tourists as well.

Being a destination preferred by foodie tourists enable to increase the number of tourists. In order to increase the number of tourists, highlighting local cuisine culture, activities related to eating and drinking and local restaurants are among the strategies that gain importance for many destinations (Yozukmaz et al., 2017). At the same time, it is necessary to understand how tourists make their decisions about purchasing and/or consuming the products related to food and beverage and to have detailed information about their decision-making processes (Mitchell and Hall, 2003). Destination management organizations and tourism planners have discovered the interaction between tourism and food and beverage production and consumption and are striving to develop the national cuisine culture. For this, they contribute to the increase of tourist experiences by determining high quality destinations (Boyne et al., 2002).

The activities that foodie tourists will participate in are important in their destination preferences. Especially foodie tourists, in their destination, take place at least one or more of the activities such as, attending cooking classes in their destination, eating out for a unique and unforgettable experience, visiting markets where organic products are sold, participating in food and drink festivals, participating in wine-related tourist activities, etc. The findings show that the foodie tourists participating in these activities are young, educated and high-income earners. In addition, these people are motivated by unique experiences in their preferred destinations (Anderson et al., 2016).

Getz and Robinson (2014) state that consuming unique and authentic food, obtaining information about food and drink and culinary culture and socializing are included among the travel experiences of foodie tourists. These experiences are thought to be enhanced by natural and cultural elements, tours or shopping. In other words, foodie tourists may attend festivals in their destination, participate in wine tours, visit farms or vineyards, and buy food and beverages from organic markets. In their study, Björk and Kauppinen-Räisänen (2014) state that while foodie tourists make their choice of destination, they pay attention to a place where they can taste local, traditional, authentic, and yet-untasted food. According to their point of view, destinations having local food are seen as major tourist attraction centers. This situation positively affects other tourists as well as foodie tourists in their destination selection. The marvelous experiences they have had, the local delicacies they have tasted, as well as their own information about food, makes the travel of foodie tourists unforgettable. Therefore, foodie tourists tend to pay higher prices during their travels.

In the light of all this information, it is possible to say that there is a positive relationship between being a foodie and travel. In other words, tasting different flavors is an important travel motivation for foodie tourists and causes foodies to visit different destinations. This makes the destinations standing out with their culinary culture, an important center of attraction for foodie tourists.

10.6 CONCLUSION

Although eating is a need, in the postmodern society, the combination of travel motivation and gastronomy brings unique experiences for tourists. People who want to live in this unique experience, are fond of eating and drinking and passionate, are called foodies. Doing research about restaurants, meeting new chefs, learning the food and drinking culture of destinations, growing vegetables and fruits to eat, participating in the adventure of a meal from soil to plate are the attitudes and behaviors which are frequently performed by foodie tourists. In addition, as gastronomy is included in cultural tourism, tourists want to get to know the eating and drinking habits that are a part of the culture. Foodie tourists, which can be considered as a niche market in destination marketing, should actually be the focal point of touristic product diversification strategies focused on gastronomy tourism. For this reason, it is necessary to follow the foodie tourists closely, to know all their characteristics, to know their typologies and to learn about the activities they perform in a destination.

Foodie tourists traveling with motivation to eat and drink are people who enjoy knowing the history, recipe and story of the food (or drink) and also putting it into action. Some foodie tourists are simply fond of coffee, wine, or any famous drink and are motivated to travel just to try it. Besides foodies are keen on organic consumption, they advocate the sustainable eating and drinking movement and adopt slow food habits. For foodie tourists who don't hesitate to pay high prices for a meal or drink, pleasure is more important than money. They consider food as a work of art and enjoy so much keeping it visually. As well as socializing in the immediate surroundings, foodie tourists, who can socialize with festivals, act with the instinct of reaching other foodies via social media, showing their talents and competencies and sharing their experiences. They internalize certain foods or drinks and consider them as works of art that symbolize themselves. They are experts in matching and harmonizing meals within the menu. They are often selective as well as curious enough to try new flavors or even adventurous enough to try food made of insects.

Knowing the foodie tourist typology in the tourism industry will positively affect the decision-making processes of the destination management organizations, locally, regionally, and nationally, regarding tourism types, such as gastronomy tourism, agricultural tourism, wine tourism, etc. In addition, within the scope of gastrodiplomacy, promotion, and marketing activities related to the cuisine culture of a destination will increase the recognition of the destination. So, it will be ensured that the destinations standing out with their culinary culture are preferred by foodie tourists. It is thought that this situation will make many contributions to the relevant destinations in economic and socio-cultural terms.

KEYWORDS

- **foodie**
- **foodie tourist behavior**
- **foodie tourist typologies**
- **travel motivations**
- **destination preferences**

REFERENCES

Ambrozas, D. *Serious Feast: Vancouver Foodies in Globalized Consumer Society*. School of Communication, Simon Fraser University, Doctoral Dissertation, Canada, 2003.

Anderson, T. D.; Getz, D.; Vujicic, S.; Robinson, R. N.; Cavicchi, A. Preferred Travel Experiences of Foodies: An Application of Photo Elicitation. *J. Vacation Market.* **2016,** *22* (1), 55–67.

Björk, P.; Kauppinen-Räisänen, H. Culinary-gastronomic Tourism—A Search for Local Food Experiences. *Nutr. Food Sci.* **2014,** *44* (4), 294–309.

Björk, P.; Kauppinen-Räisänen, H. A Destination's Gastronomy as a Means for Holiday Well-being. *Br. Food J.* **2017,** *119* (7), 1578–1591.

Boyne, S.; Williams, F.; Hall, D. On the Trail of Regional Success: Tourism, Food Production and the Isle of Arran Taste Trail. *Tour. Gastron.* **2002,** 91–114.

Collins, K. Cooking Class: The Rise of the Foodie and the Role of Mass Media. In *History of Food*; Helstosky, C., Ed.; Routledge: New York, 2015; pp 1–25.

Definition of Foodie. https://dictionary.cambridge.org/tr/ (accessed 15 Sept 2020).

Deleuze, M. A New Craze for Food: Why Is Ireland Turning into A Foodie Nation?. In *'Tickling the Palate': Gastronomy in Irish Literature and Culture*; Iomaire, M. M. C., Maher, E., Eds.; Peter Lang: Oxford, 2014; pp 143–158.

Demirel, N.; Kırıcı Tekeli, E. Profesyonel Turist Rehberlerinin Foodie Turist Tipine Bakış Açıları: Nitel Bir Araştırma. *J. Tour. Theor. Res.* **2020,** *6* (1), 35–46.

DeWitt, D. *The Founding Foodies: How Washington, Jefferson, and Franklin Revolutionized American Cuisine*; Sourcebooks: Illinois, 2010.

Foodie Tanımı. https://tureng.com/tr/turkce-ingilizce/foodie (Erişim Tarihi: 15 Sept 2020).

Getz, D.; Robinson, R. N. S. "Foodies" and Their Travel Preferences. *Tour. Analys.* **2014,** *19*, 659–672.

Getz, D.; Andersson, T.; Vujicic, S.; Robinson, R. N. Food Events in Lifestyle and Travel. *Event Manage.* **2015,** *19* (3), 407–419.

Green, E. *A Study of Travelers' Foodie Activity Dimensions, Demographic Characteristics and Trip Behaviors.* The Faculty of Sustainable Tourism, East Carolina University, Doctoral Dissertation, America, 2013.

Green, E.; Kline, C.; Hao, H.; Crawford, A. Tourist Behavior among Foodie Activity Dimensions. *J. Gastron. Tour.* **2015,** *1* (1), 33–44.

Johnston, J.; Baumann, S. *Foodies: Democracy and Distinction in the Gourmet Foodscape*; Routledge: New York, 2015.

Karim, S. A.; Chi, C. G. Q. Culinary Tourism as A Destination Attraction: An Empirical Examination of Destinations' Food Image. *J. Hosp. Market. Manage.* **2010,** *19* (6), 531–555.

Kline, C. S.; Greenwood, J.; Joyner, L. Exploring Foodie Segmentation. *J. Tour. Insight* **2015,** *6* (1), 1–35.

Megido, R. C.; Haubruge, É.; Francis, F. Insects, the Next European Foodie Craze? In *Edible Insects in Sustainable Food Systems*; Halloran, A., Flore, E., Vantomme, P., Roos, N., Eds.; Springer: Cham, 2018; pp 353–361.

Mitchell, R.; ve Hall, C. M. Consuming Tourists: Food Tourism Consumer Behaviour. In *Food Tourism around the World: Development, Management and Markets*; Hall, C. M.,

Sharples, L., Mitchell, R., Macionis, N., Cambourne, N., Eds.; Butterworth-Heinemann: Burlington, 2003; pp 60–80.

Mohsen, M. G. Foodies in the UK: A Sense of Self, Connection and Belonging beyond the Passion? . In *Creating Marketing Magic and Innovative Future Marketing Trends*; Stieler, M., Ed.; Springer: New York, 2017; pp 457–467.

Oleschuk, M. Foodies of Color: Authenticity and Exoticism in Omnivorous Food Culture. *Cult. Sociol.* **2017,** *11* (2), 217–233.

Poole, S. Let's Start the Foodie Backlash. *The Guardian* **2012,** *10*, 12.

Robinson, R. N.; Getz, D. Food Enthusiasts and Tourism: Exploring Food Involvement Dimensions. *J. Hosp. Tour. Res.* **2016,** *40* (4), 432–455.

Sloan, A. E. The Foodie Phenomenon. *Food Technol.* **2013,** *67* (2), 18.

Sogari, G.; Menozzi, D.; Mora, C. Exploring Young Foodies' Knowledge and Attitude Regarding Entomophagy: A Qualitative Study in Italy. *Int. J. Gastron. Food Sci.* **2017,** *7*, 16–19.

Türkoğlu, H.; Akoğlan Kozak, M. Türk Mutfağının Gelişiminde Gurmelerin Rollerine Yönelik Algılamalar. *Anatolia: Turizm Araştırmaları Dergisi* **2015,** *26* (2), 207–220.

Vojnoviç, N. *Contradictions, Individualism, and Conformity in the Realm of Social Media An Ethnographic Approach to the Posting Practices of the Dutch Foodie Community*. Leiden University, Doctoral Dissertation, Holland, 2017.

Watson, P. J. *Grab Your Fork: A Netnographic Study of a Foodie Blog and Its Community.* Bournemouth University, Doctoral Dissertation, England, 2013.

Yeoman, I.; Beatte, U. The Future of Food Tourism. *J. Tour. Futures* **2016,** *2* (1), 95–98.

Yozukmaz, N.; Bekar, A.; Kılıç, B. A Conceptual Review of "Foodies" in Tourism. *J. Tour. Gastron. Stud.* **2017,** 170–179.

CHAPTER 11

IMPACT OF MOBILE APPLICATIONS AND INTERNET OF THINGS (IoT) IN HALAL TOURISM

AIMAN SIRAJ[1], FAUZIA ASHRAF SHAIKH[1], VINOD KUMAR SHUKLA[1,*], IPSEETA NANDA[2], and RITU PUNHANI[3]

[1]*Amity University Dubai, UAE*

[2]*K. L. University, India*

[3]*Amity University, Noida, Uttar Pradesh, India*

**Corresponding author. E-mail: vinodkumarshukla@gmail.com*

ABSTRACT

This chapter gives a detailed insight regarding Halal tourism. This is a field in the market, which is growing tremendously where Muslim travelers all around the world seek to travel places that meet their requirements, whether it may be food, the way they dress, and the way they follow their rituals. "Halal" translates to being permissible in accordance with the teachings of Islam and to refrain from everything "Haram." The concept of Halal and Haram is briefly explained in this chapter. Halal food and how it is different from the rest is elaborated. All the Haram substances are listed out. Halal tourism and its current application are reflected upon. Research states that the religion of a person is one of the main factors

Tourist Behavior: Past, Present, and Future. Narendra Kumar, PhD, Bruno Barbosa Sousa, PhD & Swati Sharma, PhD (Eds.)

behind his/her consumer ideology. As Islam is one of the most followed religions, there is an increase in interest in this sector. Now all kinds of tour activities should comprise social and cultural interactions that should take care of the hygiene and must not harm nature.

With the increasing income of middle-class society, many countries all around the world are looking forward to expecting a huge number of tourists in the upcoming years. Muslim countries all around the globe like Malaysia and Indonesia are among the leading tourist places among Muslim countries. Other countries include Singapore, Thailand, and the United Kingdom where the tourism sector is tremendously growing. The scope and challenges of Halal tourism are discussed. It is also important to understand that non-Muslim travelers might not be willing to adhere to any type of Halal tourism services. One should comprehend if they implement Halal tourism will the tourists then visit their place. A place where Halal cuisines, edibles, and beverages should be available at all times. Islamic clothing should be recognized and respected by all people.

This chapter also proposes the impact of the Internet of things and mobile applications' role in Halal tourism. Which countries can implement this ideology and which are already implementing it. The current popular Halal tourism applications available on the Android platform are reviewed and the common features have been found. We have proposed an application that included and integrate all popular features of top-rated Halal tourism-based mobile application and also proposed to add one more feature for the ease of the user. The features that can possibly be added are mentioned and also code for one of these features has been briefed up with the help of flowcharts. An explanation of how Quick Response codes work with respect to Halal tourism is also been explained and so are the details regarding using the Global Positioning system to know the historical location of the current location of the device.

11.1 INTRODUCTION

Halal is generally translated to "lawful" or something that is allowed or permitted. This word originates from the Arabic language. Haram on the other hand is the exact antonym of Halal, which is unlawful. Both of these words are related to edibles, hygiene products, medicines, maquillage, animal meat, and any other product, ingredient, or essential that is consumed by the human body in any manner. Halal foods need to abide by the rules of the Islamic Sharia (laws) and should be made from equipment/ machine that has been cleansed properly according to the Islamic guidelines. They should abstain from using any component that is prohibited. It should also be stored in utensils that are clean and uninvolved with any haram product.

Halal finance basically avoids the usage of taking up loans and interests. Halal media is a recently trending platform that serves information regarding all Halal aspects. Halal cosmetics use only those products that come from animals that are slaughtered as per the rules of Islam. Halal clothing is a modest form of clothing. Haram foods can contain a variety of substances such as alcohol and any other intoxicants. Rennet is considered to be haram and so is lipase (from animals). Pork or any other product obtained from pigs is haram. Gelatin from any source except fish is haram. Tallow, stock, lard, and l-cysteine are haram. Enzymes, unspecified meat broth, non-Halal animal fat, and animal shortening are a few other examples. Carnivorous animals such as lions and tigers, birds of prey and all other animals that fall under this category are all considered haram. Also, any kind of food contaminated or mixed with any of the above-mentioned examples is also to be considered unlawful. It is very important for a Muslim to not eat or drink something Haram as it invalidates their worship for the next forty days. Also taking interest or loan is not allowed in Islam. As a result, anything bought with that money is also considered to be unlawful.

Now cursing and not praying on time and basically any activity that falls under the category of "sinning" is considered haram. Listening to music, indulging with the opposite gender, or committing a crime is haram. For men, any gold adornments and silk products are considered haram. Worshipping idols or anything besides God is forbidden. Wasting any resources (such as water) unnecessarily is also to be considered unlawful. Although they are plenty of other activities that are considered to be haram, these are some of the major ones. It is important for a Muslim to follow what's lawful in order to gain rewards from God.

11.2 HALAL TOURISM

This is a new concept of tourism that is focused on Muslim families and is set up according to Islamic guidelines. All the hotels and restaurants in these places do not have any kind of liquor. They have separate facilities for swimming/spa for males and females. Countries such as Turkey, Indonesia, Malaysia, and many others are doing their best in order to attract the attention of travelers who belong to Muslim families from all around the world. As of now, there aren't any internationally recognized standards regarding this though. The main idea is to offer religious convenience to such travelers. Also, this industry provides flights where only Halal food is served and they abstain from serving pork or alcohol. They also announce the timings of prayers. Various religious documentaries and podcasts are also offered to entertain people on board. So all the facilities that are offered make sure to ensure that the traveler is at their comfort level at all times (Samori et al., 2016).

Research states that a religion of a person is one of the main factors behind his/her consumer ideology. Since a lot of tourists today travel with Islamic expectations these tourists products are being created. As Islam is one of the most followed religions there is an increase in interest in this sector. Now as per research classifications there are three different kinds of concepts related to Halal hotels. The first one corresponds to those hotels where there are only separate services for men and women and do not serve alcohol. They are also called "dry" hotels. The second one in addition to this has mosques in the hotels and Qur'an and mats to pray in the hotel rooms. These are called partial Halal hotels. The third one is where all the facilities and infrastructure are designed according to the teachings of Islam and they also include Halal financing facilities. Although these are not internationally verified there is still an ongoing discussion regarding it (Ahmat et al., 2012).

This particular field of tourism is growing really fast. Many countries, travel agencies, hotels, resorts and airlines are really keen regarding this matter. Since the beginning of 2017, the number of travelers who are Muslims has increased by 30%. MasterCard and Crescent Rating is a research group that analyzes Halal-friendly tourism and has found that projects related to Halal tourism in the next decade will shoot up and contribute to the world's economy to $300 billion from $190 billion. This is very important in the demographics of the worldwide travel and tour scenario. So investing in such a business is considered to be very profitable. This is why this concept is receiving such high importance as it is highly in demand. This is also possible as the young generation is much more demanding and educated regarding their needs.

11.2.1 SCOPE OF HALAL TOURISM

All of the details about Halal tourism are given to Muslim travelers by the industrialists, policymakers, marketers, and economists who are responsible for developing Halal tourism and its facilities and infrastructure. All the travel activities and travel packages also come under this category which also assists to understand and help them develop particular ideas for marketers in order to boost communication with tourists all over the world. These people are the decision-makers in constituting and forming the backbone of the industry. Now the size of the Muslim population does not make us comprehend how much do they actually follow their religion and how active they are in terms of their religion but despite that Muslim customers constitute a major role in the rapidly developing market segments (Crouch, 2006).

This then becomes a challenge to evaluate the needs of the customers and to analyze what they would prefer. So it is important to know that what they want is to be considered really important. The state of the Global Islamic Economy report which is produced by Thomas R who has combined his work with the Dinar standard, the worldwide Muslim market travel was having a worth of over $200 billion by 2020 which represents around 14% of global expenditure and the same report states that by 2025 these values will increase tremendously. The leading three nations that lead this particular field's market are Malaysia, the United Arab Emirates and Singapore, and their main focus of attention is to be family-friendly and prove to be a great host for Muslim families all across the world (Anon, n.d.).

It is obvious to realize that there is going to be a lot of feud between the saturated marketers and the destination marketplace and that it then becomes extremely vital to understand every single factor that affects the behavior of a Muslim traveler. One should comprehend if they implement Halal tourism will the tourists then visit their place. If not then it would be foolish to carry on with it. Also, a crucial point is that if everyone would want the Halal tourism practices to be conducted or just a few people. The point being if any organization or destination experiences huge success in the Halal tourism market it will be a great encouragement for all the other marketers out there to try to get in the business. For instance destinations such as Australia and places such as Queensland can be great Muslim-friendly tourist spots. Customizing Halal travel packages is something the travel agencies are following recently where they make sure that Halal food is available, prayer timings are taken care of in the schedule, and people are taken to the mosque during that particular time and there can be Muslim tour guides.

The tourism industry has a lot of potential in areas such as Halal medical tourism, Muslim heritage souvenirs, Halal services, luxury services, business-oriented travels, and family friendly travels. There can be multiple training programs offered to Non-Muslim staff to understand in depth what all happens in this field. All in all, we need to comprehend that the scope of Halal tourism is limitless when propagated appropriately (Sharpley, 2009).

11.2.2 CHALLENGES FACED BY HALAL TOURISM

The main challenge that Halal tourism faces is that although the facilities can satisfy both Non-Muslims and Muslim travelers and proves to be an amazing business opportunity, not all Non-Muslim tourists would comply with the services. As an appropriate example would be that alcoholic drinks are not available in Halal hotels. This would not be liked by many tourists. The beaches, massage centers, and facilities are separated for males and females that would result in people not being willing to travel to these particular destinations due to all these restrictions. This is one of the main reasons that people feel discouraged to travel to Muslim countries that follow these Islamic guidelines sternly like Saudi Arabia. Wearing really short dresses such as swimsuits is not acceptable and can be viewed as a constraint by travelers though some of them can view it as a new experience (Rhama, 2017).

Separation of gender facilities can be applied if the motive is not to have a beach vacation. Maybe some non-Muslims would not like to consume Halal meat due to the way it is slaughtered and some of them might want to eat pork like they do in the home country. Only if they are bound to try to replace it with Halal meat or fish or any vegetarian cuisine then Halal tourism would be the right choice for them. However, we can come up with unique solutions like replacing and providing similar beverages without alcohol and juices or tea. We can motivate people to try other cultural cuisines and learn about the customs and traditions of that particular place (Rhama, 2017).

11.2.3 IMPLEMENTATION OF HALAL TOURISM

There are several characteristics of Halal tourism that are proposed as a tool to analyze which destinations can implement Halal tourism. Also, it is important that a place does not lose its historical identity while executing Halal tourism. Halal tourism is implemented by following all of the rules of the Islamic Sharia, including diversion between the genders and facilities to pray. Now all kinds of tour activities should comprise social and cultural interactions should take care of the hygiene and must not harm the nature. It should be the utmost interest to take care of the tourist's well-being and safety so any kind of activity that harms them or is unsecure should not take place. The end goal should be focused on the customer's satisfaction and demands (Ulfy et al., 2021).

Halal cuisines, edibles, and beverages should be available at all times. Islamic clothing should be recognized and respected by all people. The finance system should be Sharia based and should offer services for zakat provisions. Now for the holy month of Ramadan, meals should be provided for both Iftar and Suhoor that is before and after fasting time. In no case should Muslims be encouraged to break their fasts? There should be proper facilities for the Taraweeh (special) prayers. So basically a lot of care should be taken while implementing Halal tourism (Battour et al. 2010).

Countries like Malaysia, Turkey, Singapore, The Kingdom of Saudi Arabia, Indonesia, The United Arab Emirates, Japan, Taiwan, and others are implementing Halal tourism.

Countries/places where this can be implemented

- All Muslim countries can implement it.
- All Muslim-friendly countries such as Bosnia & Herzegovina, Iran, Maldives, Morocco, Kyrgyzstan, Uzbekistan, Jordan, Kazakhstan, and Sri Lanka have a chance to implement it.
- Also Muslim-friendly cities such as Hong Kong, Cape Town (South Africa), Kazan (Russia), Xi'an (China), Zanzibar (Tanzania), and Berlin & Munich (Germany) are places that can practice Halal tourism.
- There are a few Muslim-friendly islands such as Rawa Island, Tengah Island, Lankayan Island, Mataking Island, Ko Lipe, Unguja, Ko Yao Yai, Lang Tengah Island, Gemia Island, Nami Island, Boheydulang Island, Pulau Tiga, Ko Phra Thong, and Mabul Island which can be a great tourist attraction.

11.3 MOBILE APPLICATION BASED REVIEW

Following application-based review which is based on the download (Table 11.1) have been considered in this chapter to review (Download is as per Google Play Store, accessed till 2020).

TABLE 11.1 Popular Application Based on Halal Tourism.

Name	Rating (5)	Reviews	Size	Downloads
B-Halal Trip	4.8	6	24 MB	100+
Muslim Directory: Masjid, Halal	4.6	270	16 MB	10K+
Halal Gourmet Japan	4.4	195	10 MB	10K+
Halal Travel Hotels Tours Flight E-	4	0	20 MB	100+
Number Quran				
Halal Trip	3.9	2223	26 MB	100K+

11.3.1 HALAL TRIP

To use this application anyone first needs to enter a booking code. The features of this application are:

- Itinerary overview
- Itinerary details
- Qibla directions
- Location of Halal restaurants
- Reading and listening to Qur'an (Anon, n.d.).

11.3.2 MUSLIM DIRECTORY: MASJID, HALAL

Muslim directory app is the most accurate and comprehensive online listing of Masjids, Halal Restaurants, Halal Grocery Stores, Halal Markets, Islamic books, Clothing, Tax services, Dental offices, Travel Agencies (Hajj/Umrah), and many more Muslim businesses. You can also get accurate prayer timings worldwide. Find your closest Masjids and Halal restaurants and get directions.

Top reasons to download this application

- It simply has no ads at all and it's free.
- Helps you go paperless for prayer time schedule.
- Most accurate and comprehensive listing of masjids, Halal restaurants, Halal grocery stores, Halal markets, Islamic books and all professional Muslim businesses.
- Get Islamic prayer times, adhan notifications, Hijri calendar, Qibla.
- Know your local Muslim community and support (Anon, n.d.).

11.3.3 HALAL GOURMET JAPAN

It is a smartphone app from where you can receive essential details about Halal restaurants nearby.

The main facilities that this application provides are:

- You can browse the catalog of the stores and see their menu
- You can view the customer reviews
- It doesn't have any kind of ads
- You get the most accurate details and are safe from any hustling to find the perfect restaurant for yourself (Anon, n.d.).

11.3.4 HALAL TRAVEL HOTELS TOURS FLIGHT E-NUMBER QURAN

This application has a lot of advantages as mentioned below:

- Booking flights
- Phrases in different languages
- Prayer timings
- Qur'an and its translation
- E-Codes for Halal and Haram food
- Tours
- Cruises
- Hotels
- Reviews (Anon n.d.)

11.3.5 HALAL TRIP

This is the most used application of Halal tourism on the play store. It provides:

- Prayer timings
- Qibla directions
- Duas everyday
- Prayer analytics (What time have you prayed and how regular are you praying)
- Travel Etiquettes in Islam
- Travel Articles
- Details regarding Halal restaurants nearby
- Location of mosques
- It also lets you share your reviews (Anon n.d.)

11.3.6 COMMON FEATURES AVAILABLE

The most common features of these applications are:

- Displaying Prayer timings
- Qibla directions
- Halal food location
- Mosques location
- Qur'an and its translation

While it is important to note that each application has multiple different applications and has more benefits.

11.3.7 PROPOSED NEW FEATURES

It is suggested that we create an application (Fig.11.1) that includes all the features from the above mentioned five applications and in addition have more extra features which can include the following:

- Halal financing mode—Halal finance basically avoids the usage of taking up loans and interests. It includes zero percent interest, no loans, and also a zakat calculator. We can add features to locate Islamic banks nearby and also a feature to calculate zakat.

- Halal clothing stores location—We can display the data of nearby stores and their location based on the current GPS location of the device.
- QR scanner—We can add a QR code scanner to detect whether a packaged food is Halal or Haram.
- Ramadan mode—We can develop a Ramadan mode in the application where the Iftar and Suhoor timings are given and we can also get a notification or an alarm during these times.
- We can display an ayah or a hadith every day and track the Qur'an read every day.
- GPS locator—We can use the current GPS location of the device and give out the historic outline of that particular area, city, and country.
- Language Translator—It is really important to include a language translator as people are constantly traveling during a vacation and might encounter a situation where they might not be able to understand others. This will prove to be a great savior for such a situation.

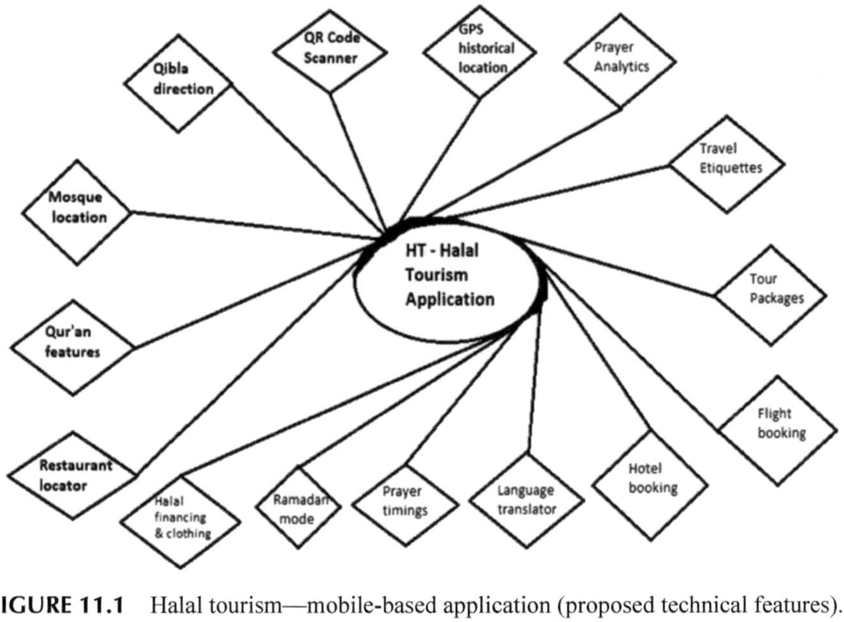

FIGURE 11.1 Halal tourism—mobile-based application (proposed technical features).

11.4 QR CODE AND SCANNERS

The quick response (QR) code was inspired by a similar product called a barcode. You can use any device to analyze and interpret the code which will then give you the necessary details about the product. This is also optically labialized and carries all details about a particular item where an item can also be an application or a web server and can track or locate it and identify all of its information. This may use numeric/alphanumeric or digital data to process the information and extend it to the device's storage in an efficient manner. You can manage all sorts of documents, identify all your products and relate to their data. It will help you track the time required to carry out the necessary information. It also proves to be a great tool to showcase your product and its design in the market. When you compare it to the previous version also called barcode you will notice that it is much faster and efficient and can store a larger amount of data. The format in which this is represented is that there is a white background which is essential so as to improve the readability of the content. There are tiny black colored squares which when scanned using an appropriate reader will provide the necessary information. Using these black squares the figures are interpreted and the different markings are found out from which the data is removed. Thus these scanners will then redirect you to an application or website or any other server where you will find the required information (Mary et al., 2016).

With respect to Halal tourism, the scanners will be able to detect whether a particular food item is Halal or not. See whether the brand name is from a reputed company or whether any ingredient used falls under the category of haram.

11.4.1 GLOBAL POSITIONING SYSTEMS

These systems have tracking units that can be carried around by any device, objects, or even humans. The systems will locate these units and interpret their current location. This is done using the object's movement which is analyzed by the systems. One can either store this data inside these units or can export it to any application or web server. Even if the device or object is moving one can see the timely updates of their location which you can view on any device that has internet enabled in it. Any smart device that has a feature of GPS can then locate this data (Mary et al., 2016).

In the Halal tourism application, we can make use of the current location of the device and display its historical outline which is basically all data reflecting back to a thousand years, the customs, the culture, the traditions, and the clothing. This will let the travelers get a deeper insight into that particular location and improve their travelling experience (Woodside and Dubelaar 2002).

11.4.2 CODE FOR THE LANGUAGE TRANSLATOR (TEXT TO SPEECH)

To code this Language translator we will use Microsoft Azure API (It supports LUIS) as the backend interface and connect it with android studio. Microsoft language translator supports at least 70 languages. The programming languages that we are going to use are C# and Java (Figs. 11.2 and 11 .3).

- Install the "microsoft-translator-java-api-0.6.1-jar-with-dependenciesjar" file
- Register yourself and you will receive a Client ID and a Client secret
- Update sting.xml with the code below and enter it and save it

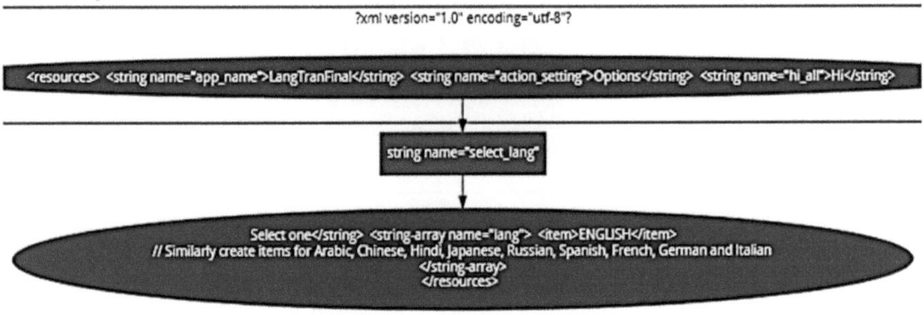

FIGURE 11.2 Text to speech conversion.

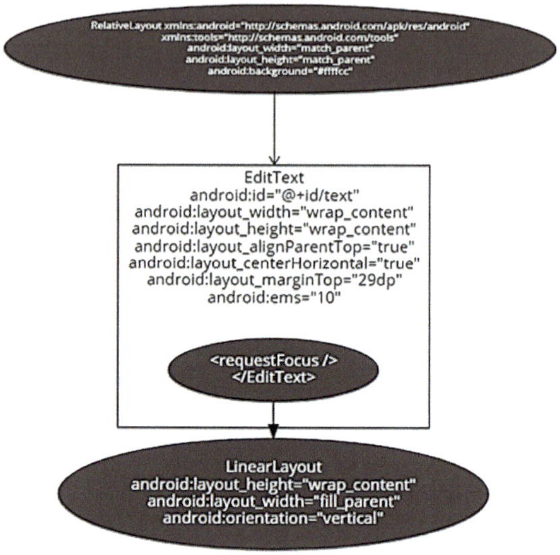

FIGURE 11.3 Text to speech.

- Find the file name main_activity and enter the code below for relative layout, text editing, and linear layout you can choose any dimensions you want (Fig. 11.4).

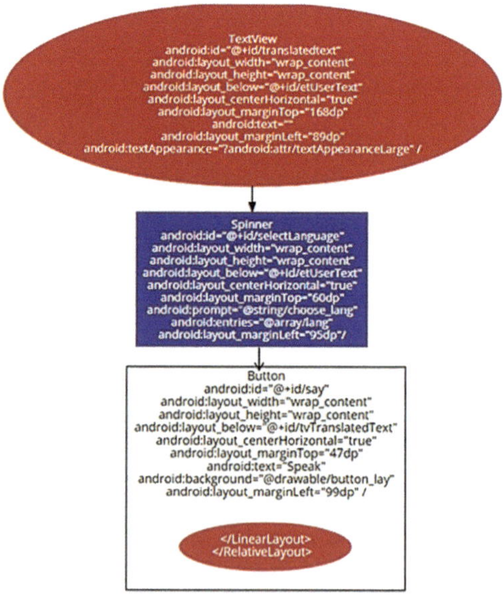

FIGURE 11.4 Layout, text editing, and liner layout.

- Then add the details regarding the text view, spinner, and button and customize it to your liking (Fig. 11.5).

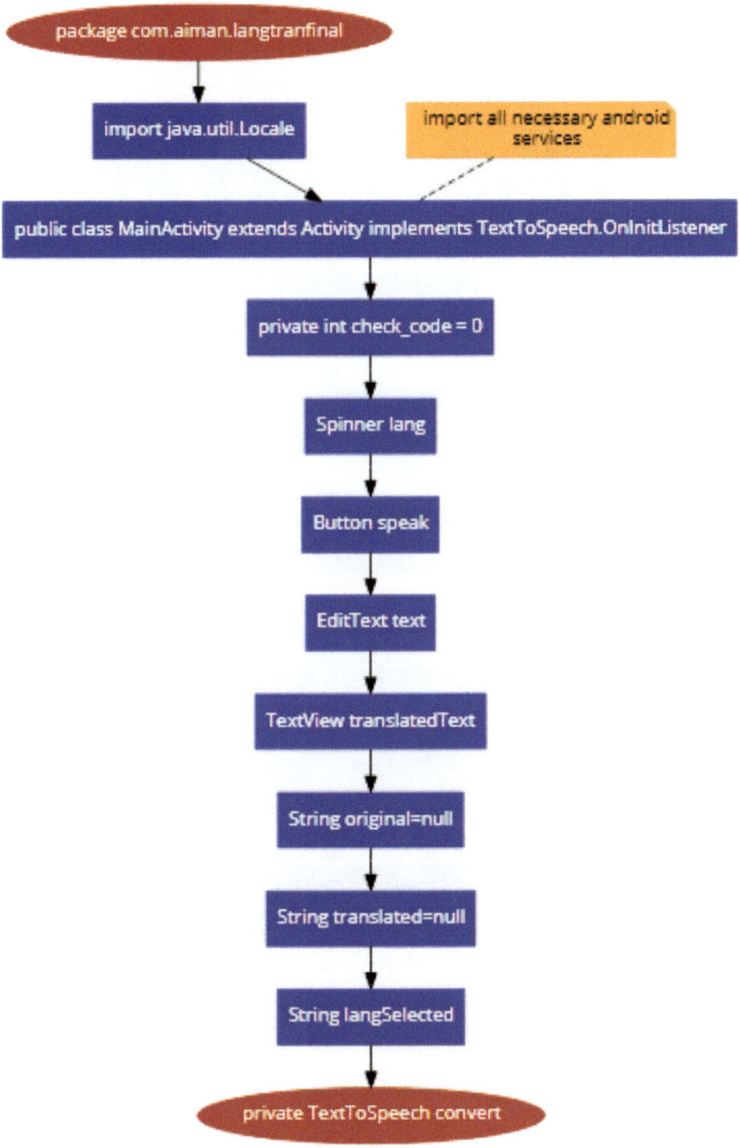

FIGURE 11.5 Text view, spinner, and customization.

- Then find main_activity and enter the code for the following into it (Fig. 11.6).

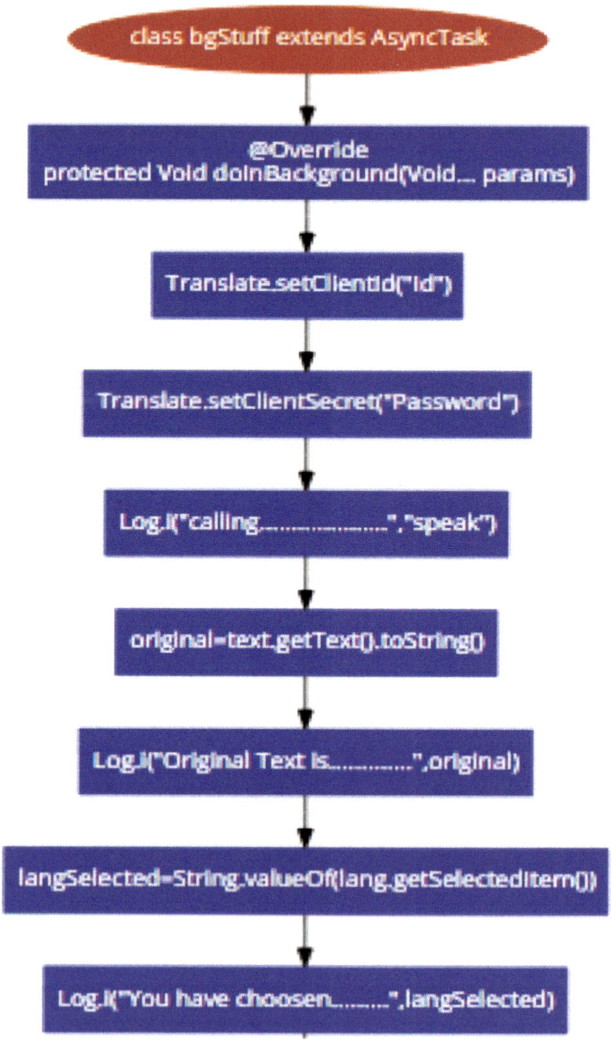

FIGURE 11.6 Flow of application.

- Then enter the code to select a particular language as shown here (Fig. 11.7).

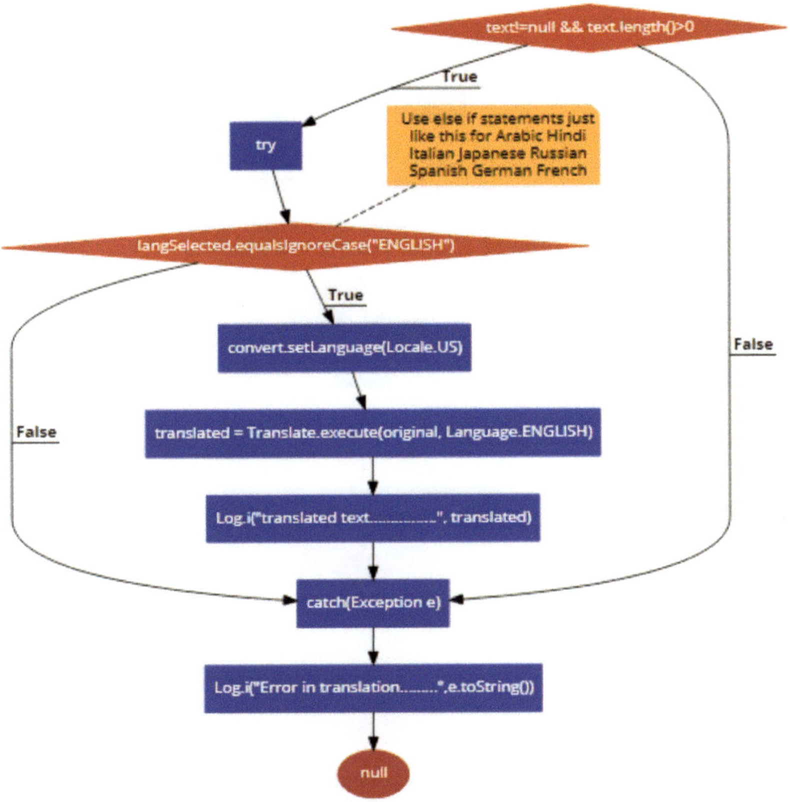

FIGURE 11.7 Code flow of the application.

- Use if and else if blocks to define languages from the Microsoft library (Fig. 11.8).
- Write a code to convert the text to speech (Fig. 11.9).
- Then write a code block to check if your text to speech engine is working (Fig.11.10).
- You then need to enter your client id and client secret when you registered yourself or else the code will not work.
- You need to then give internet permissions to the application by entering the following code inside the AndroidManifest.xml file (Fig. 11.11).
- You can then open Microsoft Azure and create a mobile service and run it on any android medium.

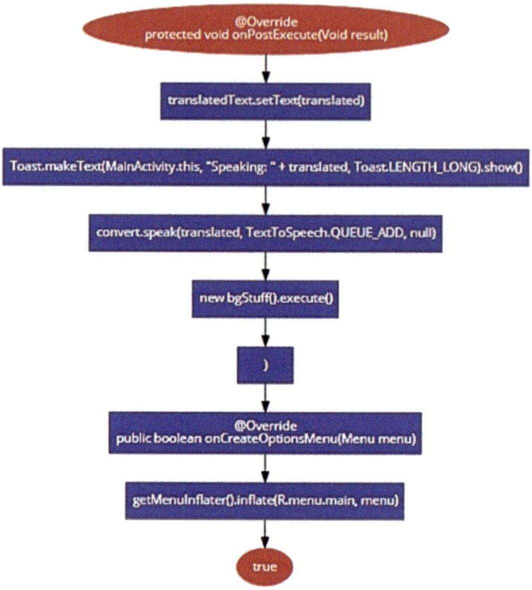

FIGURE 11.8 Multiple application an option flow.

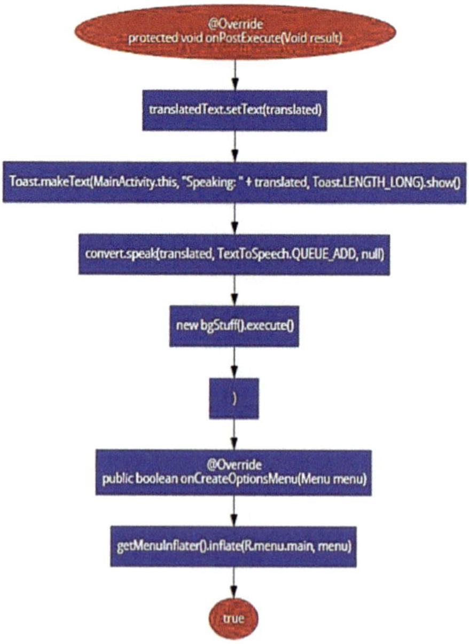

FIGURE 11.9 Text to speech.

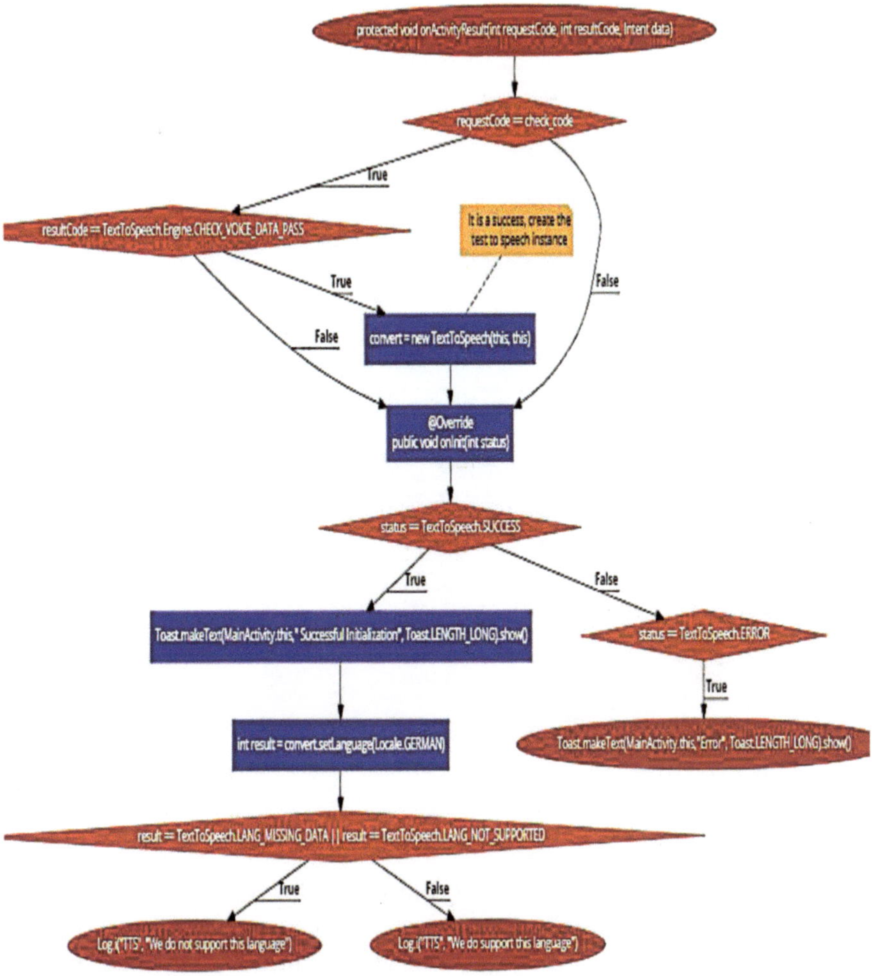

FIGURE 11.10 Validating text to speech.

11.5 THE CONTRIBUTION OF THE INTERNET OF THINGS IN HALAL TOURISM

There are a lot of ways (Fig. 11.12) in which the Internet of things can make tourism more travel friendly such as:

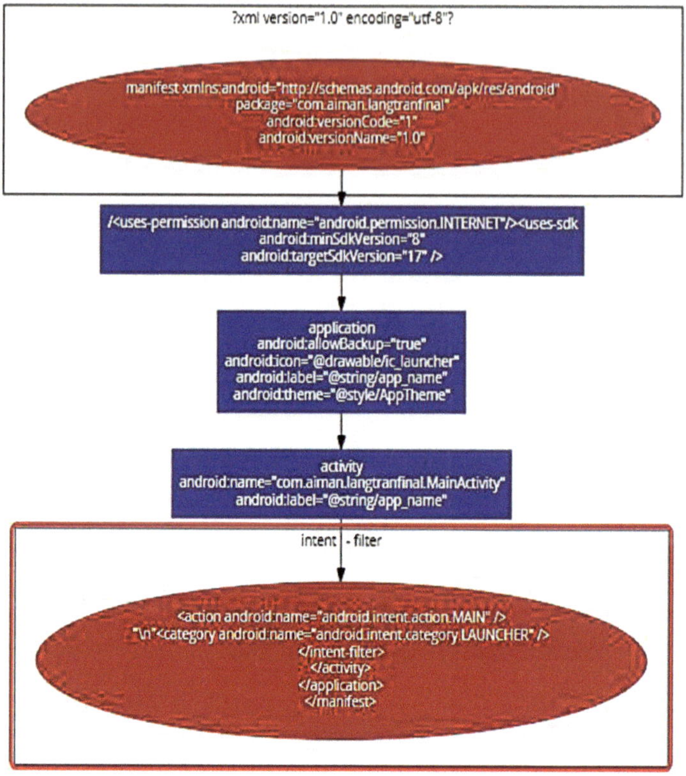

FIGURE 11.11 Managing application.

- Giving the tourists personal control—They can have a single device that can be used to control the room's temperature, lights, access to services, television by making their internet enabled so they can adjust things on their own and be more comfortable.
- Making the travel seamless—The hotel can send the tourist an electronic key card on their phone so they don't have to waste their time in the check-in process.
- Baggage arrival notification—Airports can use sensors and alert users if their baggage has arrived and sent its location to their phone saving a lot of time.
- Check for passengers—People can have RFID tag cards with them at the airport and nearby sensing devices can sense their location and send it to the central server.

- Energy Saving—The lights, heating can only be activated when a device is present in a room indicating the presence of people.
- Send information about particular locations—The tourism industry can send information to the user about the most famous places and their opening timings.
- Repair and maintenance—Any repair, for example, changing of a light bulb or any electrical issues can be reported to the hotel staff immediately that is receiving real-time information and updating it at the same time.
- Airline services—Internet of things can allow airlines to repair the parts from time to time, fuel more efficiently, and help minimize the effort and resources, and at the same time maximize and maintain the safety of the passengers.
- Achieve customer satisfaction—This can be done by conducting timely surveys in order to better a customer's experience.
- QR scanner—This can help a user scan the packaged food items and comprehend whether it is Halal or not.
- Using GPS location to its maximum—The location of a person can be found using GPS and then the facilities available nearby can be shown. Also, the historic identity of the place can also be given (Fajriyati et al. 2020).

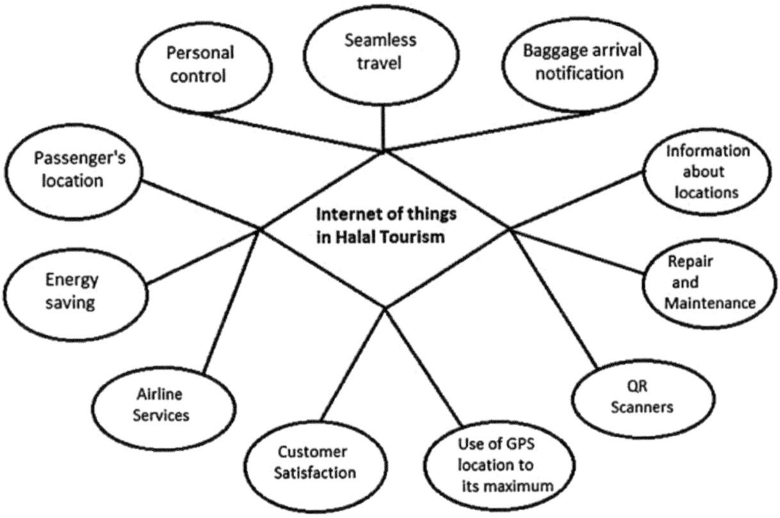

FIGURE 11.12 IoT and halal tourism.

11.6 CONCLUSION

The objective of this chapter was to highlight the importance and need for Halal tourism. This chapter covers these frequently asked questions—What is Halal? What is haram? What is Halal tourism? Which places practice and allow Halal tourism? Which is the best place for Halal tourism? How is it being implemented? What countries/places are implementing it? Which countries/places can implement it?

It is one among the recent developments that have been brought by the expansion of this market segment. With the progress of Halal within this field, countries all around the globe can try to advertise and capture the attention of these travelers by providing every kind of facility that's possible and helping them attain rewards and strengthen their faith. It is obvious to realize that there is going to be a lot of feud between the saturated marketers and the destination market place and that it then becomes extremely vital to understand every single factor that effect the behavior of a Muslim traveler. We see an increase in the revenue of Halal tourism from an industry and research standpoint. Halal tourism is usually abbreviated to evolve to Islamic teachings with anything or process which will be used or incorporated into the tourism industry. Halal foods, forbidden foods, Halal clothing with the help of Muslim guidelines, its role in Halal tourism, and Scope of Halal Tourism Challenges faced by Halal tourism have been briefly discussed. A review of the top five applications of Halal tourism in the Android Play Store is done and the common features in these applications are mentioned and suggested features that can be added are listed. Also what is the contribution of the Internet of Things in Halal Tourism is explained. It provides several samples of best practices around the world. We need to ensure that the customer's ideology is understood to achieve maximum success economically. We can surely hope and increase in demand in Halal tourism in the future and more facilities to solve any kind of problems faced and to ensure a safe, comfortable, and luxurious tour for every traveler.

KEYWORDS

- **foodie**
- **foodie tourist behavior**
- **foodie tourist typologies**
- **travel motivations**
- **destination preferences**

REFERENCES

Ahmat, N. C.; Ridzuan, A. H. A.; Zahari, M. S. S. Dry Hotel and Syariah Compliant Practices: Concepts, Challenges and Reality in Malaysia. In *ICIMTR 2012——2012 International Conference on Innovation, Management and Technology Research*, 2012.

Anon. Google Play, n.d.a. https://play.google.com/store?utm_source= emea_Med&utm_medium=hasem&utm_content=Oct1515&utm_campaign=Evergreen&pcampaignid=MKT-DR-emea-ae-all-Med-hasem-py- Evergreen-Dec0215-1 BKWS%7CONSEM_kwid_43700009235461022&gref=EkQKPAoICPDntvcFEGUSLADUibK2PGwXn49W0ucsXwfQO5E7ys3sRjqgAqFIikx_GXzg_zff99ljjvVUGgJefhAC8P8HARjj9tyaAw&gclid=CjwKCAjw57b3BRBlEiwA1ImytjxsF5-PVtLnLF8H0DuRO8rN7EY6oAKhSIpMfxl84P833_fZY471VBoCXn4QAvD_BwE&gclsrc=aw.ds (accessed 13 Mar 2021).

Anon. *State of the Global Islamic Economy 2013 Report*, n.d.b.

Battour, M. M.; Ismail, M. N.; Battor, M. Toward a Halal Tourism Market. *Tour. Analy.* **2010,** *15* (4), 461–70.

Crouch, D. Tourism, Consumption and Rurality. In *Handbook of Rural Studies*, 2006.

Fajriyati, I.; Afiff, A. Z.; Gayatri, G.; Hati, S. R. H. Generic and Islamic Attributes for Non-Muslim Majority Destinations: Application of the Three-Factor Theory of Customer Satisfaction. *Heliyon* **2020,** *6* (6), e04324.

Mary, M.; Rani, S.; Euphrasia, K. R. Data Security through QR Code Encryption and Steganography. *Adv. Comput. Int. J. (ACIJ)* **2016,** *7* (1).

Rhama, B. The Implications of the Values and Behaviours of Actors for Ecotourism Policy: A Case Study of Sebangau National Park, Central Kalimantan, Indonesia, 2017.

Samori, Z.; Salleh, N. Z. M.; Khalid, M. M. Current Trends on Halal Tourism: Cases on Selected Asian Countries. *Tour. Manage. Perspect* **2016**.

Sharpley, R. Tourism, Religion, and Spirituality." In *The Sage Handbook of Tourism Studies*; SAGE Publications Ltd: London, 2009; pp 237–254.

Ulfy, M. A.; Haque, A.; Karim, M. W.; Hossin, M. S.; Huda, M. N. Tourists Behavioral Intention to Visit Halal Tourism Destination: An Empirical Study on Muslim Tourists in Malaysia, 2021.

Woodside, A. G.; Dubelaar, C. A General Theory of Tourism Consumption Systems: A Conceptual Framework and an Empirical Exploration. *J. Travel Res.* **2002,** *41* (2), 120–132.

CHAPTER 12

A STUDY ON PERCEPTION AND REVISITING BEHAVIOR OF TOURISTS TOWARD HERITAGE TOURISM DESTINATIONS IN SIVAGANGA DISTRICT, TAMIL NADU, INDIA

A. APSARA SALETH MARY*

Centre for Tourism and Hotel Management, Madurai Kamaraj University, Madurai, India

*E-mail: *prof.apsara@gmail.com*

ABSTRACT

Heritage tourism provides opportunities for tourists to know to realize and enjoy unique features of the heritage and cultural values of those places. The exclusive features and information about heritage tourism destinations are remarkably influencing the perception of tourists on those destinations. More than two-fifth of tourists have moderate level of perception of heritage tourism destinations. A significant difference is prevailing among the profile of tourists and their perception of heritage tourism destinations. Perception of tourists on heritage tourism destinations is highly, significantly, and positively related to their revisiting behavior. To enhance the perception of tourists on heritage tourism destinations, heritage tourism destinations should keep their natural beauties and signboards must be put in appropriate positions in tourism destinations. In addition, heritage

Tourist Behavior: Past, Present, and Future. Narendra Kumar, PhD, Bruno Barbosa Sousa, PhD & Swati Sharma, PhD (Eds.)

tourism destinations should have the best hotels to offer superior quality of services to tourists and these measures would intensify revisiting behavior among tourists.

12.1 INTRODUCTION

Tourism has a major role in the development of places and people across the globe, and it is highly important for the improvement of cultural, social, and economic dimensions of people in India. In recent times, heritage tourism is emerging as one of the potential areas of tourism not only in India but also across the world. Heritage tourism is the activity to discover places that have rich cultural values and interesting historical events and it is a more striking and learning activity (Gohil, 2015). Heritage tourism provides opportunities for tourists to know, realize and enjoy unique features of heritage and cultural values of those places (Hasan and Jobaid, 2014; Surebankar and Hadli, 2010). The exclusive features and information about heritage tourism destinations are remarkably influencing the perception of tourists on those destinations. The perception of tourists about heritage tourism destinations is affecting their revisiting behavior (Rani, 2017). Therefore, it is important to study the perception and revisiting behavior of tourists toward heritage tourism destinations in the Sivaganga district.

12.2 REVIEW OF LITERATURE

Chheang (2011) found that tourists had a positive perception of cultural and heritage tour places and also of culture and cordial relation with local communities. Rahman (2012) concluded that reputation, service available, and satisfaction of tourists had a significant impact on their perception.

Rajesh (2013) revealed that cultural values, nature, accessibility, amenities, and price were generating positive perceptions on heritage tourism among tourists, and it affected their revisiting behavior. Kariru and Aloo (2014) showed that festivals, traditional food items, values of culture, amenities, and natural environment were significant features of heritage tourism places, and it affected the revisit intent of tourists.

Benur and Bramwell (2015) indicated that local culture and arts, values of culture, specific features and amenities were generating positive perceptions for heritage tourism among tourists. Praveen and Priya (2016) stated that natural beauty, transport and accommodation facilities, clean environment, and heritage values were affecting the revisit behavior of tourists.

Zeng (2017) found that nature and personal experience of culture and heritage were affecting the reputation of heritage destinations among tourists and also their revisit intends. Yeniasır and Gokbulut (2018) concluded that local communities had a positive perception of heritage tourism and specific features of destinations were influencing their perception about them.

12.2.1 RESEARCH GAP AND RESEARCH PROBLEM

Through the review of earlier studies, the research gaps are discovered and included in the present study, and it is the perception of tourists on heritage tourism destinations and its relation with their revisiting behavior of tourists. Heritage tourism destinations make attempts to preserve the heritage and cultural values, and it provides a happy and memorable experience to tourists. Promotions and programs are adopted to keep the heritage and mitigate negative impacts and improve the reputation of those destinations. Perceptions of tourists are highly important for sustainable development of heritage tourism, and it is highly vibrant and significantly influencing their revisit behavior.

12.2.2 OBJECTIVES OF THE STUDY

1. To study the perception of tourists on heritage tourism destinations
2. To scrutinize differences among profiles of tourists and their perception of heritage tourism destinations
3. To examine the relation among perception of heritage tourism destinations and revisiting behavior of tourists

12.2.3 HYPOTHESES OF THE STUDY

1. There is no significant difference in perception of heritage tourism destinations among profiles of tourists.

2. There is no significant relationship between perception of heritage tourism destinations and revisiting behavior of tourists.

12.3 RESEARCH METHODOLOGY

The present study is conducted in the Sivaganga district. The convenience sampling method is used to select tourists, and data are gathered from 300 tourists by using a structured questionnaire. Percentages are computed to know the profile of tourists and mean and standard deviation are calculated to understand the agreement level of tourists on their perception of heritage tourism destinations. *t*-test and Analysis of variance (ANOVA) test are used to scrutinize differences among profiles of tourists and their perception of heritage tourism destinations. Simple correlation analysis is applied to study the relation among perception of heritage tourism destinations and revisiting behavior of tourists.

12.4 FINDINGS

12.4.1 PROFILE OF TOURISTS

The profile of tourists is given in Table 12.1. The findings clarify that 54.67% of tourists are domestic tourists, whilst, 45.33% of them are international tourists and 57.67% of them are males, whilst, 42.33% of them are females. The findings explain that 34.00% of them are coming under the age category of 31–35 years, whilst, 12.00% of them are coming under the age category of above 40 years and 32.00% of them are possessing under graduation, whilst, 16.00% of them are possessing higher secondary. The findings disclose that 36.67% of them are getting monthly income of ₹30,001–40,000, whilst, 13.33% of them are getting monthly income of more than ₹50,000 and 74.33% of them are married, while, 25.67% of them are unmarried.

TABLE 12.1 Profile of Tourists.

Profile	Number of tourists	Percentage
Tourist type		
Domestic	164	54.67

TABLE 12.1 *(Continued)*

Profile	Number of tourists	Percentage
International	136	45.33
Gender		
Male	173	57.67
Female	127	42.33
Age category		
Below 25 years	44	14.67
26–30 years	71	23.66
31–35 years	102	34.00
36–40 years	47	15.67
Above 40 years	36	12.00
Education		
Higher secondary	48	16.00
Diploma	75	25.00
Under Graduation	96	32.00
Post Graduation	81	27.00
Monthly income		
Less than ₹30,000	52	17.33
₹30,001–40,000	110	36.67
₹40,001–50,000	98	32.67
More than ₹50,000	40	13.33
Marital status		
Married	223	74.33
Unmarried	77	25.67

12.4.2 PERCEPTION OF TOURISTS ON HERITAGE TOURISM DESTINATIONS

The perception of tourists on heritage tourism destinations is given in Table 12.2.

TABLE 12.2 Perception of Tourists on Heritage Tourism Destinations.

Perception of heritage tourism destinations	Mean	Standard deviation
Heritage tourism destinations explain local culture	3.92	0.93
Heritage tourism destinations have cultural uniqueness	3.88	0.99
Traditional foods are available in heritage tourism destinations	3.90	0.95
Conventional handicrafts are obtainable from heritage tourism destinations	3.86	1.04
Heritage tourism destinations possess undisturbed natural beauty	3.35	1.08
Heritage tourism destinations have a pleasurable environment	3.80	1.09
Access to heritage tourism destinations is very easy	3.84	1.07
Signboards are well placed in heritage tourism destinations	3.32	1.06
Heritage tourism destinations are having best hotels	3.30	1.03
Heritage tourism destinations are having sufficient area for parking	3.74	1.12

The tourists are agreed with heritage tourism destinations explain local culture, heritage tourism destinations have cultural uniqueness, traditional foods are available in heritage tourism destinations, conventional handicrafts are obtainable from heritage tourism destinations, heritage tourism destinations have a pleasurable environment, access to heritage tourism destinations is very easy and heritage tourism destinations are having sufficient area for parking, whilst, they are neutral with heritage tourism destinations possess undisturbed natural beauty, signboards are well placed in heritage tourism destinations, and heritage tourism destinations are having best hotels.

12.4.3 PROFILE OF TOURISTS AND THEIR PERCEPTION OF HERITAGE TOURISM DESTINATIONS

The distribution of tourists based on their perception of heritage tourism destinations is given in Table 12.3. The perception of tourists on heritage tourism destinations is divided into low, moderate, and high levels on the basis of Mean ± SD (Mean = 36.90; SD = 3.47).

TABLE 12.3 Distribution of Tourists Based on Their Perception of Heritage Tourism Destinations.

Level of perception of heritage tourism destinations	Number of tourists	Percentage
Low	73	24.33
Moderate	126	42.00
High	101	33.67
Total	**300**	100.00

Among 300 tourists, 33.67% of them have a high level of perception of heritage tourism destinations, whilst, 24.33% of them have a low level of it.

12.4.4 TOURIST TYPE AND PERCEPTION OF HERITAGE TOURISM DESTINATIONS

The relation among tourist type and perception of heritage tourism destinations is given in Table 12.4.

TABLE 12.4 Tourist Type and Perception of Heritage Tourism Destinations.

Tourist type	Level of perception of heritage tourism destinations			Total	t-Value	Sig.
	Low	Moderate	High			
Domestic	50 (30.49)	56 (34.15)	58 (35.36)	164 (54.67)	4.148	.000
International	23 (16.91)	70 (51.47)	43 (31.62)	136 (45.33)		
Total	73 (24.33)	126 (42.00)	101 (33.67)	300 (100.00)	–	–

Note: The figures in the parentheses are percent to the total.

Among 164 domestic tourists, 35.36% of them have a high level of perception of heritage tourism destinations, whilst, 30.49% of them have a low level of it. Among 136 international tourists, 31.62% of them have a high level of perception of heritage tourism destinations, whilst, 16.91% of them have a low level of it.

The *t*-value of 4.148 is elucidating that a significant difference is exiting in perception of heritage tourism destinations among tourist type at 1% level.

12.4.5 GENDER AND PERCEPTION OF HERITAGE TOURISM DESTINATIONS

The relation among gender of tourists and perception of heritage tourism destinations is given in Table 12.5.

TABLE 12.5 Gender and Perception of Heritage Tourism Destinations.

Gender	Level of perception of heritage tourism destinations			Total	*t*-Value	Sig.
	Low	Moderate	High			
Male	40 (23.12)	68 (39.31)	65 (37.57)	**173 (57.67)**	2.959	.003
Female	33 (25.98)	58 (45.67)	36 (28.35)	127 (42.33)		
Total	73 (24.33)	126 (42.00)	101 (33.67)	300 (100.00)	-	-

Note: The figures in the parentheses are percent to the total.

Among 173 male tourists, 37.57% of them have a high level of perception of heritage tourism destinations, whilst, 23.12% of them have a low level of it. Among 127 female tourists, 28.35% of them have a high level of perception of heritage tourism destinations, whilst, 25.98% of them have a low level of it.

The *t*-value of 2.959 is elucidating that a significant difference is exiting in perception of heritage tourism destinations among the gender of tourists at a 1% level.

12.4.6 AGE CATEGORY AND PERCEPTION OF HERITAGE TOURISM DESTINATIONS

The relation among age category of tourists and perception of heritage tourism destinations is given in Table 12.6.

TABLE 12.6 Age Category and Perception of Heritage Tourism Destinations.

Age category	Level of perception of heritage tourism destinations			Total	*F*-value	Sig.
	Low	Moderate	High			
Below 25 years	8 (18.18)	18 (40.91)	18 (40.91)	**44** **(14.67)**	5.784	.000
26–30 years	16 (22.54)	31 (43.66)	24 (33.80)	**71** **(23.66)**		
31–35 years	29 (28.43)	41 (40.20)	32 (31.37)	**102** **(34.00)**		
36–40 years	9 (19.15)	19 (40.43)	19 (40.43)	**47** **(15.67)**		
Above 40 years	11 (30.56)	17 (47.22)	8 (22.22)	**36** **(12.00)**		
Total	**73** **(24.33)**	**126** **(42.00)**	**101** **(33.67)**	**300** **(100.00)**	-	-

Note: The figures in the parentheses are percent to the total.

Among 44 tourists coming under the age category of below 25 years, 40.91% of them have a high level of perception of heritage tourism destinations, whilst, 18.18% of them have a low level of it. Among 71 tourists coming under the age category of 26–30 years, 33.80% of them have a high level of perception of heritage tourism destinations, whilst, 22.54% of them have a low level of it. Among 102 tourists coming under the age category of 31–35 years, 31.37% of them have a high level of perception of heritage tourism destinations, whilst, 28.43% of them have a low level of it.

Among 47 tourists coming under the age category of 36–40 years, 40.43% of them have a high level of perception of heritage tourism destinations, whilst, 19.15% of them have a low level of it. Among 36 tourists coming under age category of above 40 years, 22.22% of them have high level of perception of heritage tourism destinations, whilst, 30.56% of them have low level of it.

The *F*-value of 5.784 is elucidating that a significant difference is exiting in perception of heritage tourism destinations among the age category of tourists at a 1% level.

12.4.7 EDUCATION AND PERCEPTION OF HERITAGE TOURISM DESTINATIONS

The relation among education of tourists and perception of heritage tourism destinations is given in Table 12.7.

TABLE 12.7 Education and Perception of Heritage Tourism Destinations.

Education	**Level of perception of heritage tourism destinations**			**Total**	***F*-Value**	**Sig.**
	Low	**Moderate**	**High**			
Higher Secondary	9	21	18	**48**		
	(18.75)	(43.75)	(37.50)	**(16.00)**		
Diploma	20	35	20	**75**		
	(26.67)	(46.66)	(26.67)	**(25.00)**	3.762	.011
Under Graduation	22	37	37	96		
	(22.92)	(38.54)	(38.54)	(32.00)		
Post Graduation	22	33	26	**81**		
	(27.16)	(40.74)	(32.10)	**(27.00)**		
Total	**73**	**126**	**101**	**300**		
	(24.33)	**(42.00)**	**(33.67)**	**(100.00)**	-	-

Note: The figures in the parentheses are percent to the total.

Among 48 tourists possessing higher secondary, 37.50% of them have a high level of perception of heritage tourism destinations, whilst, 18.75% of them have a low level of it. Among 75 tourists possessing a diploma, 26.67% of them have a high level of perception of heritage tourism destinations, whilst, 26.67% of them have a low level of it.

Among 96 tourists possessing under graduation, 38.54% of them have a high level of perception of heritage tourism destinations, whilst, 22.92% of them have a low level of it. Among 81 tourists possessing post-graduation, 32.10% of them have a high level of perception of heritage tourism destinations, whilst, 27.16% of them have a low level of it.

The *F*-value of 3.762 is elucidating that a significant difference is exiting in perception of heritage tourism destinations among education of tourists at 1% level.

12.4.8 MONTHLY INCOME AND PERCEPTION OF HERITAGE TOURISM DESTINATIONS

The relation among monthly income of tourists and perception of heritage tourism destinations is given in Table 12.8.

TABLE 12.8 Monthly Income and Perception of Heritage Tourism Destinations.

Monthly income	Level of perception of heritage tourism destinations			Total	*F*-Value	Sig.
	Low	Moderate	High			
Less than ₹30,000	15	17	20	**52**		
	(28.85)	(32.69)	(38.46)	**(17.33)**		
₹30,001–40,000	24	46	40	**110**		
	(21.82)	(41.82)	(36.36)	**(36.67)**	6.039	.000
₹40,001–50,000	26	39	33	**98**		
	(26.53)	(39.80)	(33.67)	**(32.67)**		
More than ₹50,000	8	24	8	**40**		
	(20.00)	(60.00	(20.00)	**(13.33)**		
Total	**73**	**126**	**101**	**300**	-	-
	(24.33)	**(42.00)**	**(33.67)**	**(100.00)**		

Notes: The figures in the parentheses are percent to the total.

Among 52 tourists getting a monthly income of less than ₹30,000, 38.46% of them have a high level of perception of heritage tourism destinations, whilst, 28.85% of them have a low level of it. Among 110 tourists getting a monthly income of ₹30,001–40,000, 36.36% of them have a high level of perception of heritage tourism destinations, whilst, 21.82% of them have a low level of it.

Among 98 tourists getting a monthly income of ₹40,001–50,000, 33.67% of them have a high level of perception of heritage tourism destinations, whilst, 26.53% of them have a low level of it. Among 40 tourists getting a monthly income of more than ₹50,000, 20.00% of them have a high level of perception of heritage tourism destinations, whilst, 20.00% of them have a low level of it.

The *F*-value of 6.039 is elucidating that a significant difference is exiting in perception of heritage tourism destinations among monthly income of tourists at 1% level.

12.4.9 MARITAL STATUS AND PERCEPTION OF HERITAGE TOURISM DESTINATIONS

The relation among the marital status of tourists and perception of heritage tourism destinations is given in Table 12.9.

TABLE 12.9 Marital Status and Perception of Heritage Tourism Destinations.

Marital status	**Level of perception of heritage tourism destinations**			**Total**	***t*-Value**	**Sig.**
	Low	**Moderate**	**High**			
Married	43	89	91	**223**	5.350	.000
	(19.28)	(39.91)	(40.81)	**(74.33)**		
Unmarried	30	37	10	**77**		
	(38.96)	(48.05)	(12.99)	**(25.67)**		
Total	**73**	**126**	**101**	**300**	-	-
	(24.33)	**(42.00)**	**(33.67)**	**(100.00)**		

Note: The figures in the parentheses are percent to the total.

Among 223 married tourists, 40.81% of them have a high level of perception of heritage tourism destinations, whilst, 19.28% of them have a low level of it.

Among 77 unmarried tourists, 12.99% of them have a high level of perception of heritage tourism destinations, whilst, 38.96% of them have a low level of it. The *t*-value of 5.350 is elucidating that a significant difference is exiting in perception of heritage tourism destinations among the marital status of tourists at 1% level.

12.4.10 RELATION AMONG PERCEPTION OF HERITAGE TOURISM DESTINATIONS AND REVISITING BEHAVIOR OF TOURISTS

The relation among perception of heritage tourism destinations and revisiting behaviour of tourists was studied by applying correlation analysis and the results are given in Table 12.10.

TABLE 12.10 Relation among Perception of Heritage Tourism Destinations and Revisiting Behavior of Tourists.

Particulars	Correlation co-efficient
Perception of heritage tourism destinations and revisiting behavior of tourists	0.63**

**Significance at 1% level.

The correlation coefficient among perception of heritage tourism destinations and revisiting behavior of tourists is 0.63, and it is revealing that both are positively and highly interrelated.

12.5 CONCLUSION

The findings of this study explicate that more than two-fifth of tourists has a moderate level of perception of heritage tourism destinations. A significant difference is prevailing among the profile of tourists and their perception of heritage tourism destinations. Perception of tourists on heritage tourism destinations is highly, significantly, and positively related to their revisiting behavior. To enhance the perception of tourists on heritage tourism destinations, heritage tourism destinations should keep their natural beauties and signboards must be put in appropriate positions in tourism destinations. In addition, heritage tourism destinations should have the best hotels to offer superior quality of services to tourists and these measures would intensify revisiting behavior among tourists.

12.6 IMPLICATIONS OF THE STUDY

The outcome of this study will be useful for local administrators, tourism authorities, hotel operations, and local people to understand insights of tourists about heritage tourism destinations, need for preserving heritage, improve amenities, and various activities associated with heritage tourism and services. The results will also be helpful to formulate and implement effective and innovative strategies for the sustainable development of heritage tourism in the Sivaganga district.

KEYWORDS

- **heritage tourism**
- **perception**
- **revisiting behavior**
- **tourists**

REFERENCES

Benur, A. M.; Bramwell, B. Tourism Product Development and Product Diversification in Destination. *Tour. Manage.* **2015,** *50*, 213–224.

Chheang, V. Angkor Heritage Tourism and Tourist Perceptions. *Tourismos Int. Multidiscipli. J. Tour.* **2011,** *6* (2), 213–240.

Gohil, N. Managing Urban Heritage Tourism: A Case Study on Heritage Potential in Jabalpur (MP). *Int. J. Market., Financ. Serv. Manage. Res.* **2015,** *4* (4), 114–125.

Hasan, M.; Jobaid, M. I. Heritage Tourism Marketing: Status, Prospects and Barriers. *IOSR J. Busi. Manage.* **2014,** *16* (5), 40–48.

Kariru, A.N.; Aloo, C. Customers' Perceptions of Service Quality in Hotels in Western Tourism Circuit, Kenya. *J. Hosp., Tour. Culture* **2014,** *2* (1), 1–12.

Praveen, M.; Manju Priya, R. Travelers Satisfaction towards Tourist Destinations (An Empirical Study with Special Reference to the Nilgiri). *Int. J. Market. Human Resour. Manage. 2016, 7* (3), 64–70.

Rahman, M. S. Exploring Tourists' Perception: The Case of Bangladesh. *Tour. Int. Multidiscipl. J. Tour.* **2012,** *7* (1), 81–98.

Rajesh, R. Impact of Tourist Perceptions, Destination Image and Tourist Satisfaction on Destination Loyalty: A Conceptual Model. *Revista de Turismo y Patrimonio Cultural* **2013,** *11* (3), 67–78.

Surebankar, S.; Hadli, M. Heritage Tourism in Information Age: Need of Historians' Active Participation. *South Asian J. Tour. Herit.* **2010,** *3* (2), 116–121.

Yavana Rani, S. An Empirical Study on the Visitors' Intention to Visit Cultural Heritage Site. *Int. J. Recent Trends Busi. Tour.* **2017,** *1* (3), 14–20.

Yeniasır, M.; Gökbulut, B. Perception and Attitudes of Local People on Sustainable Cultural Tourism on the Islands: The Case of Nicosia. *Sustainability* **2018,** *10*, 1–16.

Zeng, B. Cultural Centre, Destination Cultural Offer and Visitor Satisfaction. *Sustainability* **2017,** *9*, 1–12.

CHAPTER 13

THE RELEVANCE OF VIRTUAL REALITY AND INCREASED REALITY IN EVENTS WITH STREAMING AND LIVESTREAMING IN COVID AND POST-COVID-19 ERA: A TOURISTIC PERSPECTIVE

ANNAELISE FRITZ MACHADO[1,*], BRUNO BARBOSA SOUSA[2], MIGUEL RIBEIRO[1], MAGNO ANGELO KELMER[3], and BÁRBARA DE VASCONCELLOS[1]

[1]*Polytechnic Institute of Cávado and Ave, Portugal*

[2]*Polytechnic Institute of Cávado e Ave (IPCA), CiTUR and UNIAG Research Member, Portugal*

[3]*UFJF—Federal University of Juiz de Fora, Brazil*

[*]*Corresponding author. E-mail: annaelisefritz@yahoo.com.br*

ABSTRACT

The gradual transformation of technological means has opened up new perspectives and different ways of communicating and presenting a product/service in the events area, making it possible to attract and engage participants. The watchwords became artificial intelligence, digital currencies, advancing robotics, virtual reality (VR), and

Tourist Behavior: Past, Present, and Future. Narendra Kumar, PhD, Bruno Barbosa Sousa, PhD & Swati Sharma, PhD (Eds.)

augmented reality (AR). While VR completely immerses the user within a virtual (non-physical) environment, AR allows the user to view the real world with virtual objects in overlap with real objects; in short, VR replaces reality while AR complements it. Given this scenario, events with streaming can be understood as the transfer of data on the internet. To send multimedia information from servers to clients, as livestream, enables real-time transmission that allows users to view and direct video content using a camera and a computer over the internet. These resources started to be widely used in the COVID-19 pandemic period because the world events sector is highly affected by the demands of social isolation by the Heads of State of several countries. The recognition of this need leads the consumer to a decision-making process that determines what to buy and what to consume. The general objective of this article is to present the relevance of VR and AR used in events with streaming and livestream in times of COVID and post-COVID-19.

13.1 INTRODUCTION

Tourism is characterized as a social phenomenon, which maintains interrelation with several sectors. This activity assumes as one of the most relevant engines and with the most significant impact on the development and growth of several economies. Tourism is very vulnerable to seasonality, to moments of financial crises, to situations of calamities and pandemics. An alternative to control the seasonality of tourist destinations is the insertion of events (Barbosa Sousa and Marques Teixeira, 2020). The events segment has consolidated itself as a prosperous market in the ranking of the tourist economy. The development of a country's tourist activity is globally recognized as a generator of revenue, for boasting high growth rates in recent decades but, very sensitive to any type of market variation, economic crises, climatic effects, and epidemics/pandemics, which cause many fluctuations.

The events segment has consolidated itself as a prosperous market in the ranking of the tourist economy, which contributes to the control of seasonality in tourism. With the pandemic crisis, caused by COVID-19, one of the sectors most affected and that has not yet seen a resumption, it was the events, for gathering crowds of people, something forbidden by the Heads of State of several countries.

The global society has been going through a process of structural transformation, for two decades, associated with the emergence of this new technological paradigm, based on communication and information technologies, which emerged in the 1960s and were adapting according to the needs, values, and interests of the utilities. In this way, people start to integrate communication technologies in their lives, "linking virtual reality with real 'virtuality'" (Alves et al., 2020). In recent years, there has been a strong contribution of technology in the event segment, and reinventing becomes the main premise to take the sector out of the difficulties it has been experiencing. Given this perspective, virtual reality (VR) and augmented reality (AR) in different ways guarantee a new reality to event participants and come to be viewed with good eyes.

Virtual reality is not associated with the real world; in fact, virtual environments are simulated in which the user has a totally immersive and dynamic experience with the support of equipment such as glasses connected to a computer and headset. Augmented reality is understood as mixing real and digital elements. Users interact with overlapping images as they move around the real world. The wager is for the insertion of these VR and AR technologies. In streaming events, a system that can download images and sound instantly and quickly to be consumed immediately by the participant and livestream is even faster. It uses this same technology for live online transmissions, without the need for initial loading, just click on play and watch events, shows, and other live formats; formats that are innovative and that can generate good profits. The general objective of this article is to present the relevance of VR and AR in events with streaming and livestream in the Covid and post-Covid-19 periods. Some options of these technologies are pointed out for the tourism sector. What is the relevance of virtual reality and augmented reality in events with streaming and livestream in times of Covid and post-Covid-19? In the next sections of the investigation, some real examples will be selected, which illustrate the relevance of VR and AR in events with streaming and livestream, as well as some options in the tourism industry, in order to point out the changes in behavior and best practices for Event Tourism (in an attempt to overcome the crisis, pandemic, that is, Covid-19).

13.2 COVID-19 PANDEMIC

The emergence of a series of pneumonia cases of unknown origin led to the discovery of a new coronavirus in the city of Wuhan, with an estimated population of 11 million, a province in Hubei in China. From the Chinese government's warning in December 2019 of a virus that killed 17 people and infected 400 others, the world started to monitor its progress daily, spreading to other continents (Sousa et al., 2020).

At first, it was thought to be an outbreak that could be limited and controlled, but the number of infected people increased exponentially and cases started to be registered in Thailand, Japan, Korea, and the United States. In the first instance, the cases involved people from Wuhan traveling or people from other countries who visited the Chinese city. Between January and February 2020, China made progress in infected and dead cases. The virus gained new space and Wuhan is now treated as the epicenter of the disease and new cases around the world are no longer associated with travelers who have passed through the Chinese city. In less than two months, the virus would have spread to South Korea, Iran, and Italy (Sousa et al., 2020).

On January 24, with three confirmed cases in France, the virus enters the European continent. It quickly spread and transformed Italy into the epicenter of the "Old World," placing it as the third country in the world with the most cases of the disease, behind Korea and China. The escalation of COVID-19 was rapid in the European continent and reached several countries differently. In May 2020, Italy and France, Spain, and the United Kingdom surpassed the barrier of 25,000 deaths due to COVID-19 while adding the number of deaths in Poland, Slovakia, Czech Republic, Hungary, Austria, and Romania, the total does not reach 3000 deaths. Compared to 54.42 deaths per 100,000 inhabitants recorded in Spain, 48.12 in Italy, 37.63 in France, or 43.33 in the United Kingdom, Poland records 1.84; Slovakia 0.46; the Czech Republic 2.37; Hungary 3.59; Austria 6.78 and Romania 4.20 (Sousa et al., 2020).

In April 2020, the World Health Organization (WHO) declared a new coronavirus pandemic, called Sars-Cov-2. According to the organization's data, the number of cases of COVID-19, which is the disease caused by the virus, outside China has increased 13 times and the number of affected countries has more than tripled. An epidemic of the exaggerated size that can kill tens and even hundreds of millions of people worldwide can be called a pandemic; if a disease appears that combines a great capacity of contamination with an ability to kill a large number of infected people. For the author (Higgins-Desbiolles, 2020), the rapid transmission of the virus and the contagion of millions of people around the world is mainly due to the agglomeration of people in events such as games and concerts; in confined and overloaded spaces, for example, when using means of transport; classrooms with too many students; workers performing their routine activities indoors, among various other forms of contagion.

Although identified later, the first case of COVID-19 in the African continent happened on February 14 and called the attention of the world to an announced catastrophe because it is the second most populous continent in the world, with 1.2 billion inhabitants. This fact did not occur until the end of May, since, while Europe accounts for more than 1.5 million confirmed cases, the United States exceeds 1.3 million and Latin America is close to 250,000, and the continent African population has so far 55,000 infections. The most affected African countries, until May, are South Africa, with 10,015 cases are Egypt with 9400, Morocco with 6063, and Algeria with 5723. Together, they represent almost 50% of all infections in Africa. In Oceania, Australia and New Zealand recorded the first cases of COVID-19 at the end of January. With a strategy based on mass testing of the population, the continent has few reported cases and low death rates. As of July 9, 2020, the world had 12,014,884 infected and 549,749 killed due to COVID-19. The advance of the virus, occupying new spaces, establishes territories that start to organize themselves to face it. In the illustration above, it is possible to see the spaces territorialized by the presence of COVID-19.

13.3 REFLECTION ON TOURIST ACTIVITY

Tourism is one of the most important and current engines of growth for an economy, marked by dynamism, diversity of supply, and constant growth. As tourism is one of the sectors whose degree of involvement is necessary for its success, the diversification of tourism products and services becomes a requirement capable of leading to an increase in demand for new types of tourism needs (Ferreira et al., 2019). The management of a tourist destination requires its positioning, given that the image transmitted is an influential and determinant aspect in the purchase decision.

Currently, the tourism industry comes in the most varied forms and for the most diverse audiences, aiming to meet all types of demand and satisfy the desires and needs of tourists/visitors by offering a variety of tourist products and services (Vareiro et al., 2020).

Thus, one can understand the complexity of this activity, since it is an activity that proposes to work and move various types of services present in the world market, being considered a sector very sensitive to the impacts of some phenomena, such as variations in consumer income, negative climatic effects, political and social instability, price fluctuations in the market, inflation, exchange rates, economic crises, epidemics, and pandemics (Chang et al., 2020).

Tourism is experiencing a moment of crisis, which was initiated by the spread of the COVID-19 virus. In the midst of this pandemic, touristic destinations needed to organize themselves, in order to take information on the lockdown of cities, airports to receive tourists/travelers. A serious decision that has not only negatively impacted the economy of destinations but also put employability at risk, as well as their image (Ahmad et al., 2020).

Faced with such a world crisis scenario, due to the COVID-19 pandemic, the tourism sector, being one of the most affected, needed to reinvent itself by creating job possibilities for the most varied positions in which it operates and, regarding virtual reality and augmented reality, which had already been gaining adherents, mainly in museums, started to be heavily used in other sectors, as an option to minimize the effects of the crisis. It can be clarified: The main advantages of augmented reality for tourism are past, present, and future united with AR, history comes to life, allowing the appreciation of the past, culture, and heritage through its understanding in a more visual and real way. Some examples are tourist spots, with graphic representations of a monument currently in ruins or the simulation of a battle, or tourist information being seen through smartphones. Virtual tour, using so-called VR glasses (virtual reality), you can have an immersive experience without leaving the place, getting to know various tourist spots in the world (Sousa et al., 2020); Travel agencies using virtual reality technology to help customers choose their next destination; Museums through Image Recognition and Augmented Reality, display information about pieces of art just by looking at them. This technology is already used at the Computer History Museum in Mountain View (California, USA), at the Montmartre Museum (Paris, France), and at the Royal

Ontario Museum (Ontario, Canada); Postcards with augmented reality, in addition to viewing the three-dimensional model, the visitor will also be able to obtain touristic information on the website.

In the entertainment industry, the virtual immersion experience is possible to be lived in Animal Kingdom, at Walt Disney World, with the Avatar Flight of Passage. In a 3D simulator, the tourist, connected to an avatar, flies on the back of a banshee, a flying creature from "Pandorav." In other words, with computer graphics resources and technological support for electronic chairs with motion sensors, headphones, and 3D glasses, the tourist lives an experience in another reality, the virtual one. In this case, the physical displacement of the subject or the tourist happens, as he needs to be in the place that provides such a three-dimensional experience (Taufer and Ferreira, 2019).

The purchase of online tickets should ensure the restriction of the number of visitors to enter by timeslot, as well as the maximum time they can stay at the place they visit, the digitization of the tickets themselves presents itself as an excellent tool for reducing physical contact (Taufer and Ferreira, 2019). As for hotels, the suggestion may include automatic "check-in," thus avoiding the queues to access the rooms and also suggests "the implementation of virtual assistants to help improve visitor interaction and experience," in "Destination promotion apps," as well as virtual tours that make it possible to get to know the entire hotel, making it possible to choose the environments you want to visit and the capacity of people in each environment. In the restaurant area, you can see the available tables and the places where they are in the establishment, as well as a tour of all spaces including the kitchen.

In this context, virtual reality and augmented reality are transforming forces, capable of generating quick results, where they are inserted, as it allows: to change the way countries promote their history, culture, and their people and present them to those who visit them; it helps to create a unique and differentiating memory effect through the sensory experience that it transmits to those who interact with it; it assumes itself as a decisive factor in the creation of socioeconomic value through the more efficient provision of services to the citizen; it is a matter of differentiation with safe value on several fronts: it promotes tourism, fosters the economy, helps the population, provides better services, and is also a bet for the countries, revolutionizing processes, automating tasks, making investments more profitable, and simplifying the relationship with the citizens.

Not ending there and focusing on the events segment, there are adaptations made regarding virtual reality and augmented reality, to meet the repressed demand of participants/customers who wish to interact at events but are prevented due to the pandemic, which will be presented in the following sections.

13.4 VIRTUAL REALITY AND AUGMENTED REALITY IN EVENTS

The constant advance of technology has changed consumer behavior. With the emergence of the internet, there was a great revolution in the corporate world and especially in the way of doing business. In an era marked by information and knowledge, it is imperative to create value and differentiation. However, from 2010 to the present day, the technological revolution caused by smartphones, the power of content, entertainment, and experience has increased (Sousa et al., 2020).

The culture of convergence provided a new habit for internet users regarding media consumption, since within the same media environment it is possible to consume texts, images, audios, and videos, without the need for a different device for each format. The innumerable pages and websites on the internet have also allowed the user a type of experience where he can acquire information from anywhere on the planet, all in one device (Vareiro et al., 2020).

Internet of Things (IoT), artificial intelligence, virtual reality, and augmented reality are among the eight emerging technologies that are already changing the world of commerce and causing a profound emotional impact on consumers. Despite being adopted at different scales, each will cause changes in the value chain and in the core of companies operations. We live in front of a web 3.0 (2010–2020), which incorporates high-speed connectivity, complex interactions, intelligent or semantic content, and automated applications, which can access various sources of information (data). In this section, two forms of interaction are emphasized: VR and AR (Chaffey and Ellis-Chadwick, 2016). VR is not a recent concept; it goes back to the 1960s, where the creation of the first immersive experiences or the invention of "Sensorama," related to the film industry. However, in a more recent and technological perspective, virtual reality consists of a field of study, which aims to create a system that provides a "synthetic" experience for its users. This experience can also be called "virtual" or "illusory," since all sensory stimulation is simulated and performed by a system, which includes three types of essential tools: screens, responsible for stimulating various senses; sensors, to detect the actions of individuals; and, finally, a computer capable of processing these same actions and allowing stimulation in the correct places. The goal of virtual reality technology is to fight the barriers between what is real and what is simulated, allowing individuals not only to see or hear something but also to live the experience, as if it were true, projecting the user onto the scene, is more than the mere presentation of a 3D world (Meira et al., 2021).

Virtual reality appears as something disruptive and different from augmented reality, more linked to immersion, a way of experiencing content within a computing platform, in which the user is completely immersed in a software environment. Despite the fact that virtual reality technology is more expensive and, therefore, less accessible to the public, it is important to highlight the power that it concentrates to make the entertainment and travel market disruptive, offering users the possibility to experience something that they never thought possible.

AR is one of the types of virtual reality capable of generating a vision that combines the real scene seen by the user and a virtual scene generated by the computer, creating a moment in which the user is not able to distinguish between the real world and the virtual increase of it. Deepening the definition of AR, it is about "improving" the real world through computer-generated perceptual information, in a visual, auditory, tactile, or even olfactory, thanks to the combination of real and virtual worlds, interaction in real-time and accurate 3D recording of virtual and real objects (Meira et al., 2021).

Both virtual reality and augmented reality are still in their infancy in the field of events, taking the first steps in an increasingly technological future and showing only a fraction of the impact and disruption they can bring. If virtual reality immerses the consumer in a simulated world, augmented reality superimposes information on the real world to increase the experience. On the other hand, augmented reality may appear as something less strange to users, as the devices necessary for its execution pass through objects already considered "common" in society, such as smartphones or tablets.

One of the sectors most impacted by COVID-19 was the segment of events, which cannot yet resume its activities, due to the crowding of people, which can lead to the further spread of the virus. Regarding this aspect, events are needed to think about solutions that could minimize losses. The use of technology was the strategy adopted, being able to interact, involve, and create connections between event organizers and the audience (here understood as users). Thus, virtual reality and augmented reality are the innovative platforms used for this, which replace an in-person event with a high-impact digital solution that guarantees maximum participation among participants. It is known that, despite the gradual lifting of restrictions, social distancing will continue to be the common rule between people. In terms of the benefits of virtual reality and augmented reality for the events sector, there are brand positioning, added value, and creation of important content, in addition to more assertive communication (Alves and Sousa, 2020).

In the area of events, VR and AR are new platforms, but with a chance of great successes, since it works very well in other areas, such as:

The improved navigation systems that use augmented reality to superimpose a route over the live view of the road; During football matches, television broadcasts use AR to draw lines on the field to illustrate and analyze plays; Neurosurgeons use AR projections from a 3-D of the human brain to help them with the most complex surgeries; At Singapore airport, for example, employees wear AR glasses to view information about cargo containers, improving average loading times; Military fighter pilots see an AR projection of their altitude, speed, and other data on the display of their helmet, which means that they do not need to waste their focus when looking at the plane's instruments to see them.

The use of virtual reality and augmented reality, creating virtual learning environments for different areas of knowledge. Another interesting report on the use of augmented reality on mobile devices was the article by Pérez-Sanagustín et al. (2016) that transformed traditional field educational activities into interactive and structured activities, enriching the experience and combining different types of media. The activities carried out were: Discovering the Campus and Discovering Barcelona.

Although these are deemed as super new tools for the segment of events, the notes above comply that in several areas success was guaranteed by its use. Given this context, betting on VR and AR as an alternative to help minimize the crisis in the sector, can be a solution not only for the period of COVID-19 but also for post-COVID.

13.5 EVENTS WITH STREAMING AND LIVESTREAM IN COVID AND POST-COVID-19 TIMES

The common recommendation (still) is to stay at home as much as possible and avoid crowds. Despite the new and relative normality, many events remain canceled or have been postponed, in the hope of briefly recovering the good old habits of going to the theater, enjoying a good exhibition, or seeing a great concert without fear. During the pandemic of COVID-19, the strategies adopted to help the sector's recovery were leveraged in order to minimize the negative impacts, but due to the success that has been obtained, it is understood that post-COVID will be extended, as a plus to the sector, already that taking advantage of events that take place in other geographies, from north to south of the country and even abroad, has gained fans. Both virtual reality and augmented reality technology can and should be used effectively by professionals on the segment of events since it is able to make a conference, through its use, for example, more engaging and enjoyable for participants. In this environment, streaming and livestream are widely used.

Streaming is considered a data transfer service over the internet that allows you to listen to music and watch videos without the need to store files on your computer. Its most popular services are YouTube, Netflix, and Spotify. Livestream, on the other hand, has the service of transmitting live content. This is where you interact in real time with your listeners and guests, which can make your program much more dynamic. Livestream is more recommended for those who already have experience with live broadcasts, as it requires more technological resources, cutting-edge equipment, and a waist game from the announcer and radio professionals. Today, social networks already have services such as Facebook, Instagram, and YouTube.

Thinking about the sustainability of the events segment in COVID and post-COVID times, we refer to the use of innovative technological strategies as a resource to ensure the continuity of events in their various aspects. In this environment, the livestream + 3D and the 360 degree stream enables the interaction of participants and organizers, leaving the atmosphere relaxed, immersive, participatory, and respecting the requirements of social distancing.

Livestream + 3D is a live broadcast from a virtual setting. People are recorded in Chroma and integrated into this 3D scenario in real-time during the broadcast. The model also allows inserting all kinds of holographic or augmented reality elements and can be seen on any monitor with internet access (computer, smart TV, smartphone, or tablet); Stream 360 degree is a type of event in virtual reality that can be broadcasted. Viewers, with VR glasses, will be able to watch the entire event at home, but as if they were in the middle of the action. The scenarios designed in 3D also allow the incorporation of real people previously recorded in the studio and positioned in these virtual scenarios as if they were in fact there (Pérez-Sanagustín et al., 2016).

Whatever the solution was chosen for the event, it is possible to create several virtual scenarios; integrate screens with other content (e.g., led wall, giant screen ...) within the virtual scenario; incorporate publicity strategies to sponsors in a variety of ways; subtitled with voice translation; broadcast on open signal or in Pay per View; record speakers in the country where the event is held or abroad and have access to the event's analytical data. The interaction between participants is also guaranteed, through a "face to face check in with the promoters," information desk, sponsor stands, lounge areas, chat rooms, launching surveys, or even sharing documents.

As examples to be demonstrated, there is the ticket line "Live Stage," which undergo through a reinvention during the pandemic period as the artistic environment was experiencing difficulties. Ticket line decided to launch a streaming platform, which aims to allow the public to watch events anywhere in the world. Ticket line Live Stage debuted on the 18th of July with the transmission of the show "Desconfinando e Rindo," by Herman José. But the new schedule promises theater, dance, humor, music, and magic shows.

Tomorrowland Around the World, the well-known Belgian electronic music festival, moved to the internet. Between the 25th and 26th of July, it brought together artists like Steve Aoki and Armin van Buuren but also, Dixon or Tale Of Us, and had eight stages, specially created and designed by the creative team and the 3D artists behind Tomorrowland. In addition to the performances, it was possible to attend fireworks, laser shows, and activities such as webinars, games, workshops on lifestyle, gastronomy, and magic fashion were also planned and displayed.

Concerts are back to the Hot Club (Hot Clube), but at a distance, respecting the social distancing. The main reason that led to the foundation of Hot Clube of Portugal in 1948 was "the promotion of jazz music." As the current rules of social distance are incompatible with live concerts, the alternative found to the purpose the mission of the club is the live streaming of concerts, through the Crowd cast platform, every Thursday and Friday, between 22.00 and 23.15. Access is free for HCP members with up-to-date dues, but the remaining interested parties must pay (€ 7.50) to see the concerts.

In June, Apple held a major event dedicated to programmers, the WWDC—Worldwide Developers Conference. WWDC, Apple's annual conference for programmers, kicked off in Lisbon, with Aurora's "Daydreamer" as a soundtrack in the streaming format. Due to COVID-19, this is the first all-digital event of the technology and Tim Cook, the executive president of Apple, released the news on a stage without an audience—for the first time in history—facing a camera. The Steve Jobs Theater, at Apple's headquarters in Cupertino, was empty and the presentation went through several spaces in the company's offices. Apple has used the WWDC stage to present various products, such as Macs, iPhones, or Home Pod; however, the focus of this event was the software.

Buteco at home is a presentation only with renowned singers from Brazil, the vast majority of the styles of "sertanejo" and "pagode" who, had their various concerts canceled and, in the face of this scenario, the beer brand Bohemia and Cielo got together and started sponsoring these singers, with the hashtag #Buteco, that has already around 680,000 followers, until 12/04/2020, at 5 am, through their "lives" streaming. The idea of the Buteco at home is to encourage the sale of vouchers in boutique shops (small entrepreneurs), in the amounts of $50.00, 100.00, and 150.00 "reais," Brazilian currency, that can be used until the end of the year. In beer purchases in bars/pubs in the future, there is a loyalty advantage where the buyer will earn 20% discount in the bars/pubs that he bought the vouchers. Through this strategy, the musicians are able to maintain their lifestyles despite the lack of available work, caused by COVID-19, as well as increasing the revenue on the small boutiques In order to keep them afloat and open for business.

In response to the recent COVID-19 outbreak, the major leagues in the world have been suspended, decisive matches have been postponed and some championships have been canceled. Stadiums accustomed to receiving thousands of supporters every week were closed, and the agglomerations, which previously meant euphoria, came to be seen only as a major threat and contamination focus. After long months of waiting, while the situation seems to be being controlled in some countries, the arenas are finally reopening and sporting events are resuming, but with a significant change: without the presence of an audience. In this context, the experience of the stadium is no longer physical, in person. Fans are prevented from attending the matches of their favorite team and are forced to stay at home, cheering from the safety of their homes and on their couch. The sports stars depend on the affection of the fans, and this is where new technologies come in— a way to keep this connection alive, this collective feeling of belonging and identification. Despite the considerable drop in consumption of augmented reality and virtual reality technologies because of the recent pandemic, market research has shown that sales of AR and VR headsets will grow by 23% by the end of this year. Some of the football teams that invest the most in technology worldwide, such as Manchester City and Juventus, are already experimenting with mixed media techno-logical solutions aimed at remote experience. Professional leagues from the most varied sports are investing heavily in this direction, developing interactive applications connected to Oculus visualization headsets, for example. During the last FIFA World Cup in 2018, we were able to get an idea of how the sport's world and the user experience can benefit from AR technologies, mainly through Snapchat and Facebook. An unprecedented amount of photo filters and themes started to provide fans with a different way of displaying pride for a team. The NBA also recently launched its own version of virtual reality during the finals of last year, providing an increasingly complete experience for the remote user. The new software of the major American basketball league was developed in partnership with Intel and offers a wide range of panoramic images, statistics, and analyzes in real-time and on-demand replay, luxuries that have not yet reached the stands mainly through Snapchat and Facebook. Other events using virtual reality and augmented reality technology can be innovated by the event organizers professionals or companies both in streaming and live stream formats, during the COVID-19 pandemic period, just by being well-publicized to be successful. Post-Covid can be an alternative of choice for participants, in person or virtual, with strategies that are even differenti-ated according to the type of participation or event.

13.6 FINAL CONSIDERATIONS AND FUTURE RESEARCH

The present study highlights the role of technologies in changing consumer behavior in specific contexts regarding tourism and the events segment, in periods of the pandemic, that is, COVID-19, that is, post-Covid. In this sense, the strategy adopted to minimize the crisis reveals quite satisfactory results. It was observed that the tourism sector, as well as the events segment, has been moving, innovating, and inserting cutting-edge technologies to reach its users. With immersive environments, digital technologies are used to promote sensory experiences; tours made with the use of 3D glasses; sharing information on different websites and blogs; livestream + 3D; 360 degree stream, online shopping, among others. With greater or lesser immersion, all experiences—on which are also preferably included the ones regarding the industry of tourism—whereas currently they are aiming to an approach of other configurations, trends, and technologies of the established relationships between person–machine–environment.

In this context, since the event's segment has always leveraged the tourism sector at seasonal times, it has suffered strong economic/financial pressures, as it cannot act due to the prohibitions of the Heads of States due to the prohibition of human agglomerations in the pandemic period. Faced with this reality, the insertion of VR and AR technologies is of great value, to minimize these negative effects as well as leveraging the segment economically and financially. Within this line of thought, it is possible to premise new market and user behaviors regarding different types of technologies, which can lead to a new way of promoting tourism and the event's segment. It is relevant to think that markets and users are learning to establish themselves in these technological platforms and that post-Covid, in a near future, the reality can be even for several demands from the Heads of State regarding favorable options between the real and the virtual, since acceptability is favorable to both forms of usage. The insertion of VR and AR technologies are relevant in this context as there is: brand positioning, added value, creation of important content, in addition to more assertive communication, and growth both economically and financially. In this context, some real examples associated with the promotion of events where VR and AR were included. The results show that the technological platforms have facilitated the interaction between the citizen/tourist and the event's brands, an important aspect in a context where the main goal is to inform, advise and prevent contamination therefore ensuring the protection of people's health. The focus of this initiative is to remember the importance of social detachment and knowing "how to wait" so that (in the future) everyone can once again participate in events. Future studies should move toward corroborating the conclusions evidenced here and, it is recommended to use the quantitative methodology (i.e., a questionnaire with users) to establish causal relationships between consumer behavior variables. As small as an application of virtual reality and augmented reality may seem, event organizer professionals or companies will certainly make a big difference, by obtaining socio-economic gains from innovative and disruptive initiatives. The democratization of access to technology, ease of use, and the appetite for new experiences are the motto for starting projects focused on attracting users to events.

KEYWORDS

- **Covid-19**
- **tourism**
- **events**
- **virtual reality**
- **augmented reality**
- **streaming**

REFERENCES

Ahmad, A.; Jamaludin, A.; Zuraimi, N. S. M.; Valeri, M. Visit Intention and Destination Image in Post-Covid-19 Crisis Recovery. *Curr. Iss. Tour.* **2020,** 1–6.

Alves, G. M.; Sousa, B. M.; Machado, A. The Role of Digital Marketing and Online Relationship Quality in Social Tourism: A Tourism for All Case Study. In *Digital Marketing Strategies for Tourism, Hospitality, and Airline Industries*; Santos, J., Silva, Ó., Eds.; IGI Global: Hershey, PA, 2020; pp 49–70. doi:10.4018/978-1-5225-9783-4.ch003

Barbosa Sousa, B.; Marques Teixeira, A. Integrated Marketing Communications in Contexts of Eco Tours in Porto (Portugal): Challenges (Post) Covid-19. In *COVID-19—Reshaping Marketing and Communications*; Naumovska, L., Ed.; Proud Pen, 2020; pp 1–11. https://doi.org/10.51432/978-1-8381524-7-5_1

Chaffey, D.; Ellis-Chadwick, F. *Digital Marketing: Strategy, Implementation and Practice*; Pearson Education: UK, 2016.

Chang, C. L.; McAleer, M.; Ramos, V. A Charter for Sustainable Tourism after COVID-19, 2020.

Ferreira, J.; Sousa, B.; and Gonçalves, F. Encouraging the Subsistence Artisan Entrepreneurship in Handicraft and Creative Contexts. *J.*ournal of *Enterpris.*ing

Communities: *People* and *Places* in the *Global Econ.* om **2019**y, *13 (1/2),* 64–83. https://doi.org/10.1108/JEC-09-2018-0068

Higgins-Desbiolles, F. Socialising Tourism for Social and Ecological Justice after COVID-19. *Tour. Geo.* **2020,** *22* (3), 610–623.

Meira, C.; Martins, I. S.; Sousa, B. B. Accessible Tourism and Digital Platforms: A Study Applied to the City of Viana do Castelo. In *Handbook of Research on the Role of Tourism in Achieving Sustainable Development Goals*; Brandão, F., Breda, Z., Costa, R., Costa, C., Eds.; IGI Global: 2021; pp 235–247. http://doi:10.4018/978-1-7998-5691-7.ch014

Pérez-Sanagustín, M.; Parra, D.; Verdugo, R.; García-Galleguillos, G.; Nussbaum, M. Using QR Codes to Increase User Engagement in Museum-Like Spaces. *Comput. Hum. Behav.* **2016,** *60*, 73–85.

Sousa, B. B.; Machado, A. F.; Igreja, C. M.; Campos, J. G. As redes sociais como veículo para combater os efeitos nefastos do covid-19: um estudo exploratório no contexto turístico português. *Cambiassu: Estudos em Comunicação* **2020,** *15* (25), 21–35.

Taufer, L.; Ferreira, L. T. Virtual Reality in Tourism: Entertainment or a Paradigm Shift? *Rosa dos Ventos—Tour. Hosp.* Oct–Dec **2019,** *11* (4), 908–921. doi: http://dx.doi.org/10.18226/21789061.v11i4p908

Vareiro, L.; Sousa, B. B.; Silva, S. S. The Importance of Museums in the Tourist Development and the Motivations of Their Visitors: An Analysis of the Costume Museum in Viana do Castelo. *J. Cult. Herit. Manage. Sustaina. Dev.* **2020,** *11* (1), 39–57. https://doi.org/10.1108/JCHMSD-05-2020-0065

CHAPTER 14

TOURISM ACTIVITIES' ROLE ON DESTINATION CHOICE AND POST-VISITING BEHAVIOR: EVIDENCE FROM ISTANBUL AS A CULTURAL DESTINATION

MELTEM ALTINAY OZDEMIR[1,*] and ISMAIL KIZILIRMAK[2]

[1]*Faculty of Tourism, Mugla Sitki Kocman University, Turkey*

[2]*Faculty of Economics, Istanbul University, Turkey*

[*]*Corresponding author. E-mail: meltemozdemir@mu.edu.tr*

ABSTRACT

This study aims to (1) validate a list of tourism activities in Istanbul, (2) evaluate the importance of tourism activities on a cultural destination choice, and (3) explain the impact of tourism activities on post-visiting behavior. A preference list for activities was generated by in-depth interviews were conducted on travel agencies and tourist guides organizing cultural tours, besides cultural tour programs. The research was conducted on European tourists visiting Istanbul as a cultural destination of Turkey in 2019. A purposive sampling method was used. Selection criteria are in the top six European tourists visiting Istanbul according to their nationality and staying at least one night in the destination. Data were collected by a three-part questionnaire. The results indicated that European tourists

Tourist Behavior: Past, Present, and Future. Narendra Kumar, PhD, Bruno Barbosa Sousa, PhD & Swati Sharma, PhD (Eds.)

attach importance to tourism activities on destination choice. Hence, visiting activities are the most priority activities on the choice of Istanbul as a cultural destination while festivals and sports activities are the least. By the way, European tourists prefer different tourism activities according to their nationality. The results indicated that preference for tourism activities has a positive impact on tourists' post-visiting behaviors.

14.1 INTRODUCTION

Leisure time, which is a common concept for tourism and recreation, means "non-productive, consumption of time and time not given to work" according to Veblen (1973). Similarly, Murphy (1974) interprets leisure time as "remaining time after meeting requirements of existence and work in a cycle determined by work" (Özdemir et al., 2016). In short, leisure time is defined as the time when individuals have the right to use freely on their behalf after the mandatory responsibilities and duties related to work and life are fulfilled (Toker et al., 2003). Recreation is defined as "activities that individual voluntarily chooses within a group or individually in leisure time, outside time allocated for work and compulsory needs" (Karaküçük, 1999). Accordingly, if leisure time refers to free time, recreation means voluntary activities carried out at this time. In this regard, recreation includes various leisure activities such as games, sports, entertainment, physical activity, relaxation, hobby, and artistic and cultural activities (Karaküçük, 2016, p. 331). However, it is not accepted that all of the leisure is called recreation (Ibrahim, 1997).

Regarding the relationship between recreation and tourism, they are two intertwined concepts but are activities carried out in the framework of leisure time (Mieczkowski, 1990, p. 37). Leisure is a common point that clarifies and improves the relationship between these two concepts. McKercher (1996) explains the relationship between tourism and recreation as "they often share the same resources, use same facilities, fight each other for same consumer money, create similar effects with same activity and produce common social-psychological outcomes for participants."

Activities are defined as those individuals voluntarily participate and enjoy in both recreation and tourism (Bradley and Inglis, 2012, p. 1167). In this respect, activities are defined as "efforts to rest, relax, increase knowledge and manners, provide satisfaction for psychological and physiological tastes, contribute to social service, fulfill moral belief, and develop creative abilities" (Yetim, 2011, p. 237). For this reason, activities are considered as both recreation and leisure time activities. Tourism activities of Istanbul as a cultural destination are tried to be identified in this study. Notwithstanding, a list of tourism activities of Istanbul was drafted in the study. In this respect, the research aims (1) to validate a list of preference for tourism activities, (2) to reveal the role of preference for tourism activities on tourists' destination choice, and (3) to explain the direct impact of preference for tourism activities on post-visiting behavior. Cultural attractions and destination activities are some of the factors that affect tourists' behavior (Chen and Tsai, 2007). Activity variety based on destination character is effective in changing tourists' attitudes and perceptions toward destinations (Kani et al., 2017). Indeed, since this relationship is not dealt with independently in the studies undertaken, the aforementioned effect is not well understood.

14.2 LITERATURE REVIEW

14.2.1 TOURISM ACTIVITIES

Tourism attractions provide diversification of tourism activities. Additively, activities in a cultural attrcations that constitute cultural sources are themed excursions and sightseeing programs, special interest trips, traditional games, and sports, visiting of original art and architectural structures, local language tutorial activities, visit religious structures/places, traditional food, and beverage tasting, historic places visits (ancient monuments, museums, historic villages, and castles), visiting of of places where historical events took place, participation in festivals and special events, shopping facilities (historic bazaars), traditional craft workshops and artistic activities (theatre, art galleries). Swarbrooke (1999, p. 306) describes these activities as cultural attractions, but tourists need to be offered activities to promote these attractions. In short, cultural attractions are a source of destination for tourism, while activities can be mediators increasing the attractiveness of the destination by using these attractions.

Some researchers have explained tourism activities in the classification of activities. Bucher examined the activities as cultural under four headings: music pursuits, dance activities, arts and small hand skills, and stage works (Bucher, 1989, pp. 240–248). On the other hand, Karaküçük (2001) divided activities into six groups as music activities, sports activities, games, artistic activities, out-of-place activities, and cultural activities (Karaküçük, 2001, p. 84). These groups do not appear explanatory in determining cultural activities. Although the activities were tried to be classified, tourism activities that were closely related to each other remained uncertain. For example, Dumazedier's (1974) activity classification is in the form of artistic, intellectual, social, practical, and physical activities. Activities related to handicrafts are evaluated as practical activities. Practical activities refer to activities in which individuals actively participate. Accordingly, for this classification to be valid, individuals practically must participate in an activity. However, individuals' passive interest in handicrafts rather than actively makes it difficult to evaluate these activities practically. The classification of activities in the literature is similar. It is noteworthy that activities are especially focused on sports (Hurd and Anderson, 2017; Jennings, 2007) and these activities are emphasized as classification. Therefore, tourism activities can be given as follows (Karaküçük, 2001; McLean et al., 2008; Torkildsen, 2012): sports activities, tours, historical site visits, museum visits, monument visits, religious building visits, cinema, literature, concert events, cultural events, local games, art gallery visits, local festivals, traditional crafts, local food tastings, religious rituals, and artistic activities such as ballet, dance, and opera.

14.2.2 TOURISM ACTIVITIES AND DESTINATION CHOICE BEHAVIOR

McKercher et al. (2002) said that each destination has a unique mix of cultural tourism products that will appeal to different types of cultural tourists in different ways. As such, the cultural tourism market can be uniquely divided into each destination based on tourists' behavior in the destination. McKercher et al. (2002) researched activity-based segments for cultural tourism marketing. Generally, tourism activities were investigated for marketing strategies of destinations (e.g. Backman et al., 1995; Law et al., 2004; McKercher et al., 2002). Some researchers identified destination-specific tourism activities (e.g., Law et al., 2004; Moscardo et al., 1996). But a few activity identification lists, have been tested validity and reliability, are available in the literature (e.g., Law et al., 2004; Pizam and Fleischer, 2005; Pizam et al., 2004). Moscardo et al. (1996) are the first researchers among wondering about tourism activities' role on destination choice. Tourism activities were mentioned as motives in their study. They proposed the "Activities Model of Destination Choice." This model revealed tourists' travel benefits change their destination choice. For example, tourists traveling with escape-relaxation purposes preferred nightlife and watersports, thus they give these activities more important in destination choice. Pratiwi et al. (2019) also stated that tourism activities increase a destination's attractiveness. Besides a variety of tourism activities may use for tourist attractions in alternative tourism. The activities are attractive to tourists. However, activities should be used in the best way to increase this (Pratiwi et al., 2019). Tourism activities are also considered as travel motivation (Park et al., 2015). These activities affect satisfaction (Kim et al., 2008; Law et al., 2004). In other words, they have an important role in destination choice and tourist motivation (Kim and Lehto, 2013; Moscardo et al., 1996; Sung et al., 2000). The study of Moscardo et al. (1996), which is one of the rare researches on this subject, found out the relationship between activities and destination choice. Moreover, Pizam et al. (2004) found that participants with high-risk seeking were significantly different in their travel behaviors, destination choice, and preferred tourist activities. Therefore, the following hypotheses are formulated as follows:

Hypothesis 1: Tourism activities have an important role on destination choice.

14.2.3 TOURISM ACTIVITIES AND POST-VISITING BEHAVIOR

Behavioral intention is the action that emerges with feelings of satisfaction or dissatisfaction with consumers' product and service experiences. Therefore, satisfaction is regarded as a predictor of behavioral intention and loyalty in literature. These studies indicate that satisfied visitors have a clear tendency to recommend to others the destination and revisit (Beeho and Prentice, 1997; Bigné et al., 2001; Chen and Tsai, 2007; Hui et al., 2007; Wang and Hsu, 2010). Otherwise, tourists could never visit the destination again (Alegre and Garau, 2010). In essence, they are likely to send negative messages about the destination, because they are not satisfied (Chen and Chen, 2010). Therefore, to increase the attractiveness of the destinations, it is important to recognize the tourist behavior post-visiting. The study states that tourism activities can have a potential impact on post-visiting. It is thought that tourists' behavioral intentions can be positively affected if the tourism activities they prefer are at the destination.

Hypothesis 2: Tourism activities have a positive impact on tourists' post-visiting behaviors.

14.3 METHODOLOGY

The main aims of this study are threefold: (1) to validate the list of tourism activities in Istanbul, (2) to evaluate the importance of tourism activities in a cultural destination choice, and (3) to explain the direct effect of tourism activities on post-visiting behaviors. In this context, the study was carried out in three phases. Phase 1 includes qualitative data for the formation of an initial list to identify tourism activities in Istanbul as a cultural destination. Based on the exploratory research approach appropriate items associated with tourism activities were added to the original list. Phase 2 includes experts' opinions for content validity for the list specified in Phase 1. Phase 3 validates the list of preferences for tourism activities developed in Phase 2 and includes a field examination to assess a model of preference for tourism activities and post-visiting behaviors with regression analysis. So, the interview method and expert opinion form were used in the drafting of the preference list for tourism activities; a structured questionnaire was used in data collection.

14.3.1 PHASE 1: DETERMINATION OF TOURISM ACTIVITIES-AN EXPLORATORY RESEARCH

A tourism activity list for Istanbul was identified in Phase 1. We interviewed 10 tourist guides and 10 international travel agency managers in Istanbul. Each interview took about 30 min. An open-ended interview protocol was used (*Appendix A*). Since data losses such as not remembering, forgetting, and skipping will occur during interviews, interviewees stated that tour programs of Istanbul will be more reliable. Thus, 30 tour programs covering tours in Istanbul were examined. The interviews' records and tour programs were added to the content analysis dataset for qualitative analysis. As a result of the content analysis of both interviews and tour programs, 27 items for tourism activities of Istanbul were determined as given in *Appendix B.*

14.3.2 PHASE 2: CONTENT VALIDITY OF PREFERENCE LIST FOR TOURISM ACTIVITIES

Lawshe's (1975) technique was used to ensure content validity of the list and 20 experts were consulted twice at different times. The first application was realized for a 27-item list of both English and Turkish between December 2018 and January 2019. Since content validity ratios (CVRs) of items did not reach a sufficient level, adjustments were made to list in line with the experts' (scholars in the field of tourism, who have full knowledge of tourism activities of Istanbul) suggestions. Therefore, three items (25th, 26th, and 27th) were eliminated and items were reduced to 21 from 27. 20 experts, mostly the same experts, were re-interviewed between January and March 2019 for the second time. Then, the CVR was examined. Lawshe (1975) stated that the CVR must be looked at 0.05 level of significance depending on the experts' number (for each item with positive value). Therefore, the statistical significance of CVRs obtained is determined by comparing according to content validity criteria (CVR_t). Following 20 experts' opinions, the CVR of items was obtained with help of the formula given "Equality 1." The items have been checked for with CVR criteria 0 (zero) or negative (less than zero). Then it was decided whether items would remain on the final list based on CVR_t in Table 14.1. CVR_t was based on 0.42 for 20 experts. Since no item with less than 0.42 was found, all items were decided to remain on the scale. Hence, content validity index (CVI) was calculated as 0.895 based on CVR_t. The fact that the resulting CVI is greater than CVR_t (0.89 > 0.42) indicates that the content validity of the final list of preferences for tourism activities is statistically significant.

CVR is a substance statistic based on the content validity of whether items are in scale or not, is calculated according to the formula below (Lawshe, 1975):

$$CVR = \frac{N_e}{N/2} - 1 \qquad \textit{Equality. } 1$$

Ne: Number of experts who say "essential" to items
N: Total number of experts
CVR: Content validity ratio between −1 (absolute not essential) and +1 (absolute essential).

TABLE 14.1 The Final List of Tourism Activities.

Items	Essential	Not essential	Might essential	CVR_t
Cultural activities	19	0	1	0.90^+
Religious festivals	20	0	0	1.00^+
Arts activities	19	0	1	0.90^+
Folkloric events	17	0	3	0.70^+
Festivals (music, shopping, etc.)	19	1	0	0.90^+
Active sports activities	19	0	1	0.90^+
Passive sports activities	17	2	1	0.70^+
Religious site visits	19	0	1	0.90^+
Cultural heritage visits	19	0	1	0.90^+
Museum visits	20	0	0	1.00^+
Historical Turkish bath visits	19	0	1	0.90^+
Historical bazaar trips and shopping facilities	20	0	0	1.00^+
Thematic tours	19	1	0	0.90^+
Sightseeing tours	17	1	2	0.70^+
Golden Horn and Bosporus tours	19	0	1	0.90^+
Maiden tower's boat tour	19	0	1	0.90^+
The Princes' Islands tour	20	0	0	1.00^+
Traditional crafts experience	20	0	0	1.00^+
Gastronomy experience	19	0	1	0.90^+
Trip/photographing with historical costumes	19	0	1	0.90^+
Scenic landmarks	19	0	1	0.90^+

Content validity index (mean CVR_t): 0.895.
Total panelists: 20. Min. CVR_t: 0.42 ($p = 0.05$).

14.3.3 PHASE 3: VALIDATION AND MODEL TESTING

14.3.3.1 QUESTIONNAIRE DESIGN

The three-part questionnaire (demographics, preference list for tourism activities, and post-visiting behaviors) was used in data collection. The questionnaire consists of a total of 29 questions demographic information (5), preference for tourism activities (21), and post-visiting behaviors (3). The preference list for tourism activities was asked as on 5 points Likert-type scale (1-very unimportant; 5-very important). The behavioral intention scale (three items) was adopted from Cronin et al. (2000). A 5-points Likert-type scale (1-very low; 5-very high) was used in "*The probability that I will come to Istanbul again is very low/low/moderate/high/very high,* "*The likelihood that I would recommend Istanbul to a friend is very low/low/moderate/high/very high*" and "*If I had to do it over again, my choice would be Istanbul.*" Originally, the English questionnaire has been translated by native speakers of German and Russian. Thus, the questionnaire was conducted in English, German, and Russian.

14.3.3.2 PILOT STUDY

The initial questionnaire form, in English, was implemented in Istanbul on 100 international tourists based on the convenience sampling method. The questionnaire was conducted face to face and took about 5–10 min to complete. Ambiguous sentences are corrected by the tourists in the pilot study. In the following step, the English questionnaire was revised and translated into German and Russian. The final list of tourism activities, with 21 items, which are valid and reliable results following the pilot study was seen as appropriate (Table 14.1).

14.3.3.3 RESEARCH AREA

Istanbul, a major cultural destination of Turkey, was chosen as a research area. There are important determinants as increasing tourist numbers, cultural tourist activity variety, and cultural tourism attributes for this choosing. Moreover, the main factor is marketing and promoting strategies. Istanbul is marketed and promoted as a historical and cultural destination. Especially, the tourists who are interested in historical and cultural tours visit Istanbul. In this regard, the importance of preference for tourism activities in destination choice and post-visiting behaviors is an assumption of the research. As a symbol of its historical and cultural background, Istanbul, which is included in the UNESCO World Heritage List in four regions, is one of the cultural tourism destinations with its cultural values. This is also the number of international tourists visiting Istanbul. In the last 3 years of data (2016–2018) in Istanbul, the total of international tourists visiting Turkey has hosted 30–35%. Istanbul welcomed 13 million tourists in 2018, an increase of 24% from the previous year. This number is equivalent to one-quarter of the total number of tourists visiting Turkey (39 million). Conducting a study on international tourists, from various nationalities, in the most visiting cultural destination will ensure the most comprehensive data. Based on this perspective, researchers thought to support the originality and importance of the study.

14.3.4 DATA COLLECTION

International tourists visiting Istanbul in 2018 are the population of the research. Tourists' distribution visiting Istanbul according to nationalities was used to calculate the sample size. 2018 Tourism Statistics Report publishing by Istanbul Provincial Directorate of Culture and Tourism was used to reach annual international tourist statistics. Accordingly, Istanbul welcomed 13 million tourists in 2018. A purposive sampling method was used. Selection criteria are in the top European tourists visiting Istanbul according to their nationality and staying at least one night in the destination. The sample size is 300 in 95% confidence levels, with an error of 10 and 1.25% standard deviation for infinite population formula. Based on research, data were collected from European (*German, Russian, French, Ukrainian, and Dutch*) tourists in July and October 2019. The sample consists of travelers touring both individual and by package tour. The questionnaires, in three languages, were distributed to volunteer tourists in the most visited points in Historical Peninsula (including Sultanahmet Square, Topkapi Palace, and Blue Mosque), UNESCO registered heritage site. Moreover, questionnaires were sent to accommodation businesses, travel agencies, and tourist guides. Totally 550 questionnaires were distributed and 470 responses were received. This detailed data collection process provided a nearly 85% response rate. Omitting 28 questionnaires with missing a total of 442 European tourists have participated in the research. Thus, 442 valid data have been obtained.

14.4 DATA ANALYSIS

Descriptive statistics and multiple regression analysis were used. In regression analysis, the linear model enter method was preferred. Since multivariate statistical techniques were used in data analysis, extreme values were determined for multivariate at 0.001 significance level based on Mahalanobis distance. Accordingly, 441 data were analyzed except for an extreme value.

14.5 RESULTS

14.5.1 SAMPLE PROFILE

According to the demographic profile, female tourists (55.1%) is more than males. Tourists are mostly between 25 and 34 years of age (33.1%) and married (50.1%). The majority have a bachelor's (51.7%) and a Master's (16.6%) degree. So, overall the sample is female, young, married, and educated (Table 14.2). Most German tourists are male (69.2%), married (69.2%), bachelor (30.8%), and 18–24 (30.8%) years old. Furthermore, most of the Russian tourists are female (68.9%), single (46.6%), bachelor (71.4%), and 25–34 (26.2%) years old. British tourists are mostly female (56.5%), married (60.9%), bachelor (52.2%), and 28–24 (47.8%) years old. However, French tourists are generally male (64.7%), married (76.5%), masters (58.8%), and 25–34 (58.8%) years old. Most Ukrainian tourists are female (50.0%), bachelor (50.0%), and 45–64 (44.4%) years old but the score of their marital status is equal (33.3%). Most Dutch tourists are female (62.5%), single (87.5%), bachelor (37.5%) and Ph.D. (37.5%), and 35–44 (50.0%) years old.

TABLE 14.2 Sample Profile.

	Nationality													
	German (*n* = 78)		Russian (*n* = 206)		British (*n* = 46)		French (*n* = 51)		Ukrainian (*n* = 36)		Dutch (*n* = 24)		Total (*n* = 441)	
Gender	n	%	N	%	n	%	n	%	n	%	n	%	n	%
Male	54	69.2	54	26.2	20	43.5	33	64.7	16	44.4	9	37.5	186	42.2
Female	24	30.8	142	68.9	26	56.5	18	35.3	18	50.0	15	62.5	243	55.1
No answer	0	0	10	4.9	0	0	0	0	2	5.6	0	0	12	2.7
Age														
18–24	24	30.8	29	14.1	22	47.8	9	17.6	6	16.7	0	0	90	20.4
25–34	33	42.3	54	26.2	16	34.8	30	58.8	4	11.1	9	37.5	146	33.1
35–44	9	11.5	41	19.9	0	0	0	0	4	11.1	12	50.0	66	15.0
45–54	12	15.4	32	15.5	6	13.0	6	11.8	8	22.2	3	12.5	67	15.2
55–64	0	0	38	18.4	2	4.3	6	11.8	8	22.2	0	0	54	12.2
65–74	0	0	8	3.9	0	0	0	0	4	11.1	0	0	12	2.7
No answer	0	0	4	1.9	0	0	0	0	2	5.6	0	0	6	1.4
Marital status														
Married	54	69.2	85	41.3	28	60.9	39	76.5	12	33.3	3	12.5	221	50.1
Single	24	30.8	96	46.6	18	39.1	12	23.5	12	33.3	21	87.5	183	41.5
No answer	0	0	25	12.1	0	0	0	0	12	33.3	0	0	37	8.4
Education														
Primary school	0	0	1	0.5	0	0	0	0	0	0	3	12.5	4	0.9
High school	18	23.1	6	2.9	8	17.4	3	5.9	4	11.1	0	0	39	8.8
Associate degree	9	11.5	32	15.5	4	8.7	3	5.9	12	33.3	0	0	60	13.6
Bachelor	24	30.8	147	71.4	24	52.2	6	11.8	18	50.0	9	37.5	228	51.7

TABLE 14.2 *(Continued)*

	Nationality													
	German (n = 78)		Russian (n = 206)		British (n = 46)		French (n = 51)		Ukrainian (n = 36)		Dutch (n = 24)		Total (n = 441)	
Gender	**n**	**%**	**N**	**%**	**n**	**%**	**n**	**%**	**n**	**%**	**n**	**%**	**n**	**%**
Masters	18	23.1	10	4.9	10	21.7	30	58.8	2	5.6	3	12.5	73	16.6
Ph.D.	9	11.5	4	1.9	0	0	17.6	17.6	0	0	9	37.5	31	7.0
No answer	0	0	6	2.9	0	0	0	0	0	0	0	0	6	1.4

14.5.2 RELIABILITY

Internal consistency of the scale was evaluated by Cronbach's alpha (α), a widely used measure of reliability. All global scores were assessed using this approach, and Cronbach's α values >0.70 were considered satisfactory (Nunnally, 1994). Cronbach's α values were above the threshold of 0.70 (Table 14.3). This coefficient means a high level of internal consistency of the scale (Özdamar, 2001). Item-total correlations were evaluated by Pearson's product-moment correlation coefficient (r). The correlation between each item and scores was assessed; correlation values >0.20 were considered satisfactory (Ebel, 1965; Erkuş, 2016; Wang et al., 2017). Tourism activities showed a satisfactory correlation with the index score (Pearson's $r > 0.2$) and ranged from 0.32 to 0.68. Thus, it is stated that tourism activities are valid and reliable in explaining the importance of preference for tourism activities in destination choice. The behavior intention scale's reliability is 0.83, thus a reliable scale was used for measuring post-visiting tourist behavior.

14.5.3 VALIDITY

14.5.3.1 CONSTRUCT VALIDITY

An exploratory factor analysis (EFA) was performed on 441 cases to explore the underlying structure of tourism activities. In the factor analysis, it is expected that the Bartlett test result will be in significance level and data will come from multiple normal distributions since a high correlation relationship is searched between the variables. According to EFA results, the Bartlett test was found as significant (3990, 385; $p < 0.001$), and the sample size was sufficient since Kaiser–Meyer–Olkin (KMO) ranges from 0 to 1, with 0.83 considered highly suitable for factor analysis (Gorsuch, 1988; Hair et al., 1998). According to these results, data were accepted as suitable following factor analysis.

The total variance and eigen value of the *Tourism Activities* list described initial and after rotation are exported in Table 14.3. Common variance refers to the amount of variance that one item shares with other items. It is expected to be at least 0.30 (Tabachnick and Fidell, 2007). Similarly, it is recommended that cross-loadings of items should be 0.50 or higher (Karasar, 1999, p. 140). By these values, the common variance sub-limit is 0.49; the lower limit of factor loading was found as 0.50. All factor loadings were retained more than 0.50.

TABLE 14.3 EFA Summaries of Tourism Activities.

Components	M	Common variance	Factor load	Eigen value	EV (%)	Cronbach's α
City tours and landscape activities	3.80			3.472	16.5	0.811
Sightseeing tours	3.91	0.671	0.727			
Maiden Tower's boat tour	3.81	0.579	0.711			
Thematic tours	3.64	0.648	0.750			
The Princes' Islands tour	3.84	0.598	0.699			
Golden Horn and Bosporus tours	4.14	0.664	0.665			
Scenic landmarks	4.06	0.570	0.501			
Festivals and sports activities	2.98			2.895	13.7	0.756
Passive sports activities	2.83	0.738	0.754			
Active sports activities	3.05	0.671	0.763			
Festivals	3.05	0.573	0.624			
Artistic and religious activities	3.13			2.424	11.5	0.793
Arts activities	3.30	0.647	0.695			
Folkloric events	3.16	0.653	0.506			
Religious festivals	3.02	0.532	0.584			
Trip/photographing with historical costumes	2.95	0.490	0.503			

TABLE 14.3 *(Continued)*

Components	M	Common variance	Factor load	Eigen value	EV (%)	Cronbach's α
Traditional crafts experience	3.24	0.694	0.670			
Visiting activities	4.08			2.244	10.6	0.706
Cultural heritage visits	4.35	0.722	0.740			
Museum visits	4.11	0.694	0.692			
Religious site visits	4.08	0.567	0.653			
Historical Turkish bath (Hamam) visits	3.90	0.549	0.570			
Authentic experience activities	3.73			2.124	10.1	0.703
Gastronomy experience	3.94	0.657	0.761			
Historical Bazaar trips and shopping facilities	4.04	0.536	0.625			
Cultural activities	3.65	0.708	0.667			

EV, explained variance by each factor; Cα, Cronbach's α; M, mean.

Cronbach's coefficient alpha = 0 887.

Total variance explained (%) 62.6; M: 3.57; SD: 0.586 ; *n*: 441.

KMO: 0.830; Barlett test: 3990.385; $p < 0.001$.

Extraction method: Principal component analysis.

Rotation method: Varimax with Kaiser normalization.

As a result of the EFA principal component analysis and varimax rotation method, a construct of 21 items and four dimensions were obtained, explaining 62.6% of the total variance. Total variance explained is considered sufficient because it is located at the upper limit of the range of 40–60% for social sciences (Karagöz, 2017, p. 404). Furthermore, in social science, explained variance is commonly as low as 50–60% (Pett et al., 2003). So, Table 14.5 demonstrates a cumulative percentage of the variance of 62.6% and a total of five dimensions having an eigenvalue (λ) ≥1—*city tours and landscape activities* (six items), *festivals and sports activities* and (three items), *artistic and religious activities* (five items), *visiting activities* (four items), and *authentic experience activities* (three items).

According to the results obtained from EFA, factor loadings of *city tours and landscape activities* ranged from 0.50 to 0.75 and explained 16.5% of the total variance. *Festivals and sports activities* factor loadings ranged from 0.62 to 0.75 and explained 13.7% of the total variance. The factor loading of *artistic and religious activities* was between 0.50 and 0.69, so this component explained 11.5% of the total variance. *Visiting activities* factor loading ranged from 0.57 to 0.74 and 10.6% of the total variance was explained. Finally, *authentic experience activities* factor loadings were between 0.62 and 0.76 with 10.1% of the total variance. The factor loadings as well as factor means, common variances, eigenvalues, explained variance, and reliability values are shown in Table 14.3. The results indicated approximately 62.6% of the importance given to tourism activities can be determined with the tourism activities list. Cronbach's alpha's internal consistency coefficient for the preference list for tourism activities remained at 0.887. The Cronbach's alpha scores for components (0.811, 0.756, 0.793, 0.706, and 0.703) were exceeded the recommended criterion of 0.70, pointing reliability.

14.5.3.2 DISCRIMINANT VALIDITY

Discriminant validity is based on the principle that similar scales aiming to measure different trends have a negative correlation, low level of relationship, or no relation at all. Discriminant validity reveals the discrimination of factors obtained after explanatory factor analysis and supports the construct validity of the scales. It is defined as a low-level correlation from 0.0 to 0.3, medium correlation from 0.3 to 0.7, and a high-level correlation value from 0.7 to 1.0 according to levels (Büyüköztürk, 2002). In this context, to provide discriminant validity, correlation coefficients between factors are expected to be low or medium (Bearden and Netemeyer, 1999).

TABLE 14.4 Discriminant Validity of Tourism Activity List.

Components	**M**	**SD**	**1**	**2**	**3**	**4**	**5**
City tours and landscape activities	3.80	0.70	1				
Festivals and sports activities	2.98	0.92	0.391*	1			
Artistic and religious activities	3.13	0.85	0.464*	0.633*	1		
Visiting activities	4.08	0.69	0.400*	0.289*	0.423*	1	
Authentic experience activities	3.73	0.85	0.259*	0.308*	0.529*	0.476*	1

*$p < 0.01$.

Pearson correlations were used in evaluating discriminant validity in the study (Table 14.4). Accordingly, the relationship between the five components of tourism activities has been examined. It was determined that there are significant positive correlations between factors ($p < 0.01$). However, since these relationships show a medium and low correlation between the factors (from 0.25 to 0.63), the discriminant validity of the tourism activities list is accepted given Bearden and Netemeyer (1999).

These activities have a mean of 3.57 for European tourists in choosing Istanbul as a cultural destination. When differences between groups are examined, the importance of preference for tourism activities in destination choice differs significantly ($p < 0.05$; $p < 0.10$). Table 14.5 reveals the European tourists' preferences for tourism activities on destination choice. The tourism activities have importance for European tourists on destination choice, respectively, are visiting activities (4.08), city tours and landscape activities (3.80), authentic experience activities (3.73), artistic and religious activities (3.13), and festivals and sports activities (2.98). There is a significant difference in importance given to the activities according to tourists' nationalities. Visiting activities are the most important activities for German, Russian, Ukrainian, and Dutch tourists on their destination choice whereas authentic experience activities are the most priority activities for British and French tourists. Particularly, German tourists give more importance to tourism activities than Dutch tourists on destination choice. However, it is stated that British tourists need tourism activities more than Russian, Ukrainian, and Dutch on destination choice.

The difference between groups in city tours and landscape activities is the following ($F = 7.603$; $p < 0.05$). These activities are more important for German and Russian according to Dutch tourists. However, Russian, British, and Ukrainian tourists pointed out that the activities are more priority for them according to French and Dutch tourists. There is no significant difference between groups for festivals and sports activities. Besides, artistic and religious activities have a significant difference between European tourist groups ($F = 4.523$; $p < 0.05$). Especially, it is noticed that these activities have less importance on destination choice for Ukrainian tourists than German, British, and French tourists. Similar to Ukrainian, Russian tourists prefer artistic and religious activities more than German and British ($F = 1.955$; $p < 0.05$). Moreover, German tourists' preferences for visiting activities are more than Russian and Dutch. It is noteworthy that tourists' preferences of authentic experience activities

TABLE 14.5 European Tourists' Preferences of Tourism Activities on Destination Choice.

Tourism activities	Nationality														F	p	Dif.
	1		2		3		4		5		6						
	German (n = 78)		Russian (n = 206)		British (n = 46)		French (n = 51)		Ukrainian (n = 36)		Dutch (n = 24)		Total (n = 441)				
	M	SD	M	SD	M	SD	M	SD	M	SD	M	SD	M	SD			
City tours and landscape activities	3.69	0.72	3.97	0.60	3.37	0.63	3.50	0.85	3.82	0.74	3.31	0.61	3.80	0.70	7.603	0.00**	**6 < 1 1; 4; 6 < 2 4; 6 < 3 4; 6 < 5**
Festivals and sports activities	3.08	0.96	2.29	0.83	3.21	0.93	2.90	1.10	3.07	1.08	2.70	0.73	2.98	0.92	1.539	0.17	--
Artistic and religious activities	3.34	0.87	3.01	0.83	3.46	0.72	3.25	0.87	2.82	1.01	3.12	0.11	3.13	0.85	4.523	0.00**	**2; 5 < 1 2; 5 < 3 5 < 4**
Visiting activities	4.25	0.47	4.01	0.71	3.85	0.58	4.05	0.79	3.93	0.48	3.93	1.16	4.08	0.69	1.955	0.08*	**2; 6 < 1**
Authentic experience activities	4.02	0.69	3.44	0.80	4.20	0.65	4.54	0.47	3.66	0.85	3.66	1.19	3.73	0.85	22.122	0.00**	**2; 5; 6 < 1 2; 5 < 3 1; 2; 3; 5; 6 < 4**
Tourism activities (whole)	3.67	0.56	3.52	0.03	3.75	0.54	3.61	0.67	3.46	0.62	3.35	0.59	**3.57**	0.58	2.713	0.02**	**6 < 1 2; 5; 6 < 3**

M, Mean; SD, Standard Deviation; Dif, Difference between groups

$^{*}p < 0.10$; $^{**}p < 0.05$.

Total M: 3.57; SD: 0.58.

POSTHOC = LSD ALPHA (0.05).

differ significantly between groups ($F = 22.122$; $p < 0.05$). German, British, and French tourists prefer these activities more than Ukrainian and Russian. Especially, these activities are more significant on the destination choice of French tourists than the other five tourist groups. Findings showed that visiting activities are the most priority activities on the choice of Istanbul as a cultural destination while festivals and sports activities are the least. Therefore, findings supported the hypothesis of "*H1: Tourism activities have an important role on destination choice.*"

TABLE 14.6 The Impact of Tourism Activities on Post-Visiting Behaviors.

	Tourism activities	**Post-visiting behaviors**				
		β	**t**	**p (t)**		
Model-1	City tours and landscape activities	0.047	0.632	0.35	R	0.443
	Festivals and sports activities	−0.003	−0.049	0.96	ΔR^2	0.187
	Artistic and religious activities	0.042	0.641	0.52	F	21.241
	Visiting activities	0.186	0.943	0.00*	*P*	0.00*
	Authentic experience activities	0.275	3.585	0.00*	0	
Model-2	Tourism activities (whole)	0.375	8.482	0.00*	R	0.375
					ΔR^2	0.139
					F	71.978
					P	0.00*

*$p < 0.05$.

The model established in Table 14.6 was found significance in the relationship between preference for tourism activities and post-visiting behaviors ($F = 21.241$; $p < 0.05$). It was revealed that there was a moderately significant positive relationship between tourism activities and post-visiting behaviors ($R = 0.443$). 18.7% of the change in post-visiting behaviors was explained by tourism activities components ($\Delta R^2 = 0.187$). Considering the effect of the independent variable – *preference for tourism activities* – on the dependent variable, *post-visiting behaviors*, it was revealed that visiting activities and authentic experience activities have a significant positive impact on post-visiting behaviors ($\beta_4 = 0.186$; $\beta_5 = 0.275$). The activities showing the highest impact on post-visiting behaviors are, respectively, "authentic experience activities ($\beta_5 = 0.275$)." Overall tourism activities explain 13.9% of post-visiting behavior ($\Delta R^2 =$

0.139). This result indicates that tourism activities have a positive impact on post-visiting behaviors ($p < 0.05$). Meanwhile, findings supported the hypothesis of "*H2: Tourism activities have a positive impact on post-visiting behaviors.*"

14.6 DISCUSSION AND CONCLUSION

Tourism activities are significant on the choice of Istanbul as a cultural tourism destination, and tourism activities impact tourists' post-visiting behaviors. Results noticed that visiting activities are of higher importance than other activities on destination choice for European tourists while festivals and sports activities are of lower importance. The result looks similar to previous studies (Law et al., 2004; Park et al., 2015) in literature. Law et al. (2004) mentioned that Hong Kong travelers mostly preferred activities about local food, relaxing, and sightseeing, visiting scenic landmarks, and shopping. Also, they stated that tourists preferred sports activities at least. In this respect, these results are so similar to the study of Law et al. (2004). Tourism activities have an important role respectively on British, German, French, Russian, Ukrainian, and Dutch tourists' choice of Istanbul. Visiting activities have higher importance for German, Russian, Ukrainian, and Dutch tourists although authentic experience activities are more priority for British and French on destination choice. According to a study by Park et al. (2015), European tourists prefer historical and cultural activities while Asian tourists are interested in local product shopping. Similarly, the results of this study underlined that European tourists mostly prefer visiting activities, city tours and landscape activities, authentic experience activities, and artistic and religious activities rather than festivals and sports activities in the destination they want to go to.

Tourists' preferences show that these activities are an impact on tourists' destination choices as Timothy (1997) mentioned. However, as van der Ark and Richards (2006) stated in their study that tourists chose destinations especially for participating in activities, this study also shows that visiting activities and city tours and landscape activities are among tourism activities that are considered important on destination choice. According to Moscardo et al. (1996), tourists, who obtain information about their destination through information sources before coming, make their destination choices in line with this information. Accordingly, it can be said that tourism activities that tourists are informed about before coming to a destination are an important factor on destination choice. However, tourism activities' preference differs according to culture and interests of tourists. For example, Honkanen (2002) claims that many tourists visit cultural attractions for other reasons than their cultural enlightenment. McKercher and Chow So-Ming (2001) defined cultural distance as a discriminator. Their research suggests that the greater difference between tourists' own culture and destination culture, the greater motivation of international tourists to learn more about the culture of the destination. When looking at the nationality of tourists, it is seen that especially British tourists give more importance to tourism activities than tourists (German, French, Russian, Ukrainian, and Dutch) from a closer distance. It is assumed that Russian and Ukrainian tourists give less importance to tourism activities than other tourists (except Dutch) on the choice of Istanbul because Turkey is closer to Russia and Ukraine than Britain, Germany, France, and Netherlands. This result partly supports McKercher and Chow So-Ming (2001). Timothy (1997) says that the search for cultural connections is less important for culturally close tourists because they already feel connected. Therefore, they visit destinations as recreational rather than cultural. As Timothy (1997) said, tourists are now motivated for recreational travel as well as their cultural connectivity. Tourists want to learn the culture of the destination in the fastest and most fun way, regardless of their purpose. At this point tourism activities have key role.

Tourists' tourism activities preferences vary depending on their characteristics. Honkanen (2002) emphasized that Finland is appropriate for nature-based tourism as well as cultural attractiveness and cultural tourism, which explores Finland's tourism characteristics; festivals, cultural heritage, and museum visits by tourists are the most interesting charms. According to this research, the cultural destination identity of Istanbul is dominant compared to other tourism types and at the same time, similar to the result of Honkanen (2002), cultural heritage areas, museums, religious sites, and historical Turkish Baths are seen to be the most popular activities in Istanbul. Although there are few studies in the literature explaining the relationship between tourism activities and post-visiting behaviors, this research confirms the opinion of Kani et al. (2017) "The diversity of activities according to the character of each destination is effective in changing attitudes and perceptions of tourists toward destinations." However, this research scientifically reveals the relationship between tourism activities and post-visiting behaviors, and the impact of tourism activities on post-visiting behaviors. Results showed that there is a significant positive relationship between tourism activities and post-visiting behaviors. Overall tourism activities have a positive impact on tourists' post-visiting behaviors. The positive impact of tourism activities on post-visiting behaviors partially supports the study done by Kani et al. (2017).

14.7 MANAGERIAL IMPLICATIONS

The study revealed that tourism activities are important to the tourists' destination choice. First of all, local governments, tourism professional organizations, and tourism enterprises should know that activities are important for the destination choice of potential tourists. So, they should ensure the effective marketing. In package tour programs, it is thought that visits and recommendations of the destination will increase by introducing and drawing attention to little or no known activities specific to a destination. The low importance of festivals and sports activities on choosing Istanbul as a destination can be explained by the fact that tourists do not have enough information about these activities or by focusing on certain activities in tour programs. Istanbul, which has potential especially with sports events and festivals, is expected to work more effectively on the promotion of these activities.

The interest and importance of the activities differ according to nationality in the study allows reviewing tour programs especially in Istanbul. Especially, the travel agencies that organize nationality-oriented package tours are expected to provide tourist satisfaction by knowing the activities that tourists are interested in. According to the variety of existing resources of destinations, tourism activities of destinations are adapted to the types of tourism. Istanbul's more adaptation to cultural tourism with its resources has enabled cultural activities to come to the fore. However, it should be known by the authorities that such activities have a positive impact on tourists' post-visiting intentions. Among the activity dimensions, visiting activities, city tours, and landscape activities, and authentic experience activities are the most popular activities in Istanbul for all four international tourist groups. They are mostly cultural activities. Thus, tourism businesses in Istanbul should try to reveal cultural activities that are behind Istanbul through tours. The Provincial Culture and Tourism Directorates should research to define tourists' profile at certain intervals and tourists' activity preference with the field expert. This will provide that destinations are better recognized and promoted.

14.8 LIMITATIONS AND FUTURE RESEARCH

A limitation of this research is that it is difficult to make broad generalizations about tourist behavior for all destinations. Repeating this work elsewhere will undoubtedly produce quite different segments. Therefore, the results should be evaluated under some limitations. Firstly, the sample was collected from European tourists. Similar research could be implemented on local tourists. Secondly, this study was carried out only in one country and one city. So, it would be interesting to carry out such studies in different destinations using more representative samples. More locations need to be involved to enhance the scope and generalizability of the results. Thirdly, the activities included in this study are relevant to Istanbul only; hence to apply this list of preferences in other cultures or locations requires further refining of the list by including activities related to that particular location or culture. This study will assist future research in measuring the importance of activity choice in destinations. The research has contributed to defining the list of tourism activities preferences which represents a way to look at the role of activities on cultural destination choice. It should be considered as a complement for measuring the importance of tourism activities in destinations and the impact on tourists' post-visiting intentions. In the future, the appropriateness of tourism activity preferences can be tested across different cultures.

ACKNOWLEDGMENT

This study is a section of Ph.D. dissertation completed in the department of Tourism Management at Istanbul University, Istanbul, Turkey.

KEYWORDS

- **tourism activities**
- **destination choice**
- **behavioral intention**
- **destination activities**

REFERENCES

Alegre, J.; Garau, J. Tourist Satisfaction and Dissatisfaction. *Ann. Tour. Res.* **2010,** *37* (1), 52–73. https://doi.org/10.1016/j.annals.2009.07.001

Backman, K.; Backman, S.; Uysal, M.; Sunshine, K. M. Event Tourism: An Examination of Motivations and Activities. *Fest. Manage. Event Tour.* **1995,** *3* (1), 15–24.

Bearden, W. O.; Netemeyer, R. G. *Handbook of Marketing Scales: Multi-Item Measures for Marketing and Consumer Behavior Research*; Sage: Thousand Oaks, CA, 1999.

Beeho, A. J.; Prentice, R. C. Conceptualizing the Experiences of Heritage Tourists: A Case Study of New Lanark World Heritage Village. *Tour. Manage.* **1997,** *18* (2), 75–87. https://doi.org/10.1016/S0261-5177(96)00103-3

Bigné, J. E.; Sánchez, M. I.; Sánchez, J. Tourism Image, Evaluation Variables and after Purhase Behaviour: Inter-Relationship. *Tour. Manage.* **2001,** *22* (6), 607–616. https://doi.org/10.1016/S0261-5177(01)00035-8

Bradley, G. L.; Inglis, B. C. Adolescent Leisure Dimensions, Psychosocial Adjustment, and Gender Effects. *J. Adolesc.* **2012,** *35* (5), 1167–1176.

Bucher, A. C. *Foundations of Physical Education*; Mosby: Saint Louis, 1989.

Büyüköztürk, Ş. *Sosyal bilimler için veri analizi el kitabı: İstatistik Araştırma Deseni, SPSS Uygulamaları ve Yorum*; Pegem Akademi: Ankara, 2002.

Chen, C.-F.; Chen, F.-S. Experience Quality, Perceived Value, Satisfaction and Behavioral Intentions for Heritage Tourists. *Tour. Manage.* **2010,** *31* (1), 29–35. https://doi.org/10.1016/j.tourman.2009.02.008

Chen, C. F.; Tsai, D. C. How Destination Image and Evaluative Factors Affect Behavioral Intentions? *Tour. Manage.* **2007,** *28* (4), 1115–1122. https://doi.org/10.1016/j.tourman.2006.07.007

Cronin, J. J.; Brady, M. K.; Hult, G. T. M. Assessing the Effects of Quality, Value, and Customer Satisfaction on Consumer Behavioral Intentions in Service Environments. *J. Retail.* **2000,** *76* (2), 193–218. https://doi.org/10.1016/S0022-4359(00)00028-2

Dumazedier, J. *Sociology of Leisure*; Elsevier Scientific Publ. Co., 1974.

Ebel, R. L. *Measuring Educational Achievement*; Englewood Cliffs: New Jersey, 1965.

Erkuş, A. *Psikolojide ölçme ve ölçek geliştirme-I: temel kavramlar ve işlemler*, 3rd ed.; Pegem Akademi: Ankara, 2016.

Gorsuch, R. L. Exploratory Factor Analysis. In *Handbook of Multivariate Experimental Psychology*; Nesselroade, J. R., Cattell, R. B., Eds.; Springer: Berlin, 1988; pp 231–258.

Hair, J. F.; Black, W. C.; Babin, B. J.; Anderson, R. E.; Tatham, R. L. *Multivariate Data Analysis*, Vol. 5; Prentice Hall: Upper Saddle River, NJ, 1998.

Honkanen, A. Churches and Statues: Cultural Tourism in Finland. *Tour. Hosp. Res.* **2002,** *3* (4), 371–379.

Hui, T. K.; Wan, D.; Ho, A. Tourists' Satisfaction, Recommendation and Revisiting Singapore. *Tour. Manage.* **2007,** *28* (4), 965–975. https://doi.org/10.1016/j.tourman.2006.08.008

Hurd, A.; ve Anderson, D. M. *Kraus' Recreation ve Leisure in Modern Society*; Jones ve Bartlett Learning, 2017.

Ibrahim, H. Atipik boş zaman faaliyetleri. *İstanbul: Cogito Dergisi, Yapı Kredi Yayıncılık* **1997,** *12* (10), 181–189.

Jennings, G. Water-based Tourism, Sport, Leisure, and Recreation Experiences. In *Water-Based Tourism, Sport, Leisure, and Recreation Experiences*; Jennings, G., Ed.; Routledge: London, UK, 2007; pp 19–38.

Kani, Y.; Aziz, Y. A.; Sambasivan, M.; Bojei, J. Antecedents and Outcomes of Destination Image of Malaysia. *J. Hosp. Tour. Manage.* **2017,** *32*, 89–98. https://doi.org/10.1016/j.jhtm.2017.05.001

Karagöz, Y. *Spss ve amos uygulamaları nitel-nicel karma bilimsel araştırma yöntemleri ve yayın etiği*; Nobel Yayın-Dağıtım: Ankara, 2017.

Karaküçük, S. *Rekreasyon boş zamanları değerlendirme*, 3rd ed.; Gazi Kitapevi: Ankara, 1999.

Karaküçük, S. *Rekreasyon boş zamanları değerlendirme*, 4th ed.; Gazi Kitabevi: Ankara, 2001.

Karaküçük, S. *Rekreasyon bilimi*, 3rd ed.; Gazi Kitapevi: Ankara, 2016.

Karasar, N. *Bilimsel araştırma yöntemi*; Nobel: Ankara, 1999.

Kim, S.; Lehto, X. Y. Projected and Perceived Destination Brand Personalities: The Case of South Korea. *J. Travel Res.* **2013,** *52* (1), 117–130. https://doi.org/10.1177/0047287512457259

Law, R.; Cheung, C.; Lo, A. The Relevance of Profiling Travel Activities for Improving Destination Marketing Strategies. *Int. J. Contemp. Hosp. Manage.* **2004,** *16* (6), 355–362. https://doi.org/10.1108/09596110410550798

Lawshe, C. H. A Quantitative Approach to Content Validity. *Personnel Psychol.* **1975,** *28* (4), 563–575. https://doi.org/10.1111/j.1744-6570.1975.tb01393.x

McKercher, B. Differences between Tourism and Recreation in Parks. *Ann. Tour. Res.* **1996,** *23* (3), 563–575.

McKercher, B.; Chow So-Ming, B. Cultural Distance and Participation in Cultural Tourism. *Pacific Tour. Rev.* **2001,** *5* (1–2), 23–32.

McKercher, B.; PhD, P. S. Y. H.; Cros, H. Du; So-Ming, B. C. Activities-Based Segmentation of the Cultural Tourism Market. *J. Travel Tour. Market.* **2002,** *12* (1), 23–46. https://doi.org/10.1300/J073v12n01_02

McLean, D. D.; Hurd, A.; Rogers, N. *Kraus' Recreation and Leisure in Modern Society*; Jones & Barlett Publishers: Sudbury, MA, 2008.

Mieczkowski, Z. *World Trends in Tourism and Recreation*; Peter Lang Publishing: New York, 1990.

Moscardo, G.; Morrison, A. M.; Pearce, P. L.; Lang, C.-T.; O'Leary, J. T. Understanding Vacation Destination Choice through Travel Motivation and Activities. *J. Vacation Market.* **1996,** *2* (2), 109–122. https://doi.org/10.1177/135676679600200202

Murphy, J. F. Leisure Determinants of Life Style. *J. Health Phys. Educ. Recreat.* **1974,** *45* (9), 35–37.

Nunnally, J. C. *Psychometric Theory 3E*; Tata McGraw-Hill Education, 1994.

Özdamar, K. *Spss ile biyoistatistik*, 4th ed.; Eskişehir: Kaan Kitapevi, 2001.

Özdemir, A.; Güçer, E.; Karaküçük, S. Rekreasyon ve turizm. In *Rekreasyon bilimi*; Karaküçük, S., Ed.; Gazi Kitapevi: Ankara, 2016; pp 313–392.

Park, S. H.; Lee, C. K.; Miller, J. C. A Comparative Study of the Motivations, Activities, Overall Satisfaction, and Post-Trip Behaviors of International Tourists in Macau: Mainland Chinese, Hongkongese, Taiwanese, and Westerners. *Asia Pacific J. Tour. Res.* **2015,** *20* (10), 1174–1193. https://doi.org/10.1080/10941665.2014.965184

Pett, M. A.; Lackey, N. R.; Sullivan, J. J. *Making Sense of Factor Analysis: The Use of Factor Analysis for Instrument Development in Health Care Research*; Sage Publications Inc: Thousand Oaks, CA, 2003.

Pizam, A.; Fleischer, A. The Relationship between Cultural Characteristics and Preference for Active vs. Passive tourist activities. *J. Hosp. Leisure Market.* **2005,** *12* (4), 5–25. https://doi.org/10.1300/J150v12n04_02

Pizam, A.; Jeong, G. H.; Reichel, A.; van Boemmel, H.; Lusson, J. M.; Steynberg, L.; Montmany, N. The Relationship between Risk-Taking, Sensation-Seeking, and the Tourist Behavior of Young Adults: A Cross-Cultural Study. *J. Travel Res.* **2004,** *42* (3), 251–260. https://doi.org/10.1177/0047287503258837

Pratiwi, Y. P.; Sutono, A.; Suherlan, H. Tourism Activities Development Based on Sustainable Livelihood Approach in the Tourism Villages of Purwakarta Regency. *J. Adv. Soc. Sci. Hum.* **2019,** *5* (2), 588–591.

Sung, H. Y.; Morrison, A. M.; O'leary, J. T. Segmenting the Adventure Travel Market by Activities: From the North American Industry Providers' Perspective. *J. Travel Tour. Market.* **2000,** *9* (4), 1–20. https://doi.org/10.1300/J073v09n04_01

Swarbrooke, J. *Sustainable Tourism Management*; CABI Pub, 1999. Retrieved from https://books.google.com.tr/books?id=p_NOAAAAMAAJ

Tabachnick, B. G.; Fidell, L. S. *Using Multivariatestatistics*, 5th ed.; Allynand Bacon: Boston, 2007.

Timothy, D. J. Tourism and the Personal Heritage Experience. *Ann. Tour. Res.* **1997,** *24* (3), 751–754. https://doi.org/10.1016/s0160-7383(97)00006-6

Toker, F.; Çankaya, C.; Doğan, M. Uludağ Üniversitesinde Öğrenim Gören Yabancı Uyruklu Öğrencilerin Boş Zaman Etkinliklerine Katılımlarının Araştırılması. *Uludağ Üniversitesi Eğitim Fakültesi Dergisi* **2003,** *17* (1), 219–233.

Torkildsen, G. *Leisure and Recreation Management*; Routledge: London, New York, 2012.

Van Der Ark, L. A.; Richards, G. Attractiveness of Cultural Activities in European Cities: A Latent Class Approach. *Tour. Manage.* **2006,** *27* (6), 1408–1413.

Veblen, T. *The Theory of the Leisure Class*; Houghton Mifflin: Boston, 1973.

Wang, C. Y.; Hsu, M. K. The Relationships of Destination Image, Satisfaction, and Behavioral Intentions: An Integrated Model. *J. Travel Tour. Market.* **2010,** *27* (8), 829–843. https://doi.org/10.1080/10548408.2010.527249

Wang, M.; Batt, K.; Kessler, C.; Neff, A.; Iyer, N. N.; Cooper, D. L.; Kempton, C. L. Internal Consistency and Item-Total Correlation of Patient-Reported Outcome Instruments and Hemophilia Joint Health Score v2. 1 in US Adult People with Hemophilia: Results from the Pain, Functional Impairment, and Quality of Life (P-FiQ) Study. *Patient Pref. Adher.* **2017,** *11*, 1831.
Yetim, A. *Sosyoloji ve Spor*; Berikan Yayınevi: Ankara, 2011.

APPENDIX A: INTERVIEW PROTOCOL

The 20 individual interviews (10 managers of international travel agency, 10 professional tourist guides in Istanbul) were conducted to explore the importance of tourism activities on destination choice (Istanbul). Each interview began with an introduction to the background of the study. The interview questions were open-ended. Two questions were asked the destination stakeholders:

1. What do you think about tourism activities in Istanbul as a cultural destination?
2. Which tourism activities do international tourists prefer in Istanbul the most? So, are these activities enough for the cultural importance of Istanbul?

APPENDIX B: INITIAL LIST OF PREFERENCE FOR TOURISM ACTIVITIES

Tourism activities
1. Spiritual events (Whirling Dervish performances, Ramadan events such as Karagoz, iftar)
2. Artistic and cultural activities (Harem Dance, Folk Dance, Janissaries shows, etc.)
3. Cultural tours (Historical peninsula and historical buildings: Hagia Sophia, Blue Mosque, Little Hagia Sophia, etc.)
4. Various sportive and physical activities (running, walking, etc.)
5. The opportunity to watch various sporting events
6. Activities based on amusement parks
7. City tours by bus
8. Museum and exhibition tours
9. Historic Peninsula tours by walking

10. Health and wellness activities (Turkish bath, massage, etc.)
11. Golden Horn and Bosporus boat tours
12. Historical Bazaar trips (Grand Bazaar and Spice Bazaar)
13. Shopping facilities (shopping malls, souvenir shops, etc.)
14. Photo facilities
15. Trip to the Prince Islands
16. Phaeton trips
17. Trip/photographing with historical costumes
18. Traditional crafts experience (Ebru/calligraphy/stained glass)
19. Gastronomy experience (local cooking/tasting)
20. Bosporus tours by helicopter
21. Landscape monitoring facilities (Camlica Hill, Pierre Loti Hill, etc.)
22. Cycling island tour
23. Bosporus tours with dinner and concept (Ottoman night, etc.)
24. Maiden's tower cruise
25. Swimming
26. Private excursions (VIP)
27. Doing nothing (passive holiday)

INDEX